Springer Series in
MATERIALS SCIENCE 37

Springer
*Berlin
Heidelberg
New York
Barcelona
Hong Kong
London
Milan
Paris
Singapore
Tokyo*

Springer Series in
MATERIALS SCIENCE

Editors: R. Hull · R. M. Osgood, Jr. · H. Sakaki · A. Zunger

Springer Series in Materials Science covers the complete spectrum of materials physics, including fundamental principles, physical properties, materials theory and design. Recognizing the increasing importance of materials science in future device technologies, the book titles in this series reflect the state-of-the-art in understanding and controlling the structure and properties of all important classes of materials.

31 **Nanostructures and Quantum Effects**
By H. Sakaki and H. Noge

32 **Nitride Semiconductors and Devices**
By H. Morkoç

33 **Supercarbon**
Synthesis, Properties and Applications
Editors: S. Yoshimura and R. P. H. Chang

34 **Computational Materials Design**
Editor: T. Saito

35 **Macromolecular Science
and Engineering**
New Aspects
Editor: Y. Tanabe

36 **Ceramics**
Mechanical Properties, Failure
Behaviour, Materials Selection
By D. Munz and T. Fett

37 **Technology and Applications
of Amorphous Silicon**
Editor: R. A. Street

38 **Fullerene Polymers
and Fullerene Polymer Composites**
Editors: P. C. Eklund and A. M. Rao

Volumes 1–30 are listed at the end of the book.

Robert A. Street (Ed.)

Technology and Applications of Amorphous Silicon

With 279 Figures and 20 Tables

 Springer

Dr. Robert A. Street
Xerox PARC
3333 Coyote Hill Road
CA 94304 Palo Alto
USA

Series Editors:

Prof. Robert Hull
University of Virginia
Dept. of Materials Science and Engineering
Thornton Hall
Charlottesville, VA 22903-2442, USA

Prof. H. Sakaki
Institute of Industrial Science
University of Tokyo
7-22-1 Roppongi, Minato-ku
Tokyo 106, Japan

Prof. R. M. Osgood, Jr.
Microelectronics Science Laboratory
Department of Electrical Engineering
Columbia University
Seeley W. Mudd Building
New York, NY 10027, USA

Prof. Alex Zunger
NREL
National Renewable Energy Laboratory
1617 Cole Boulevard
Golden Colorado 80401-3393, USA

ISSN 0933-033x

ISBN 3-540-65714-2 Springer-Verlag Berlin Heidelberg New York

Library of Congress Cataloging-in-Publication Data.
Technology and applications of amorphous silicon / R. A. Street (ed.). p.cm. – (Springer series in materials science; v. 37). Includes bibliographical references and index. ISBN 3-540-65714-2 (hardcover). 1. Amorphous semiconductors. 2. Silicon. 3. Optoelectronic devices – Materials. 4. Liquid crystal displays. I. Street, R.A. II. Series: Springer series in materials science; v. 37. TK7871.99.A45T38 2000 621.3815'2–dc21 99-23864

This work is subject to copyright. All rights are reserved, whether the whole or part of the material is concerned, specifically the rights of translation, reprinting, reuse of illustrations, recitation, broadcasting, reproduction on microfilm or in any other way, and storage in data banks. Duplication of this publication or parts thereof is permitted only under the provisions of the German Copyright Law of September 9, 1965, in its current version, and permission for use must always be obtained from Springer-Verlag. Violations are liable for prosecution under the German Copyright Law.

© Springer-Verlag Berlin Heidelberg 2000
Printed in Germany

The use of general descriptive names, registered names, trademarks, etc. in this publication does not imply, even in the absence of a specific statement, that such names are exempt from the relevant protective laws and regulations and therefore free for general use.

Typesetting: Data conversion by Steingraeber Satztechnik GmbH, Heidelberg
Cover concept: eStudio Calamar Steinen
Cover production: *design & production* GmbH, Heidelberg

SPIN: 10719334   57/3144/tr - 5 4 3 2 1 0 – Printed on acid-free paper

Preface

Hydrogenated amorphous silicon (a-Si:H) has become an established material in semiconductor technology, led by photovoltaic and active matrix display applications. The primary attribute of the technology is its large area capability, which provides applications that are otherwise unavailable. The extraordinary ability to make devices on one-meter glass plates or long rolls of metal foil is outside the scope of traditional semiconductor manufacturing. A-Si:H exhibits the full range of semiconducting properties, although with lower speed and current compared with single crystal silicon, and the most important devices are thin film transistors and photodiodes. The plasma deposition technology along with the amorphous structure provide a wide set of compatible materials that allows diversity in device design and considerable band gap engineering. For example, a triple solar cell structure (Chap. 6) has 10–12 distinct layers, each specifically optimized to its role.

The book describes both established and emerging applications. Active matrix addressing is ideally suited to a-Si:H thin film transistors, and is applied to liquid crystal displays (Chap. 2), and image sensor arrays (Chap. 4). Such arrays are made with > 10 million distinct a-Si:H devices, a number which compares respectably with any silicon IC. The arrays are embedded in electronic and optical systems, and the chapters describe how the devices interact with the external systems to fulfill their role.

One focus of the emerging technology is the development of novel devices with complex layered structures. Examples are integrated color sensors (Chap. 7) and large area position sensors (Chap. 8). Another focus is the integration of laser recrystallized polycrystalline silicon into the technology (Chap. 3). Polysilicon adds flexibility to the large area technology since, in addition to co-existing with a-Si:H devices on the same array, parts of an individual a-Si:H device, such as the TFT contacts, can be selectively recrystallized. Finally, some revolutionary approaches to the manufacture of devices using printing technology are described in Chap. 5. The expanding range of device options makes the future of a-Si:H and large area electronics look very promising.

As editor I would like to thank the authors for their excellent contributions to this book. I am grateful to the Xerox Palo Alto Research Center for their support of the project, and to many colleagues with whom it has been a pleasure to work on this subject.

Palo Alto, 1999 *Bob Street*

Contents

1 Introduction 1
Robert Street
1.1 Overview of the Book 1
1.2 Development of Amorphous Silicon 2
1.3 Basic Properties of Amorphous Silicon 3
References 5

2 Active-Matrix Liquid-Crystal Displays 7
Toshihisa Tsukada
2.1 Introduction 7
2.2 TFT LCD 9
    2.2.1 TFT LCD Configuration 9
    2.2.2 Pixel Design 14
    2.2.3 Design Analysis 17
    2.2.4 Scaling Theory of TFT LCD 26
    2.2.5 Fabrication of TFT Panels 34
2.3 Thin-Film Transistors 36
    2.3.1 Hydrogenated Amorphous Silicon Thin-Film Transistors ... 39
    2.3.2 TFT Characteristics 41
    2.3.3 Threshold Voltage Shift 49
    2.3.4 Simulation of TFT Behavior 53
    2.3.5 Two-Terminal Devices 60
2.4 Liquid Crystal 61
    2.4.1 Physical Constants of Liquid Crystal 64
    2.4.2 Twisted-Nematic Cell 69
    2.4.3 In-Plane-Switching Cell 81
    2.4.4 Super-Twisted Nematic (STN) Cell 87
References 89

3 Laser Crystallization
for Polycrystalline Silicon Device Applications 94
James B. Boyce and Ping Mei
3.1 Introduction 94
3.2 Laser Processing of Polysilicon 96

    3.2.1 Polysilicon... 96
    3.2.2 Laser Crystallization................................. 101
    3.2.3 Grain Growth... 105
    3.2.4 Surface Roughening................................. 110
    3.2.5 Laser Doping.. 111
3.3 Low-Temperature Poly-Si Devices............................ 117
    3.3.1 Device Fabrication.................................. 118
    3.3.2 CMOS Device Performance........................ 121
    3.3.3 Device Leakage Currents........................... 126
    3.3.4 Device Stability.................................... 130
3.4 Integration of a-Si and Poly-Si TFTs........................ 132
    3.4.1 Development of Hybrid a-Si and Poly-Si Devices.... 133
    3.4.2 Hybrid Materials Processing........................ 135
    3.4.3 Device Fabrication and Performance............... 138
3.5 Conclusion... 142
References... 143

4 Large Area Image Sensor Arrays............................ 147
Robert Street
4.1 Introduction... 147
4.2 Devices.. 148
    4.2.1 P-i-n Photodiodes................................... 148
    4.2.2 Thin Film Transistors............................... 157
4.3 Sensor Array Designs...................................... 160
    4.3.1 Matrix Addressed Readout......................... 161
    4.3.2 TFT Addressed, p-i-n Photodiode Arrays........... 161
    4.3.3 High Fill Factor Array Designs..................... 171
    4.3.4 TFT Addressed, X-Ray Photoconductor Arrays..... 172
    4.3.5 Diode Addressed Arrays............................ 175
    4.3.6 CMOS Sensors..................................... 178
4.4 Imaging Systems and Their Performance..................... 178
    4.4.1 Electronics... 179
    4.4.2 Electronic Noise.................................... 185
    4.4.3 X-Ray Detection.................................... 191
    4.4.4 The Performance of X-Ray Detectors............... 194
4.5 Applications of Large Area Image Sensors................... 204
    4.5.1 Medical X-Ray Imaging............................. 204
    4.5.2 Other Radiation Imaging Applications.............. 211
    4.5.3 Document Scanning................................ 214
4.6 Future Developments....................................... 216
References... 217

5 Novel Processing Technology for Macroelectronics......... 222
S. Wagner, H. Gleskova, J.C. Sturm, and Z. Suo
5.1 Introduction... 222

5.2 Resolution and Registration:
  The Density of Functions Achievable by Printing................ 225
5.3 Printed Toner Masks for Etching and Liftoff.................... 228
  5.3.1 Toner Masks via Paper Transfer: TFTs on Glass Foil 228
  5.3.2 All Masks Printed Directly: TFTs on Steel Foil 230
5.4 Printing Active Materials: Jetting Doped Polymers
  for Organic Light Emitting Devices............................ 232
5.5 Substrates and Encapsulation for Macroelectronic Circuits 236
5.6 Plastic Substrate Foil: TFT on Polyimide 244
5.7 3-D Integration on a Foil Substrate:
  OLED/TFT Pixel Elements on Steel 246
5.8 Outlook 249
References 250

6 Multijunction Solar Cells and Modules 252
Subhendu Guha
6.1 Introduction... 252
6.2 Deposition Methods 254
  6.2.1 Glow-Discharge Deposition Technique...................... 254
  6.2.2 Plasma Chemistry and the Growth Process 254
  6.2.3 Factors that Influence Film and Cell Quality 256
6.3 Single-Junction Cells 258
  6.3.1 Cell Structure... 258
  6.3.2 Cell Characteristics...................................... 259
  6.3.3 Numerical Modeling 261
  6.3.4 Light-Induced Degradation 264
6.4 High Efficiency Cells 268
  6.4.1 Introduction 268
  6.4.2 Multijunction Cell....................................... 269
  6.4.3 Key Requirements for Obtaining High Efficiency 270
  6.4.4 Back Reflector 270
  6.4.5 Doped Layer... 275
  6.4.6 Intrinsic Layers 277
  6.4.7 Optimization of the Component Cells
       and Current Matching 281
  6.4.8 Tunnel Junction.. 282
  6.4.9 Top Conducting Oxide 285
  6.4.10 Cell and Module Performance 285
6.5 Manufacturing Technology 287
  6.5.1 Manufacturing Process 287
  6.5.2 Production Status and Product Advantage 293
6.6 Alternative Technologies and Future Trends 295
References 299

7 Multilayer Color Detectors 306
Fabrizio Palma
7.1 Introduction 306
    7.1.1 Applications of a-Si:H Color Sensors 307
7.2 Optical Properties of Amorphous Silicon 308
    7.2.1 Optical Properties of Amorphous Silicon Alloys 310
    7.2.2 Optical Design of Layered a-Si:H Structures 312
7.3 Two-Color Sensors 315
    7.3.1 Steady State and Transient Operation 317
    7.3.2 SPICE Model of the Two Color Detector 319
7.4 Three-Color Sensors 320
    7.4.1 Three Color Discrimination with Two Electrical Terminals 320
    7.4.2 Adjustable Threshold Three Color Detector (ATCD) 323
    7.4.3 Three Color Detectors in the Time Integration Regime 327
    7.4.4 Mechanism of Autopolarization of the Stacked Cells 328
7.5 a-Si:H Based UV Sensors 332
    7.5.1 Structure and Operation of the UV Detector 333
7.6 a-Si:H Based IR Sensors 334
    7.6.1 IR Detection by Differential Photo-Capacitance 336
References 338

8 Thin Film Position Sensitive Detectors: From 1D to 3D Applications 342
Rodrigo Martins and Elvira Fortunato
8.1 Introduction and Historical Background 342
    8.1.1 Why Use Amorphous Silicon to Produce Position Sensitive Detectors? 343
8.2 Principles of Operation of 1D and 2D PSD 346
    8.2.1 The Different Types of PSD Devices That Can Be Produced 348
    8.2.2 Different Types of a-Si:H TFPSD and the Production Processes Used 349
8.3 Physical Model for the Lateral Photo-effect in a-Si:H p-i-n 1D and 2D TFPSD 358
    8.3.1 Introduction 358
    8.3.2 General Description of the 1D Theoretical Model 359
    8.3.3 Role of the Recombination Losses for the Fall-Off Parameter 362
    8.3.4 Static Behaviour of E_y and ϕ_y 364
    8.3.5 Role of ρ_s and ρ_{sd} for the Device Detection Limits, Linearity, and Spatial Resolution 365
    8.3.6 Static Distribution of the Lateral Current 365
    8.3.7 Extension of the Theoretical 1D Model to the 2D Case 367
    8.3.8 Determination of the Transient Response Time of the TFPSD 368
8.4 Static and Dynamic Detection Limits 371

    8.4.1 Static Detection Limits of 1D TFPSD 371
    8.4.2 Linearity and Spatial Resolution of 1D TFPSD 372
    8.4.3 Position Response to Multiple Light Beams 374
    8.4.4 Static Predicted and Experimental Performance
        of the 2D TFPSD Device 376
8.5 Dynamic Performance of the 1D and 2D TFPSD 376
    8.5.1 Response Time of the TFPSD 379
    8.5.2 Detection of Light Signals with Different Wavelengths 381
8.6 Characteristics of the a-Si:H p-i-n Structures
    Used to Produce the TFPSD 383
    8.6.1 J–V Curves 383
    8.6.2 Dependence of the Saturation Current of the Device on T 385
    8.6.3 Spectral Response and Detectivity 386
8.7 Peripherals for 1D
    and 2D TFPSD Signal Processing 387
    8.7.1 Optical Methods 387
    8.7.2 Peripherals for Signal Processing 389
8.8 Simulated and Experimental Data in 2D Optical Inspection Systems
    with TFPSD Detector 392
8.9 Linear Array of Thin Film Position Sensitive Detector (LTFPSD) .. 393
    8.9.1 Principles of the Optical Methods Used 394
    8.9.2 Positional Resolution of the Array 395
    8.9.3 Hardware to Control Arrays of Multiple 1D Sensors 396
    8.9.4 Bandwidth Requirements for the Preamplifiers
        Used in the Hardware Control Unit of the LTFPSD 398
8.10 Summary and Future Outlook 399
References 400

Symbols and Abbreviations 404

Subject Index 411

1 Introduction

Robert Street

Xerox Palo Alto Research Center, 3333 Coyote Hill Road, Palo Alto, CA 94304, USA; E-mail: street@parc.xerox.com

1.1 Overview of the Book

We are all familiar with the extraordinary progress of microelectronics, in which continued miniaturization has produced increasingly powerful computation at ever decreasing cost. Despite the ubiquitous silicon chip, there is another class of electronic devices which need a certain size to be useful and therefore require a different approach to fabrication. The displays and scanners that form the interface between people and the electronic world, as well as solar cells and X-ray imagers, are examples of electronic devices for which a large size is essential. This book describes the technology of such devices, in which glass sheets or metal foils form the substrate and the active materials are deposited thin films. The emphasis is on substrate sizes measured in square meters, rather than sub-micron feature sizes.

Hydrogenated amorphous silicon (a-Si:H) has proved to be the material of choice for large area electronics principally for four reasons. Amorphous silicon has all the requisite semiconducting properties of doping, photoconductivity, junction formation, and so on. The plasma deposition process lends itself to large area, and indeed the material tends to have higher purity and greater uniformity as the reactor size gets larger. Being a form of silicon with the same chemical properties as its crystalline cousin, device fabrication takes advantage of much of the knowledge about processing crystalline silicon that has been gained through the microelectronics industry. Finally, the same plasma deposition process allows the formation of a diverse set of alloy materials which provide the dielectrics and passivation layers needed for electronic devices as well as semiconductors with a range of band gaps.

The book focuses on the application of a-Si:H – liquid crystal displays, position sensors, medical imaging and solar cells, and the transistor and photodiode devices that they are made from. Polycrystalline silicon is also included in the book, since this material belongs in the family of thin film semiconductor and insulator materials that can be selected for large-area application. The matrix-addressed arrays which presently dominate the market for large-area electronics use the same processing technology as for the silicon IC industry, but on a larger scale. We discuss new approaches to processing which may take advantage of the particular large area approach.

1.2 Development of Amorphous Silicon

Studies of amorphous silicon began in the 1960s when physicists turned their attention to understanding the differences between the electronic states of amorphous and crystalline semiconductors. However, evaporated or sputtered pure silicon has a very large defect density, which precludes most semiconducting properties, and the research was largely confined to studies of the disordered structure and the hopping conduction of electrons in the defect states [1.1]. The first great breakthrough came with the growth of amorphous silicon from a silane (SiH_4) plasma by Chittick et al. [1.2]. Although that project was not continued for long, the approach was taken up by Spear at the University of Dundee, who confirmed the early measurements and made extensive further studies which showed that the plasma deposited material had greatly improved semiconducting properties, in particular, good photoconductivity [1.3]. Although not clear at first, subsequent work by the Harvard and Chicago groups, showed that the improved properties were the result of hydrogen incorporation into the material [1.4, 5]. The hydrogen binds to dangling bond defects and removes the corresponding electronic states in the band gap, thus eliminating most of the trapping and recombination centers.

The second major breakthrough was the 1975 discovery of substitutional doping [1.6]. The addition of phosphine (PH_3) or diborane (B_2H_6) to the plasma during growth gave n-type and p-type conduction and the ability to control the position of the Fermi energy across most of the band gap. Previously it had been widely believed that amorphous semiconductors could not be doped because of their different impurity chemistry compared to the crystalline materials, so that the observation of doping was a fortunate surprise.

Doping quickly led to the solar cell [1.7], which was the principal application of a-Si:H for the decade from 1975 to 1985, and sustained the extensive research effort into the basic physical properties of a-Si:H. The attraction of a-Si:H for solar cells is that a thickness of less than 1 micron absorbs a large fraction of sunlight, and the plasma deposition is easily scalable to provide the very large area that are needed for significant power generation. Amorphous silicon has found its way into many solar cell applications, but the holy grail of the industry, large scale grid power generation, has so far not proved economically competitive with oil and coal generation. It seems only a matter of time that renewable sources will become the only feasible method of power generation, and then a-Si:H can compete for its share of a trillion dollar business.

The first a-Si:H field effect transistor, announced in 1978 [1.8], had a silicon nitride gate dielectric, deposited in the same plasma reactor by combining the gases SiH_4 and NH_3. The plasma nitride was an important innovation which allowed these thin film transistors (TFT) to be fabricated on glass, at low temperature and with a very high on/off ratio. By this time it was realized that a great variety of silicon alloys could be fabricated by plasma deposition so that an extensive family of materials was available for the technology. The modern solar cells described in Chap. 6 and the color sensors in Chap. 7 make particularly effective use of this range of materials. The a-Si:H TFT emerged as the best solution to the fabrication of matrix addressed arrays for displays and image sensors [1.9]. The first such devices were linear sensor arrays for FAX machines, which appeared in the mid-1980s [1.10]. Active matrix liquid crystal displays (Chap. 2) followed and found a major application in lap-top computers. This market, which continues to grow rapidly and is presently about $20 billion per year, has firmly established the technology for large area processing. In the last decade, the display industry has moved through four generations of processing technology, each of which is characterized by a larger substrate size. The fourth generation uses glass of 80×80 to 100×100 cm. This manufacturing base has opened the way for other electronic devices, the most significant of which are the image sensor arrays for X-ray medical diagnostic imaging described in Chap. 4.

1.3 Basic Properties of Amorphous Silicon

Although amorphous silicon is disordered at the atomic scale, it retains the same local chemical bonding as in crystalline silicon. The silicon atoms are 4-fold coordinated in a tetrahedral bonding symmetry, but with a significant distribution in bond lengths and bond angles. The structure is a random covalent network, which is connected in rings of different sizes, unlike the ordered 6-fold rings of the crystal. The 5–10 atomic percent of hydrogen is bonded to the silicon atoms, mostly in the form of Si-H bonds, either isolated or on the surface of small voids. A model for the atomic structure is shown in Fig. 1.1.

The bonding disorder determines most of the electronic properties of a-Si:H and strongly influences the design of the devices [1.1, 11]. Free electrons

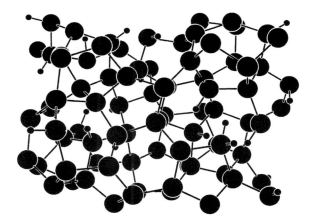

Fig. 1.1. Model for the atomic structure of a-Si:H showing the random covalent network and the Si-H bonds

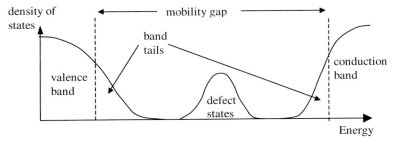

Fig. 1.2. Schematic density of states of a-Si:H showing the localized band tails and defect states

and holes have a scattering length of about an interatomic spacing, and consequently a free carrier mobility of only about 10–20 cm^2/Vs compared to about 500 cm^2/Vs in crystalline silicon. Perhaps more significantly, the disorder causes an exponential tail of localized states at the band edges extending into the forbidden gap. The energy dividing the extended and localized states is known as the mobility edge. The mobility gap separating the valence and conduction band mobility edges is approximately the optical band gap. An illustration of the electronic structure of a-Si:H is shown in Fig. 1.2. Conduction of both electrons and holes occurs near the mobility edges, but involves frequent trapping and release from these localized states and consequently, the effective carrier mobility is further reduced and is also thermally activated. It turns out that the conduction band tail is narrower than the valence band tail, so that electrons have a higher mobility than holes. Some of the key electronic parameters are listed in Table 1.1.

The disorder also affects the optical absorption, which is discussed further in Chap. 7. The rapid scattering of free carriers causes a large uncertainty in the momentum of the electron and hole. Conservation of momentum no longer applies to electronic transitions and the distinction between a direct and indirect optical transition disappears. Essentially all the optical tran-

Table 1.1. Some typical material parameters for good electronic quality a-SiH [1.11]. The exact values depend on the details of the deposition conditions

Electron drift mobility	1 cm^2/Vs
Hole drift mobility	0.003 cm^2/Vs
Optical band gap	1.7 eV
300 K conductivity (undoped)	$10^{-11}\,\Omega^{-1}\,\mathrm{cm}^{-1}$
300 K conductivity (n+)	$10^{-2}\,\Omega^{-1}\,\mathrm{cm}^{-1}$
300 K conductivity (p+)	$10^{-3}\,\Omega^{-1}\,\mathrm{cm}^{-1}$
Defect density	$10^{-15}\,\mathrm{cm}^{-1}$
Diffusion length	3000 Å
Hydrogen concentration	10 at.%

sitions are allowed and the optical absorption follows the density of states. There is an exponential region below that band gap energy, which reflects the exponential band tails, and an approximately parabolic region above the band gap. Even though it has the larger band gap, the optical absorption of a-Si:H is actually larger than crystalline silicon in the region of the band edge because of the indirect gap in the crystal. It has never been completely clear whether hydrogen or the disorder plays a greater role in the larger band gap compared to crystalline silicon.

Despite the large amount of bonded hydrogen, undoped a-Si:H retains about 10^{15} cm^{-3} dangling bond defects which form electronic states near the middle of the band gap. These defects control the trapping and recombination of carriers, and hence determine the carrier lifetimes, photoconductivity and depletion layer width of Schottky barriers and p-i-n junctions. Many years of experimenting with the plasma deposition conditions have not succeeded in further reducing either the defect density or the band tail widths, which suggests that fundamental limits may have been reached. In doped material the defect density increases by 2–3 orders of magnitude, which greatly reduces the minority carrier lifetimes. For this reason, doped layers are primarily used for junctions rather than active layers. Hence, the preferred a-Si:H sensor is a p-i-n structure (see Chaps. 4, 7 and 8), and the channel of the TFT is undoped.

Hydrogen is essential for the good electronic properties of a-Si:H, but does have some deleterious effects. Light induced defect generation was discovered early in the studies of a-Si:H solar cells [1.12], and causes the solar cell efficiency to decrease slowly by 10–20% over many days exposure. The mechanism of defect generation has to do with the breaking of Si-H bonds by the energy of electron-hole recombination. The simplest version of the reaction is,

$$\text{Si-H} \leftrightarrow \text{Si}_{\text{dangling bond}} + \text{H} \,. \tag{1.1}$$

The movement of the free hydrogen and the nature of its subsequent metastable state are the subject of on-going research. Elevated temperatures of about 200°C reverse the metastable changes, but also speeds up the defect generation. Metastability is also induced by current flow and charge accumulation, and so can influence the performance of a-Si:H TFTs and sensors. Much study has gone into minimizing the effects of metastability in devices by careful design, particularly in solar cells (see Chap. 6). TFTs and sensors are stable at room temperature but are increasingly unstable at elevated temperature.

References

1.1 See for example, N. F. Mott and E. A. Davis, Electronic Processes in Non-Crystalline Materials, (Oxford Press) p. 345ff, 1979.

1.2 R. C. Chittick, J. H. Alexander and H. F. Sterling, J. Electrochem. Soc. **116**, 77 (1969).
1.3 W. E. Spear, Proc. 5th Int. Conf. on Amorphous and Liquid Semiconductors, Taylor and Francis, p. 1 (1974).
1.4 A. J. Lewis, G. A. N. Connell, W. Paul, W. Pawlik and R. Temkin, AIP Conf. Proc. **20**, 27 (1974).
1.5 H. Fritzsche, Proc. 7th. Int. Conf. on Amorphous and Liquid Semiconductors, (CICL, Edinburgh), 3 (1977).
1.6 W. E. Spear and P. G. LeComber, Solid State Commun. **17**, 1193 (1975).
1.7 D. E. Carlson and C. R. Wronski, Appl. Phys. Lett **28**, 671 (1976).
1.8 A. J. Snell, K. D. Mackenzie, W. E. Spear, P. G. LeComber and A. J. Hughes, Appl. Phys. **24**, 357 (1981).
1.9 P. G. LeComber, W. E. Spear and A. Ghaith, Electron. Lett. **15**, 179 (1979).
1.10 F. Okamura and S.Kaneko, MRS Symp. Proc. **33**, 275 (1983).
1.11 See, for example, R. A. Street, Hydrogenated Amorphous Silicon (Cambridge University Press), 1991.
1.12 D. L. Staebler and C. R. Wronski, Appl. Phys. Lett. **31**, 292 (1977).

2 Active-Matrix Liquid-Crystal Displays

Toshihisa Tsukada

Central Research Laboratory, Hitachi, Ltd., Kokubunji, Tokyo 185-8601, Japan
E-mail: tsukada@crl.hitachi.co.jp

2.1 Introduction

The active-matrix liquid-crystal display (AMLCD) is a flat-panel display in which the display medium is liquid crystal and each picture element (pixel) is driven by such active devices as diodes or transistors. These active devices are arranged in rows and columns on a glass substrate to control each pixel, and hence the name of active matrix. Before the AMLCDs were introduced, liquid-crystal displays were operated on a basis of simple matrix or passive-matrix. Passive-matrix liquid-crystal displays feature flatness, lightweight, and low-power consumption. Due to these features, they have been first installed in such devices as wrist watches or calculators. Then, their application fields have expanded to pocket TVs, word processors, and factory automation machines. There was a constant demand for larger sizes and higher resolutions.

As the number of pixels got larger, however, a crosstalk problem arose limiting further increase of pixels. This limitation appeared apparent when the pixel count exceeds one hundred by one hundred. One solution to overcome this problem is to use a different liquid-crystal display mode. Conventional twisted-nematic (TN) display mode had been replaced by super-twisted-nematic (STN) display mode in which the threshold slope of transmittance is steeper than that of TN mode. As the resolution of display panels were increased up to a few hundred by a few hundred, however, even the STN mode could not be operated without suffering from a crosstalk problem. In order to further increase the number of pixels, each pixel has to be isolated from others with regard to applied voltage. The active device works to apply the signal voltage to the right pixel and to hold the voltage without being affected by false signals.

The most commonly used active device is a thin-film transistor (TFT). The original form of TFTs can be traced as far back as 1961 [2.1]. It was a thin-film version of single-crystalline transistors. The target of their application is thin-film logic circuitry for computers. Their intention had not resulted in a great success due to the development of MOS transistors of silicon crystal. In 1966, Weimer described the concept of TFT LCD [2.2] in which TFTs are used as display switches. A more detailed concept was described by Lechner et al. [2.3] in 1971, where the use of diodes or triodes (transistors) was discussed as switches for active-matrix liquid crystal displays. Use of storage capacitors implemented in parallel with the liquid crystal cell capacitor was also mentioned. Preceding this discussion, Heilmeier et al. [2.4] proposed nematic liquid crystal as a material for a flat-panel display. A sandwich cell consisting of a transparent front electrode, a reflecting back electrode, and nematic liquid crystal in between was prepared (reflective mode). When there was no applied field, the cell appeared black. When a dc voltage was applied, the liquid crystal became turbulent and scattered light: the cell appeared white. This phenomenon was termed "dynamic scattering", and was applied to demonstrate the first liquid crystal display both in reflective and transmissive mode operations. A 3.5-by-4-inch alphanumeric display was fabricated and a maximum contrast ratio higher than 20 to 1, was demonstrated.

Brody et al. applied the CdSe TFT to the active-matrix liquid crystal panel [2.5]. This display panel consisted of 14 000 transistors, storage capacitors and the twisted-nematic (TN) liquid crystal cell. Although TFTs were made of CdSe rather than a-Si:H, the configuration is essentially the same as today's TFT LCD panels. The TN liquid crystal cell was first proposed by Schadt and Helfrich [2.6] and featured low-voltage operation, low power consumption, and a fast response time. In the TN cell, the average direction (director) of the liquid crystal molecules is twisted 90° as they go from the back to the front glass substrate. The polarization of the light is also rotated 90° as the light passes though the liquid crystal cell. When the front polarizer is set parallel to the rear polarizer, the light is not transmitted when there is no applied voltage. When the ac voltage is applied to the cell, the director of liquid crystal molecules becomes perpendicular to the substrate. In this state, the TN cell becomes optically inactive and linearly polarized light travels through the cell without any rotation of polarization. Therefore, the light is transmitted through the front polarizer. Since its development, the TN cell has played a very important role as the display medium for TFT LCDs.

Hydrogenated amorphous silicon (a-Si:H) was a late arrival in TFT technologies. However, it had a great impact on achieving practical TFT LCDs. Since the first report by the Dundee group [2.7], a-Si:H TFT has been recognized as the most suitable device for TFT LCDs. The mobility of a-Si:H has proven to be just enough to charge the liquid crystal capacitance and storage capacitor. The off-current of a-Si:H TFT has turned out to be on a sufficiently low level not to discharge the pixel capacitance during the frame time, owing to the high resistivity of undoped a-Si:H. Another important feature of a-Si:H is that it can be deposited over a large area. Uniformity over the area, and reproducibility from run to run can be obtained relatively easily. Also important is that it can be deposited at low temperatures, so glass substrates can be used. Moreover, good interface properties between a-Si:H and other thin films, like metals, insulators, and semiconductors can be obtained. Due to these properties, the hydrogenated amorphous silicon TFT has acquired wide acceptance among the many candidates to make TFT LCDs.

As for liquid crystal, more than one hundred years have passed since its discovery. In 1888, an Austrian botanist, F. Reinitzer reported that cholesteric benzoate he purified showed a strong birefringence at a temperature range between 145.5°C and 178.5°C. He sent the sample, which became a turbulent liquid in this temperature range, to a German physicist O. Lehmann. Lehmann found that this material showed an optical anisotropy when observed under crossed polarizers and coined the term "liquid crystal" for the first time.

In the following chapters, AM LCDs are described focusing on panels based on a-Si:H TFT technologies.

2.2 TFT LCD

The thin-film-transistor-addressed liquid-crystal display (TFT LCD) is a flat-panel display which is used in such fields as consumer electronics, computers, and communication terminals. TFT LCDs are generally characterized by the diagonal length of the panel and their resolution. Figure 2.1 shows the correspondence between display area and the number of pixels (picture elements). In computer display applications each pixel generally consists of three colored stripes (red, green, and blue). The inserts in this figure show the relative sizes of this configuration, assuming the pixels are square. A resolution of 640 (horizontal) by 480 (vertical) pixels corresponds to a video-graphics-array (VGA) monitor, and if a pixel size of $0.33\,\mathrm{mm}^2$ is assumed, the diagonal size of VGA monitors becomes 10.4 inch or 26.4 cm. In panels with a resolution of 1280 (H) by 1024 (V) which is called super XGA (extended graphics array), the number of pixels exceeds one million, and there are more than three million RGB dots or subpixels. Thus, more than three million TFTs must be fabricated on these panels.

2.2.1 TFT LCD Configuration

The basic configuration of a TFT LCD is shown in Fig. 2.2 [2.8]. Liquid crystal is encapsulated between two glass substrates, a TFT substrate and a color-filter substrate. In order to obtain good display quality, the cell gap of the liquid crystal (i. e., the spacing between the two glass substrates) has to be precisely controlled to a specific value, e.g., 5 µm. This gap has to be uniform over the whole display area and reproducible from run to run. Therefore, transparent spacers such as plastic beads are placed on the surface of the glass substrate.

The liquid crystal cells generally used are twisted-nematic type (see Sect. 2.4) in which the *director* (orientation) of the liquid crystal molecules is twisted 90° between the TFT substrate and the color-filter substrate. In

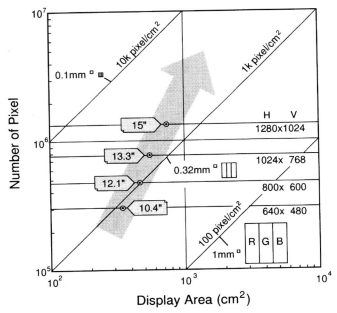

Fig. 2.1. Number of pixel vs. display area of TFT LCD. Numbers shown by H and V correspond to row and column pixel number of computer terminal displays. Typical diagonal panel sizes for each resolution are shown in inserts. Shaded arrow represents the relation of (number of pixel) $\propto$ (display area)^2 [2.2], indicating the trend of the panel size and resolution

Fig. 2.2, the crossed-polarizer system is shown, in which the first polarizer works as a backlight polarizer and the other acts as an analyzer. In this system, light passes through the analyzer when there is no applied voltage on the cell, and is blocked when the applied voltage is high enough to align the liquid crystal molecules vertically. The liquid crystal molecules are anchored on the surface of the glass substrates so that they are oriented to a proper direction. In order to set the anchoring direction, the glass substrate is coated with an organic film such as a polyimide film and the surface of the film is rubbed with a fabric in a specific direction. The liquid crystal molecules are tilted several degrees with respect to the glass surface. This tilt angle is called the pretilt angle and plays an important role in determining the electrical and optical characteristics of the TFT LCD.

The TFT substrate consists of a TFT array and an array of external terminal pads on which LSIs are bonded to drive the TFT panel. The driver LSIs are essentially scan generators for the horizontal and vertical buslines. These LSIs are directly bonded to the glass with TAB (tape-automated bonding) connectors, and they provide each pixel of the panel with video signals that

2 Active-Matrix Liquid-Crystal Displays 11

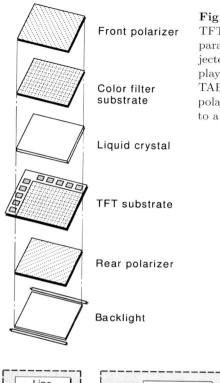

Fig. 2.2. Configuration of a TFT LCD. The TFT and the color filter substrates are two parallel sheets of glass with liquid crystal injected between them. LSIs for driving the display are bonded to the TFT substrate via TAB (tape-automated bonding). A crossed-polarizer system is shown here, corresponding to a normally-white display

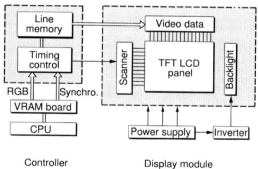

Fig. 2.3. Schematic diagram of the TFT LCD module and controllers

are transferred to the panel via a video signal processor and controller. A schematic diagram of TFT LCD module and controllers is shown in Fig. 2.3.

The backlight system can be either direct or indirect. With direct lighting, one or more fluorescent lamps are positioned directly beneath the rear polarizer, and with indirect lighting, a light-guide is used to guide the light from lamp(s) situated beside it. Three-wavelength-type illumination is generally used as a backlight.

The backlight illumination is attenuated as it passes through the display module as shown in Fig. 2.4. The maximum transmittances of the polarizer and color filter are one half and one third, respectively, resulting in a utility factor of one sixth. The aperture ratio of the pixels further reduces this factor. If an aperture ratio of 50% is assumed the utility factor will be 8%. However, this is only the upper limit of the transmittance of the total system. In a practical system, the total utility factor is 3–6%.

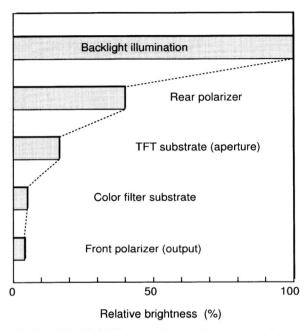

Fig. 2.4. Backlight illumination is attenuated as it passes through the display module. Only several percent of the original illumination is output from the front polarizer

Figure 2.5 shows a schematic diagram of a TFT LCD. There are two sets of buslines, i.e., horizontal gate buslines and vertical data buslines. A TFT is formed at each intersection of these buslines to turn on and off the voltage applied to the liquid crystal cell. This cell is represented by an equivalent capacitance and in parallel with this capacitor a storage capacitor or additional capacitor (C_{st} or C_{add}) is formed to improve the retention characteristics of the signal charge. The color filter is formed in a striped R, G, and B configuration, three stripes (dots) forming one pixel. The digital signal is fed to each dot, and it is possible to display $(2^n)^3$ colors for n bits per color channel. For $n = 2$, the panel displays 64 colors, and 4096 colors for $n = 4$. The 24 bit color graphics represent $16\,777\,216$ colors $[= (2^8)^3]$.

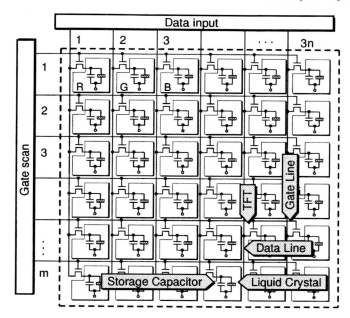

Fig. 2.5. Schematic diagram of a TFT LCD. This shows a display for personal computers. Each pixel consists of three RGB subpixels or dots and is designed to be square-shaped

The display operates one line at a time – video signals are fed to the data buslines simultaneously through a data buffer during the gate turn-on time. The scan gate voltage pulse applied to a certain (say, i-th) gate busline opens the gates of the TFTs connected to this busline. The signal voltage is then applied through the data buslines to the pixel electrode of each dot on this gate busline. The maximum time t_{ON} allotted to turn on the TFT is given by

$$t_{ON} = (m\, f_F)^{-1}\ , \tag{2.1}$$

where m is the number of gatelines and f_F is the frame frequency. If the frame frequency is 60 Hz and if there are 480 gate buslines, t_{ON} is 34.7 µs. During this time period, the charging of the capacitance (the liquid crystal cell and the storage capacitor) has to be completed.

After this charging period, the liquid crystal cells on the i-th gate line are cut off from the data lines and the cells connected to the $(i+1)$-th gate lines are charged. The cell cutoff has to be perfect, i. e., the cutoff cell has to keep its charged voltage until the next charging step takes place. If, for some reason, there is an increase of the off-current of the TFT, the signal voltage will discharge causing crosstalk and degraded display quality. In fact, the most frequent cause of crosstalk is an increase in leakage current due to the photocurrent induced by intense illumination from the backlight.

2.2.2 Pixel Design

Twisted-Nematic Cell. There are many variations in the pixel design layout. First, the designer must choose whether the a-Si:H TFTs will be configured as back-channel-etched TFTs or channel-passivated TFTs (see Sect. 2.3.1). Next, the TFT layout and the pixel electrode design must be considered. Then, the configuration of the storage capacitance has to be decided. Two examples of possible configurations are shown in Fig. 2.6. In the additional capacitance (C_{add}) scheme (Fig. 2.6a), the pattern of the gate busline is formed to overlap the transparent pixel electrode of ITO (Indium Tin Oxide), so that no extra buslines are needed. In the storage capacitance design of C_{st} (Fig. 2.6b), an independent electrode is provided for the storage capacitance. The C_{add} design is simple to fabricate, but it increases the gate busline capacitance. On the other hand, the C_{st} design has an increased number of cross-over points between the C_{st} busline and the data busline; possible shorts between these buslines can lower the production yield of TFT panels, and the panel processing becomes more complex.

The design rule defines the minimum pattern size and the gap or spacing between various patterns. For example, it sets the minimum size for the spacing between pixel electrodes and data buslines, and also governs the channel length of a-Si:H TFTs. A design rule or minimum design size of 10 µm is generally adopted for panels with a diagonal size of around 10 in. and video-graphic-array (VGA) resolution (640 × 480). For higher resolution panels, sub 10 µm technology is used for the design and processing. Effects of modifying the design rule are discussed in Sect. 2.2.4.

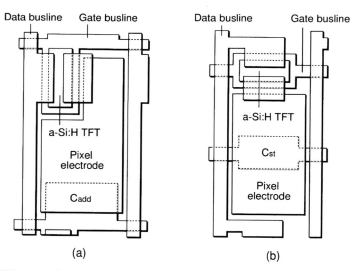

Fig. 2.6a,b. Examples of pixel layout of a TN cell: (**a**) the additional capacitance case, C_{add}, and (**b**) the storage capacitance case, C_{st}

As described before (Fig. 2.4), the utility factor of the backlight illumination is limited by the aperture ratio of the pixels. The aperture is defined as the area ratio of the transparent electrode to the pixel. Since light can only be modulated at the transparent pixel electrodes, light passing through the gaps between the pixel electrodes and the metallization patterns degrades the display quality and should be blocked. Therefore, the gap between metallizations is covered by an opaque material; this is the so-called "black matrix" design. The black matrix is usually formed on the color-filter substrate. Since the two glass substrates, i.e., the color-filter and TFT substrates, are assembled and aligned after processing each substrate separately, the alignment cannot be as precise as can be achieved on a single substrate. Therefore, the black matrix must have a large margin, resulting in a reduced aperture ratio. This large margin can be decreased by fabricating the black matrix on the TFT substrate. Precise alignment can be achieved in this scheme, allowing the aperture and brightness to be increased. However, since the materials for the black matrix are metals like chromium, capacitive coupling between the black matrix and the electrodes can be a problem. Therefore, the black matrix is usually fabricated on the color-filter substrate. When chromium is used for the black matrix, it reflects light from outside, degrading the display quality of the panel. This effect can be lessened by the use of a chromium oxide. Another role of the black matrix is to shield the a-Si:H TFTs from incident light. Since a-Si:H is very sensitive to visible radiation, the visible light from the backlight incident upon the TFTs generates a photocurrent which increases the TFT off-current. Thus, the black matrix is designed to cover the whole TFT area.

The aperture ratio is also affected by the disclination of the liquid crystal. The disclination line corresponds to the boundary between regions of the liquid crystal that have directors with opposite orientations. In a twisted-nematic TFT LCD, the electric field is generally applied vertically or perpendicular to the glass substrate. However, the electrode configuration on the TFT substrate gives rise to a local horizontal electric field. This local field, in turn, induces reverse tilt domains or disclinations. This phenomenon is also related to the rubbing direction of the alignment layer. Generally, the rubbing direction is at 45° to the busline direction and the disclination tends to occur at the corner of each pixel electrode corresponding to the tail of the rubbing direction. The reverse tilt introduces transmittance irregularities and decreases the contrast ratio in that region. Therefore, in order to avoid undesirable effects, this area has to be covered by the black matrix, even though reverse tilt only occurs occasionally.

In-Plane-Switching Mode. In the preceding section, the pixel design based on the twisted-nematic (TN) liquid-crystal cell was discussed. These TN-cell-based displays suffer from the narrow viewing-angle characteristics. One of the most promising cell modes to overcome this situation is the in-

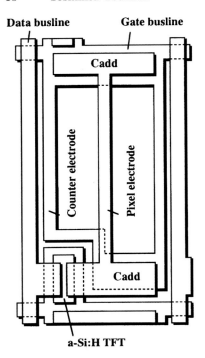

Fig. 2.7. Pixel layout example of the in-plane-switching mode display

plane-switching (IPS) mode (see Sect. 2.4). In this mode, the electric field is applied in parallel to glass substrates in contrast to the TN mode where the field is applied in perpendicular to the substrates.

The pixel design of the IPS mode is shown in Fig. 2.7 [2.9]. The pixel electrode is designed in a striped pattern and is made of such metals as chromium or aluminum. The signal voltage is applied between this pixel electrode and the counter electrode extending from the gate busline. The transparent electrode of ITO used in the TN version as shown in Fig. 2.6, is not used in this IPS scheme. This considerably simplifies the process steps to fabricate the panel. Since the typical number of masks used in the TFT fabrication is five, reduction of one mask (ITO) is a substantial improvement.

The separation between pixel electrode and counter electrode, s, is directly related to the threshold voltage. In the IPS mode, V_{th} is proportional to (s/d) (see Sect. 2.4). When the panel size becomes large and/or the resolution of the panel is not so high, this separation could be large resulting in a high V_{th}. In this case, interdigital electrode formation will be favored for reducing driving voltages. Since the cell gap d of liquid crystal ranges from 5 to 7 μm, the separation s is chosen to be between 10 to 30 μm.

2.2.3 Design Analysis

This section discusses three major issues of the TFT LCD design: (1) the charging behavior of the pixel capacitance by the a-Si:H TFT, (2) the dc voltage offset generated by the parasitic capacitance, C_{gs}, and (3) the delay and distortion of the gate pulse voltage, i. e., the gate delay.

Figure 2.8 shows an equivalent circuit representation of a pixel to be used in the analysis. The distinction between the source and the drain electrodes of the TFTs is usually quite clear. In an n-channel TFT such as an a-Si:H TFT, the drain electrode is biased with a higher potential than the source electrode which is usually grounded. In the TFT LCD panel, however, the signal voltage fed through the data busline has an alternating polarity as shown in Fig. 2.8. Therefore, during a positive cycle of the signal voltage, the TFT electrode on the data busline corresponds to the drain electrode. However, during a negative cycle of the signal voltage, the situation is reversed, and the data busline electrode is biased to a lower potential with respect to the pixel electrode and thus corresponds to the source electrode in its usual notation. For convenience, however, we call the electrode connected to the data busline the *drain* and the electrode connected to the pixel electrode the *source*. According to this notation, the parasitic capacitance between the gate and the source is C_{gs}, which corresponds to the overlapping capacitance between the source and the gate electrodes.

Charging of Pixel Capacitance As shown in Fig. 2.8, the signal voltage is applied to the liquid crystal cell through a TFT which acts as a voltage switch, and this voltage controls the intensity of the illumination from the

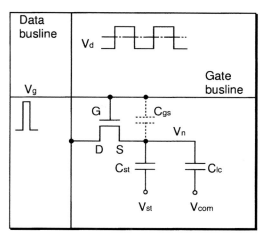

Fig. 2.8. Equivalent circuit of a pixel, showing the drain and source electrode notations. The node potential V_n represents the potential of the pixel electrode

backlight. Following the notation of Fig. 2.8, we derive a formula for the charging behavior of a liquid crystal cell. The stored charge Q_n at a node or pixel electrode is given by

$$Q_n = C_{gs}(V_n - V_g) + C_{st}(V_n - V_{st}) + C_{lc}(V_n - V_{com}), \qquad (2.2)$$

where V_n is the node potential, V_g is the gate voltage, C_{lc} is the capacitance of the liquid crystal, and the charging current or drain current of the a-Si:H TFT, I_d, is given by

$$I_d = dQ_n/dt. \qquad (2.3)$$

Since the TFT is operated in the linear region of its I_d-V_d characteristics to charge the liquid crystal cell, the drain current is given by the formula (gradual-channel approximation),

$$I_d = \beta_0 \left[(V_g - V_t - V_n)(V_d - V_n) - (V_d - V_n)^2/2 \right], \qquad (2.4)$$
$$\beta_0 = \mu_n C_i (W/L), \qquad (2.5)$$

where μ_n is the electron mobility in the channel, C_i is the gate insulator capacitance per unit area, V_d is the drain voltage, V_t is the threshold voltage of the TFT, and W and L are the width and length of the TFT, respectively.

Under the assumption of constant parameters we obtain from (2.2–5),

$$dV_n/dt = (1/2)(\beta_0/C_{px}) \left[(V_g - V_t - V_n)^2 - (V_g - V_t - V_d)^2 \right], \qquad (2.6)$$

where $C_{px} = C_{lc} + C_{st} + C_{gs}$. As will be discussed in Sects. 2.3.2 and 2.4.2, both C_{lc} and C_{gs} depend on voltage and thus time. However, these variations have little effect on the charging behavior. Therefore, the solution to (2.6) provides us with a good description of the charging behavior. If we set an initial value as follows (Fig. 2.9)

$$V_n = V_{n0} \quad \text{at} \quad t = 0, \qquad (2.7)$$

the node potential V_n is given by the following equation for the case of charging from a negative cycle to a positive cycle:

$$V_n = \frac{1 - a\exp(-t/\tau)}{1 - b\exp(-t/\tau)} V_d, \qquad (2.8)$$

where τ is a time constant [2.10] and a and b are dimensionless constants [2.11]. These constants are given by

$$\tau = \frac{C_{px}}{\beta_0 (V_g - V_t - V_d)}, \qquad (2.9)$$

$$a = \frac{V_d - V_{n0}}{V_d} \cdot \frac{2(V_g - V_t - V_d) + V_d}{2(V_g - V_t - V_d) + V_d - V_{n0}}, \qquad (2.10)$$

and

$$b = \frac{V_d - V_{n0}}{2(V_g - V_t - V_d) + V_d - V_{n0}}. \qquad (2.11)$$

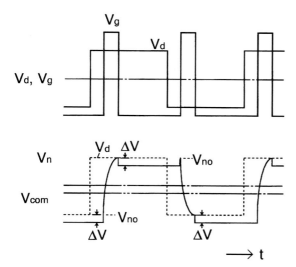

Fig. 2.9. Voltage waveforms of TFT LCD. The signal voltage is ac with positive and negative cycles. A voltage jump occurs due to the feed-through voltage from the gate pulse which results in a dc voltage offset. The common electrode voltage V_{com} is adjusted to this offset voltage

Since Equation (2.8) corresponds to the transition from a negative to a positive cycle, V_d in this equation is positive and V_{n0} is negative as shown in Fig. 2.9. The initial value, V_{n0} is defined as the node potential at the leading edge of a gate pulse. At this point, the node potential V_n jumps by ΔV, which is given by

$$\Delta V = V_a C_{gs}/C_{px}, \qquad (2.12)$$

where V_a is the pulse amplitude of the gate voltage. V_{n0} is the value of V_n after this jump takes place. V_{n0} is therefore nearly equal to V_d in the preceding cycle, or $-V_d$, and in this negative cycle, the node potential is equal to $-V_d - \Delta V$. The time constant of the pixel is given by

$$\tau_d = \frac{R_{lc} R_{off}}{R_{lc} + R_{off}} C_{px}, \qquad (2.13)$$

where R_{lc} is the resistance of the liquid crystal cell and R_{off} is the off-resistance of the a-Si:H TFT. If τ_d is large enough, the node potential will not decay from its initially charged value. In this case, the initial node potential is

$$V_{n0} = -V_d, \qquad (2.14)$$

and the constants of (2.10) and (2.11) are simplified to

$$\begin{aligned} a &= 2 - V_d/(V_g - V_t), \\ b &= V_d/(V_g - V_t). \end{aligned} \qquad (2.15)$$

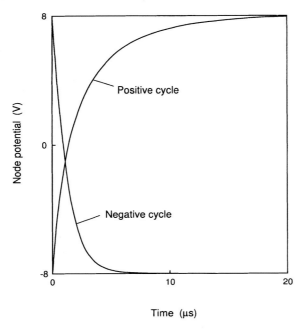

Fig. 2.10. The charging characteristics of the node. The positive cycle corresponds to the transition from a negative to a positive cycle ($V_d = 8V$), and the negative cycle corresponds to the reverse case ($V_d = -8V$). The numerical values used in the calculation are as follows: $C_{px} = 1\,\text{pF}$, $\mu_n = 0.5\,\text{cm}^2/\text{V s}$

For the transition from a positive to a negative cycle, the same formula as above can be used with $V_d < 0$, and $V_{n0} > 0$. In this case, V_{n0} is again approximately equal to V_d including the effect of the voltage step ΔV due to C_{gs}, if τ_d of (2.13) is larger than the frame time.

The simulated charging behavior of the node potential is shown in Fig. 2.10. In this example, the time constant of the charging process τ (2.9) is calculated to be $\tau = 5\,\mu\text{s}$ and $1\,\mu\text{s}$ for positive and negative cycles, respectively, corresponding to charging times (95%) of $13.3\,\mu\text{s}$ and $4.3\,\mu\text{s}$, respectively.

From (2.8), when $t \ll \tau$ this formula reduces to

$$V_n = V_{n0}[1 + \alpha(t/\tau)], \qquad (2.16)$$

where the coefficient α is given by

$$\alpha = \frac{V_d - V_{n0}}{V_{n0}} \frac{2(V_g - V_t - V_d) + V_d - V_{n0}}{2(V_g - V_t - V_d)}. \qquad (2.17)$$

Substituting the numerical values shown in the caption of Fig. 2.10 into (2.17) yields $\alpha = -6$ and -1.2 for the positive and negative cycles, respectively.

One of the most important parameters in designing a TFT is the ratio of the width to the length, W/L, of the transistor. It is directly related to the time t_r required to charge the pixel capacitance. If the ratio of the charged node potential to the data voltage is defined as r_c, we obtain from (2.8)

$$t_r = k_t \tau, \tag{2.18}$$

where k_t is given by

$$k_t = \ln\left[(a - br_c)/(1 - r_c)\right]. \tag{2.19}$$

This must be shorter than the scan time, which is the frame time t_f divided by the number of scan lines, m, and this condition is given by

$$t_r < t_f/m. \tag{2.20}$$

From (2.9, 18–20), we obtain

$$W/L > k_t(m/t_f)(C_{px}/\mu_n C_i)/(V_g - V_t - V_d). \tag{2.21}$$

With a frame rate of 60 Hz, 480 scan lines, and $r_c = 0.95$, (2.21) reduces to

$$W/L > 1.9. \tag{2.22}$$

Here, we assumed the same figures as described above. This value for W/L is almost ideal from the viewpoint of TFT design, since $W/L = 1$ provides the minimum TFT cell area. If we assume $r_c = 0.99$, however, (2.21) reduces to $W/L > 3.0$.

In the derivation of (2.22), we assumed the gate line delay associated with voltage pulses traveling along the gate busline was negligible. However, the gate delay can represent a considerable part of the scan time as TFT/LCD panels get larger and their resolution gets higher. When the gate delay is included, (2.20) becomes

$$t_r < (t_f/m) - t_d, \tag{2.23}$$

and numerical evaluation of this equation with $r_c = 0.99$ and $t_d = 10\,\mu\text{s}$ gives

$$W/L > 4.3, \tag{2.24}$$

which is still a reasonable value.

Another basic criterion can be deduced from the risetime formula of (2.20). The TFT must have sufficient retention capability so as not to discharge the stored charge during the frame time. The retention ratio r is written as

$$r = \exp(-t_f/\tau_d). \tag{2.25}$$

If τ_d of (2.13) is assumed to be much larger than t_f, we obtain from (2.18–20, 25):

$$\tau_d(1 - r) > k_t\, m\, \tau. \tag{2.26}$$

If the on-resistance of the TFT R_{on} is defined as

$$R_{on}^{-1} = \beta_0(V_g - V_t - V_d), \tag{2.27}$$

the ratio of the on- and off-resistance of the TFT is obtained from (2.9, 13, 26, 27):

$$R_{off}/R_{on} > k_t m[1 + (R_{off}/R_{lc})]/(1 - r). \tag{2.28}$$

For 480 scan lines, a retention ratio of 99% ($r = 0.99$), $r_c = 0.99$, and $R_{lc} \approx R_{off}$, the on/off ratio of the TFT is given by

$$R_{off}/R_{on} > 4 \times 10^5, \tag{2.29}$$

where we assumed the same voltage figures as in Fig. 2.10. This criterion sets a minimum level of the on/off current ratio of the TFTs. Although the transfer characteristics of the a-Si:H TFT seem to satisfy this condition as will be seen in Sect. 2.3.2, the numerical check will be of value. By substituting $\mu_n = 0.5 \, \text{cm}^2/\text{V s}$, $C_i = 20 \, \text{nF/cm}^2$, $W/L = 5$, $V_g = 13 \, \text{V}$, $V_t = 1 \, \text{V}$, and $V_d = 8 \, \text{V}$ into (2.27), we obtain an on-resistance of $5 \times 10^6 \, \Omega$. With this R_{on}, R_{off} is calculated to be larger than $2 \times 10^{12} \, \Omega$ from (2.29). These resistance values correspond to on- and off-current of $2 \, \mu\text{A}$ and $5 \, \text{pA}$ for $V_d = 10 \, \text{V}$, respectively, and this current range can well be covered by a-Si:H TFTs. From an assumption of $R_{lc} \approx R_{off}$, the resistivity of liquid crystal is calculated to be $5 \times 10^{11} \, \Omega \cdot \text{cm}$ for a cell gap of $5 \, \mu\text{m}$ and an aperture of $250 \times 50 \, \mu\text{m}^2$ in a pixel area of $300 \times 100 \, \mu\text{m}^2$. Since the bulk resistivity of liquid crystal is 10^{12}–$10^{14} \, \Omega \cdot \text{cm}$, there is a wide margin. The time constant of the pixel is calculated from (2.13) to be one second, which is also consistent with the assumption made in deriving (2.26) from (2.25).

Voltage Offset. Liquid crystal displays do not work properly under dc bias, so they are always operated in ac mode as shown by V_d in Fig. 2.9. Due to this ac operation and the parasitic capacitance formed between the gate and the source electrode of the TFT, however, a dc voltage offset, ΔV, appears in the node potential, V_n [2.12]. This offset is given by

$$\Delta V = V_a C_{gs}/(C_{gs} + C_{st} + C_{lc}), \tag{2.30}$$

where V_a is the amplitude of the voltage pulse applied to the gate busline to turn on the TFTs connected to it. When the gate voltage appears on the line, C_{gs} is charged at the leading edge and discharged at the trailing edge of the voltage pulse. Under normal conditions, the charging and discharging cancel each other out, and no effect is observed. However, in TFT LCD panels, the ac signal is applied to the liquid crystal through a TFT, and only the discharging effect remains on the pixel electrode as shown in Fig. 2.9.

The resultant dc voltage offset causes undesirable effects on the performance of the liquid crystal display such as flicker [2.13], image sticking [2.14,

15] and permanent brightness nonuniformities on the panel. Flicker noise appears as low-frequency brightness variation. The voltage offset causes asymmetry in the ac node potential from frame to frame. The liquid crystal cell reacts to this asymmetrical signal voltage, resulting in fluctuation of its transmittance. When the frame frequency is 60 Hz, this fluctuation occurs at a frequency of 30 Hz, and is recognizable.

The storage capacitance is included in the design of the pixel electrode to reduce this level shift. Without C_{st}, (2.27) reduces to

$$\Delta V = V_a/(C_{gs} + C_{lc}). \tag{2.31}$$

For $V_a = 22$ V (assumed base level of gate voltage is -9 V), $C_{gs} = 0.1$ pF, and $C_{lc} = 0.3$ pF, ΔV is as high as 5.5 V. Since the threshold voltage, V_{th}, of a TN liquid crystal is 2–3 V, this value of ΔV is about twice that of V_{th}. With C_{st}, this voltage shift takes a lower value. If the same figures are substituted as described above with $C_{st} = 0.6$ pF, ΔV becomes

$$\Delta V = 22 \times 0.1/1.0 = 2.2 \text{ V}. \tag{2.32}$$

This represents a considerable improvement, but the dc bias is still rather high. This is compensated by adjusting the voltage of the common electrode on the color filter substrate, V_{com}, so as to cancel out the voltage shift. However, this canceling is not perfect since both C_{lc} and C_{gs} of (2.30) are voltage dependent. The effects of this voltage dependence are discussed below.

The parasitic capacitance C_{gs} depends on V_g as will be discussed in Sect. 2.3.2. However, since the transient of the gate pulse ($\sim$ ns) is much faster than the response time of a-Si:H TFTs ($\sim$ μs), C_{gs} can be assumed to be constant and be approximated by the value corresponding to the peak gate voltage. Equation (2.30) then reduces to

$$\Delta V = V_a C_{gs}/\left[C_{gs} + C_{st} + C_{lc}(V)\right]. \tag{2.33}$$

As will be described in Sect. 2.4.2, the liquid crystal capacitance varies as a function of applied voltage. This variation is due to the anisotropic dielectric constant of the liquid crystal. In a TN liquid crystal, the dielectric constant is low ($\varepsilon = \varepsilon_\perp$) when there is no applied voltage and approaches $\varepsilon_\parallel$ as the voltage applied to the cell is increased.

The difference in ΔV between two pixels, dV, is given by [2.14]

$$dV = \frac{V_a C_{gs}}{C_{gs} + C_{st} + C_{lc}(V_1)} - \frac{V_a C_{gs}}{C_{gs} + C_{st} + C_{lc}(V_2)}, \tag{2.34}$$

where V_1 and V_2 are the applied rms voltages to these pixels. The maximum value of (2.34), is obtained by substituting $C_{lc}(V_1)$ and $C_{lc}(V_2)$ with $C_\perp$ and $C_\parallel$, which correspond to $\varepsilon = \varepsilon_\perp$ and $\varepsilon = \varepsilon_\parallel$, respectively. When C_{lc} is equal to $C_\perp$ ($= 0.3$ pF) and $C_\parallel$ ($= 0.6$ pF), ΔV becomes 2.2 V and 1.7 V, respectively. Therefore dV is 500 mV, and V_{com} is set to -1.95 V in this case. Then, the maximum offset voltage δV is given by $\delta V = \pm 250$ mV, where

$$\delta V = \Delta V + V_{com}. \tag{2.35}$$

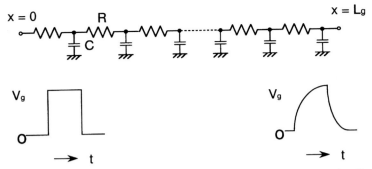

Fig. 2.11. Distributed-constant representation of the gate busline. The voltage pulse is delayed and distorted by the series resistance and parallel capacitance of the line as it propagates along the gate busline. R and C correspond to the resistance and capacitance per unit length, respectively

If there is no parallel (storage) capacitance, $\delta V = \pm 1.2\,\text{V}$ when $V_{\text{com}} = -4.3\,\text{V}$. Therefore, in this case the effect of the storage capacitance is to decrease δV by a factor of 4.8.

Although the display quality is improved considerably by the addition of a storage capacitance (C_{st} or C_{add}), there still remains a residual dc voltage of hundreds of mV as discussed above. Increasing the storage capacitance is the simplest way to lower ΔV, but this reduces the aperture ratio since the area taken up by the storage capacitance will increase. Another approach is to decrease C_{gs} by decreasing the overlap area between the source and the gate electrode of the TFT. This can be achieved with self-aligning technology [2.16, 17], in which gate electrode is used as a mask to define the amorphous silicon island and the edge of the passivation layer. However, if the a-Si:H TFT is fabricated with a perfect self-alignment process, i.e., with little overlap between the gate and source electrodes, it does not work well. A small gap between the source and gate electrodes impedes the injection of carriers, resulting in a high-resistance region that impairs the TFT operation; an overlap width of a few µm is necessary to achieve sufficient injection.

Gate Delay. TFT LCD panels like the one shown in Fig. 2.5 are operated on a line-at-a-time basis. Each gate busline is selected sequentially and a voltage pulse is applied to the selected gate busline. This pulse propagates down the busline which, we assume, is represented by the CR distributed-constant circuit shown in Fig. 2.11, where C and R are the capacitance and resistance per unit length. The rectangular gate pulse at the input is delayed and distorted as it travels along the gate busline. The gate delay is defined as the time between the onset of the pulse at the starting point of the busline and the time when the pulse height at the end of the busline reaches 90% of the pulse amplitude V_{g}.

The transmission of pulses through the distributed-constant RC line is governed by the following equations:

$$-\partial v/\partial x = Ri, \tag{2.36}$$
$$-\partial i/\partial x = C\partial v/\partial t, \tag{2.37}$$

where the voltage v and the current i are both functions of x and t. This is the so-called Thomson cable equation, which has been analyzed for the cases of an infinite line and finite-length lines terminated with open-circuits and short-circuits. The TFT LCD gate busline corresponds to the case of a finite-length RC line terminated with an open circuit. In this case, (2.36) and (2.37) are reduced to

$$\partial^2 v/\partial x^2 = CR\partial v/\partial t, \tag{2.38}$$
$$\partial^2 i/\partial x^2 = CR\partial i/\partial t. \tag{2.39}$$

The Laplace transform is a convenient tool for analyzing this problem. The solution of $v(x,t)$ and the corresponding $i(x,t)$ are given as follows:

$$\frac{v(x,t)}{V_g} = 1 - \frac{4}{\pi}\sum_{n=1}^{\infty}\frac{1}{2n-1}\exp\left(-\frac{(2n-1)^2\pi^2}{4RCL_g^2}t\right)\sin\left(\frac{(2n-1)\pi}{2L_g}x\right), \tag{2.40}$$

$$\frac{i(x,t)}{(2V_g/RL_g)} = \sum_{n=1}^{\infty}\exp\left(-\frac{(2n-1)^2\pi^2}{4RCL_g^2}t\right)\cos\left(\frac{(2n-1)\pi}{2L_g}x\right). \tag{2.41}$$

The gate delay is estimated by putting $x = L_g$ in (2.40), and is given by

$$\frac{v(L_g,t)}{V_g} = 1 + \frac{4}{\pi}\sum_{n=1}^{\infty}\frac{(-1)^n}{2n-1}\exp\left(-\frac{\pi^2(2n-1)^2}{4RCL_g^2}t\right). \tag{2.42}$$

When $t/RCL_g^2 > 1$, (2.42) is approximated as

$$\frac{v(L_g,t)}{V_g} \approx 1 - \frac{4}{\pi}\exp\left(-\frac{\pi^2 t}{4RCL_g^2}\right). \tag{2.43}$$

The gate delay t_d of TFT/LCD panels can be estimated from (2.43):

$$t_d = 1.03\,RCL_g^2, \tag{2.44}$$

where t_d is defined as the time at which the voltage reaches 90% of the pulse voltage. Therefore, t_d can be closely approximated by a simple expression of RCL_g^2. Another good approximation for t_d is $2\,RCL_g^2$, which corresponds to the time for 99% charging. And more practical expression for t_d (90%) is $R_p C_p n^2$ where R_p and C_p are the resistance and capacitance per pixel, respectively, and n is the number of pixel.

In (2.44), R is the resistance per unit length of the gate metallization. In small- and medium-sized panels, Cr is the most widely used metal. With

Cr's specific resistivity of 55 µΩ cm and SiN gate insulators with a dielectric constant of 6.9, t_d becomes as high as 20 µs. This is adequate for low- and medium-resolution panels in which the gate selection time is $(1/60) \times (1/480) = 34.7$ µs or longer. However, in high-resolution panels with over a thousand gate lines, the gate address time is 16.7 µs or less. In this case the gate metallization must be made with lower resistivity metal. Al gate metallization is suitable for this purpose. The resistivity of Al is 3–6 µΩ cm, and from (2.44) the value of t_d is 1–2 µs for the case of 10-in. (25 cm) TFT LCD panels. Since the charging time of a unit cell of liquid crystal is approximately 13 µs, as discussed in Sect. 2.3.1, t_d of 1–2 µs is more than reasonable.

The diagonal size is limited by the resistivity of the metals used for the gate line metallization. The lower the resistivity, the larger the panels that can be fabricated and the higher their resolution. The calculated results are shown in Figs. 2.12a and b [2.18]. Larger panels can be designed with C_{st}, since the dotwise capacitance is larger in C_{add} case than in C_{st} case due to a large overlap area of C_{add} between the gate busline and the pixel electrode. The maximum diagonal size can reach 30 in. or more for Al metallization. According to these results, a-Si:H TFTs are applicable to a wide range of TFT LCD panels. Figure 2.12 shows the results calculated with a rather conservative design rule, i.e., assuming a gate busline width of 10 µm. Different designs including wider gate buslines and larger W/L of a-Si:H TFTs allow panels larger than one meter (40 in.) to be constructed.

2.2.4 Scaling Theory of TFT LCD

As TFT LCDs get large in their sizes and high in their resolutions, the pixel sizes become small as shown in Fig. 2.1. When the pixel size gets small, the aperture ratio in each pixel becomes low if the design rule for fabricating TFT panels is the same. This urges us to modify the design rule. A decrease in linewidth makes possible obtaining small TFTs and narrow buslines. This in turn results in a large aperture ratio. Therefore, it is essential to reduce the minimum size for fabricating TFT panels. In the following section, the effects of shrunk design rule are described with a discussion of the performance improvements to be expected from the TFT LCD panels fabricated under this modified design rule.

TN Panels. In this section a scaling theory is developed for the TN cell where the panel size and resolution are kept the same. The effects of design rule shrinkage are discussed for this cell. The assumptions made in developing the scaling theory are constant field and constant voltage [2.19]. The first assumption is the same as for MOS LSIs [2.20]. The electrical field is kept constant before and after scaling in semiconductors, insulator films, and liquid crystals. The second assumption comes from the fact that the operational voltage is determined from the liquid-crystal-cell configuration. In the TN-

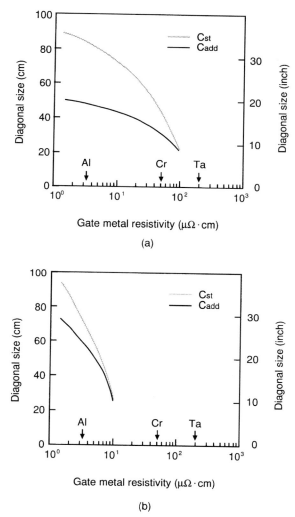

Fig. 2.12a,b. Relation between the gate metal resistivity and the diagonal size of the display panel: (**a**) the panel size of VGA displays where the number of pixels is 640×480; (**b**) the size of higher-resolution panels (1920×1120). When aluminum is used for the gate busline metallization, panels larger than 30 inch can be fabricated, even those of the high-resolution type

mode operation of liquid-crystal displays, the threshold voltage of a cell is given by

$$V_{\text{th}} = \pi \left\{ [k_{11} + (k_{33} - 2k_{22})/4] / \Delta\varepsilon \right\}^{1/2}, \tag{2.45}$$

where k_{ii} ($i = 1, 2,$ and 3) are the elastic constants of splay, twist, and bend stress, respectively, and $\Delta\varepsilon$ is the anisotropic dielectric permittivity (see

Table 2.1. Scaling of TFT/LCDs

	Physical parameter	Expression	Scaling factor
TFT	Channel length	L	1
	Channel width	W	$1/k$
	Insulator thickness	t_i	1
	Drain voltage	V_d	1
	Gate voltage	V_g	1
	Threshold voltage	V_t	1
	Gain factor, β_0	$\mu C_i W/L$	$1/k$
	Drain current, I_d	$\beta_0[(V_g - V_t)V_d - V_d^2/2]$	$1/k$
	Stray capacitance	C_{gs}	$1/k^2$
LC	Treshold voltage (TN), V_{th}	$\pi\{[k_{11} + (k_{33} - 2k_{22})/4]/\Delta\varepsilon\}^{1/2}$	1
	Cell gap	d	1
	Cell capacitance, C_{lc}	$\varepsilon A/d$	$\gtrsim 1$
	Cell resistance, R_{lc}	$\rho d/A$	$\lesssim 1$
Panel	Pixel capacitance, C_{px}	$C_{gs} + C_{st} + C_{lc}$	$1/k$
	Busline width	W_B	$1/k$
Panel characteristics	Gate delay, t_d	$C_p R_p N^2$	$1/k$
	Offset voltage, ΔV	$V_a C_{gs}/C_{px}$	$1/k$
	Time constant (charge), τ	$C_{px}/[\beta_0(V_g - V_t - V_d)]$	1
	Time constant (discharge), τ_d	$R_{lc} R_{off} C_{px}/(R_{lc} + R_{off})$	$2/(1+k)$
	Aperture ratio, γ	$(1 - \gamma_1)/(1 - \gamma_0)$	$< 1/k$

Sect. 2.4.2). Since the V_{th} is solely determined by the physical parameters of the liquid crystal, the scaling (reduction) of the threshold voltage is not very realistic. If the V_{th} is not scaled, then the relevant voltages are not scaled either. The TFT-related voltages are kept as they are. Then, TFT channel length L is not scaled, and the film thicknesses, including the a-Si:H and insulators, are not scaled either due to a constant-field assumption. However, TFT channel width W is scaled to $1/k$, where k (> 1) is the scaling factor. The busline width is also scaled to $1/k$.

Table 2.1 shows the physical parameters and their scaling factors. Stray capacitance, C_{gs}, is scaled to $1/k^2$ since the overlapping area of source and gate electrodes are scaled to $1/k^2$. Pixel capacitance, C_{px}, has to be considered separately from discussions described above. The C_{px} is the sum of the stray capacitance C_{gs} of the TFT, cell capacitance, C_{lc}, and storage capacitance C_{st}. Generally speaking, C_{px} is not a parameter to be scaled. Since C_{st} is large compared to other capacitances, however, C_{px} can be reduced by decreasing C_{st}. The C_{px} is reduced (scaled in a sense) by decreasing the overlapping area between the pixel electrode and the storage capacitance electrode. The C_{px}

is scaled to $1/k$ by putting the storage capacitance after scaling, $C'_{\text{st}}(>0)$, to satisfy

$$C'_{\text{st}} = \frac{C_{\text{st}}}{k} - \left(1 - \frac{1}{k}\right)C_{\text{lc}} + \frac{C_{\text{gs}}(k-1)}{k^2}. \tag{2.46}$$

Now we are ready to discuss the panel characteristics, and the results are also shown in Table 2.1. Most notable in this list is the aperture ratio improvement. The design rule change narrows the busline width, narrows the gap between the pixel electrode and the buslines, shrinks the TFT and reduces the storage capacitance area. All of these contribute to the improved aperture ratio. We estimate the aperture ratio, γ by assuming as follows: the pixel is square and each pixel is composed of three RGB subpixels which measures L_p by $L_\text{p}/3$, the busline widths are scaled to $1/k$, the TFT area is scaled to $1/k$ (W is scaled to $1/k$), and the C_{px} is scaled to $1/k$. From (2.46) the storage capacitance C'_{st} is seen to be less than C_{st}/k. Since there is no scaling for the gap insulator of the storage capacitance, the area A'_{st} is smaller than A_{st}/k, where A_{st} and A'_{st} are the areas of storage capacitors before and after scaling, respectively. Then, the aperture ratio before and after scaling, γ_0 and γ_1, are written as follows:

$$1 - \gamma_0 = [(4/3)L_\text{p}W_\text{B} + LW + A_{\text{st}}]/\left(L_\text{p}^2/3\right), \tag{2.47}$$

$$1 - \gamma_1 = [(4/3)L_\text{p}W_\text{B}/k + LW/k + A_{\text{st}}/k]/\left(L_\text{p}^2/3\right). \tag{2.48}$$

From (2.47) and (2.48), γ_1 is calculated as

$$\gamma_1 > 1 - (1-\gamma_0)/k. \tag{2.49}$$

Substitution of $\gamma_0 = 50\%$ and $k = 2$ into (2.49), gives a value larger than 75%. Accordingly, the improvement is substantial.

The dc offset voltage ΔV is also improved by scaling. Since the ΔV is written as

$$\Delta V = V_\text{a}C_{\text{gs}}/C_{\text{px}}; \tag{2.50}$$

it is scaled to $1/k$. The C_{gs} is scaled to $1/k^2$ and the pixel capacitance is scaled to $1/k$. The offset voltage characteristics are improved in spite of the reduced storage capacitance.

The other parameters in Table 2.1 are related to the signal charging and its retention characteristics. In order to fully charge the pixel capacitance to the signal voltage, the charging time constant should be kept as low as possible. On the other hand, the discharging time constant should be kept high. The charging time constant is given by

$$\tau = C_{\text{px}}/\left[\beta_0(V_\text{g} - V_\text{t} - V_\text{d})\right]. \tag{2.51}$$

Since both C_{px} and β_0 are scaled to $1/k$, the time constant τ for charging the pixel capacitance is not scaled.

After the pixel is charged to the signal voltage, the TFT is turned off till the next charging step takes place. Since both the resistance of liquid crystal cell, R_{lc} and the off-resistance of a-Si:H TFT, R_{off} are finite, the charged potential at the node decays to some extent. The decay constant or the time constant for discharging is written as

$$\tau_d = \frac{R_{lc} R_{off}}{R_{lc} + R_{off}} C_{px}. \tag{2.52}$$

The off-resistance of the TFT is scaled to k (channel width W is scaled to $1/k$) and the pixel capacitance is scaled to $1/k$. Therefore, the product of R_{off} and C_{px} remains the same. The effect of the scaling appears in the factor of $R_{lc}/(R_{lc} + R_{off})$. Putting the decay constant after scaling τ'_d, we obtain

$$\frac{\tau'_d}{\tau_d} = \frac{R_{lc} + R_{off}}{R_{lc} + kR_{off}}. \tag{2.53}$$

This ratio is reduced to

$$\tau'_d/\tau_d = 1 \quad (R_{lc} \gg R_{off}), \tag{2.54}$$

$$\tau'_d/\tau_d = 1/k \quad (R_{off} \gg R_{lc}). \tag{2.55}$$

When R_{lc} is equal to R_{off}, this ratio becomes

$$\tau'_d/\tau_d = 2/(1+k). \tag{2.56}$$

Since the decay constant becomes smaller after scaling, the retention characteristics is degraded to some extent. However, its effect is not so large. If $k = 2$ is substituted into (2.56), τ'_d/τ_d becomes $2/3$. If it is further assumed that the frame frequency is 60 Hz and the decay constant before scaling is 1 s, scaling can be estimated to be $\exp[-(1/60)/1]$ and $\exp[-(1/60)/(2/3)]$, respectively. These exponential factors are calculated to be 0.983 and 0.975, respectively. Therefore, the voltage decay difference is less than 1%, which is fairly reasonable and acceptable. Equation (2.54) corresponds to the best case where the decay constant is kept the same before and after scaling. There is no degradation in the retention characteristics. This implies that the resistivity of liquid crystal should be kept as high as possible.

To take a full usage of the scan time for charging, the gate delay should be minimized. The gate delay, t_d, discussed in the previous section is given by $R_p C_p n^2$, where C_p is the busline capacitance per pixel, R_p is the busline resistance, and n is the number of pixels in the horizontal direction of the panel. C_p and R_p are expressed as

$$C_p/3 = C_c + C_{gs} + C_{gd}, \tag{2.57}$$

and

$$R_p = \rho L_p/(W_B t_G), \tag{2.58}$$

where C_c is the capacitance of crossover between the gate busline and the data busline, ρ is the specific resistivity of the busline metallization, L_p is the horizontal length of the pixel, W_B is the width of the busline and t_G is the film thickness of the gate metallization.

Since we are discussing the case that the resolution of the panel remains the same, the effect of scaling on the gate delay can be summarized as $1/k^2$ multiplied by k. The $1/k^2$ comes from C_p, which is proportional to the overlapping area, and k comes from R_p which is inversely proportional to W_B. Because the other parameters are kept constant, the gate delay is scaled to $1/k$.

Larger, Higher-Resolution Panels. So far, the panel characteristics of TFT LCDs were discussed without changing the size or resolution of the panels. In this section, the characteristics for larger, higher-resolution panels are discussed. We assume here that the panel is widened from H to H' and its resolution is increased from n to n' (horizontal). Since we are talking about displays for PCs, such as for VGA (video graphics array) and XGA (extented graphics array) formats, we can reasonably assume that the pixels are square and consists of three striped RGB dots.

The offset voltage is given by (2.30). C_{gs} is scaled to $1/k^2$, as it was discussed for the simple scaling. The C_{px} needs to be estimated more carefully. In larger, higher-resolution panels, the pixel area is reduced by a ratio of r_p. This ratio is given as

$$r_p = (H'n/Hn')^2. \tag{2.59}$$

The general trend in TFT LCD panels is for this ratio to decrease as panel-size and resolution increase. We can reasonably assume that the C_{lc} is reduced in proportion to this ratio. However, if this effect is included in the C_{px} scaling, the discussion will be unnecessarily complicated. Therefore, we assume here that C_{px} is still scaled to $1/k$. This is equal to saying that the reduction of C_{lc} is compensated for by an adjustment of C_{st}, which is easily achieved by modifying the area of C_{st}. Thus, the offset voltage is scaled to $1/k$, since the C_{gs} is scaled to $1/k^2$ and the C_{px} is scaled to $1/k$. This is the same result as for simple scaling.

The same discussion as for the offset voltage applies to the charging time constant. The τ remains unchanged since C_{px} is reduced to $1/k$ and β_0 is scaled to $1/k$ in (2.9). The decay time constant, τ_d'' of larger, higher-resolution panels with scaling is estimated from (2.13). The resistance of liquid-crystal cell increases by a factor of $(1/r_p)$ due to cell-area shrinkage, the off-resistance of the TFT increases by k, and C_{px} is reduced by $1/k$, as before. Therefore, the ratio of τ_d'' to τ_d is written as

$$\frac{\tau_d''}{\tau_d} = \frac{R_{lc} + R_{off}}{R_{lc} + kr_p R_{off}}. \tag{2.60}$$

Table 2.2. Scaling of larger, higher resolution TFT/LCDs

	Physical parameter	Expression	Scaling factor
Panel charac- teristic	Gate delay, t_d	$C_p R_p n^2$	r_g
	Offset voltage, ΔV	$V_a C_{gs}/C_{px}$	$1/k$
	Time constant (charge), τ	$C_{px}/[\beta_0(V_g - V_t - V_d)]$	1
	Time constant (discharge), τ_d	$R_{lc} R_{off} C_{px}/(R_{lc} + R_{off})$	$2/(1 + kr_p)$

This is further reduced to

$$\tau_d''/\tau_d = 1 \quad (\text{for } R_{lc} \gg R_{off}), \tag{2.61}$$

$$\tau_d''/\tau_d = 1/(kr_p) \quad (\text{for } R_{off} \gg R_{lc}). \tag{2.62}$$

The decrease of decay time is rather relaxed as compared to the case of simple scaling. The ratio, r_p is calculated to be 0.639 for the case of panels enlarged from 10.4″ VGA ($H = 211$ mm, $n = 640$) to 13.3″ XGA ($H' = 270$ mm, $n' = 1024$). Then, the ratio τ_d''/τ_d becomes 0.78. This means that even the worst case estimate of (2.62) is better than the numerical example described before for the simple scaling. Therefore, the effect of the decreased decay constant on the retention characteristics is negligible.

The last item to be discussed is the gate delay. The t_d is modified for C_p' and R_p':

$$C_p' = C_p, \tag{2.63}$$

$$R_p' = (H'n/Hn')R_p. \tag{2.64}$$

The modified gate delay t_d' is thus given by

$$t_d' = r_g t_d, \tag{2.65}$$

where r_g is expressed as (scaling included)

$$r_g = (1/k)(H'n'/Hn). \tag{2.66}$$

The gate delay is longer in larger, higher-resolution panels by a factor of $(H'n'/Hn)$ and is reduced by a scaling factor of $(1/k)$. Numerical evaluation of r_g for the above case of 10.4″ VGA and 13.3″ XGA with $k = 2$ yields $r_g = 1.0$. The gate delay increases as the panel gets larger and its resolution becomes higher, and this increase is just canceled out when $k = 2$. These results are summarized in Table 2.2.

The time allocated for signal-voltage charging is proportional to the inverse of the number of gate busline, i. e., $1/m$. The maximum allocated time for VGA panels is 34.7 μs, but it becomes as short as 16.3 μs in SXGA panels (assuming a frame rate of 60 Hz). Because the charging-time constant remains the same after scaling, it becomes severe for higher-resolution panels to be fully charged.

Table 2.3. Scaling of TFT/LCDs by using unloaded-pixel scheme

	Physical parameter	Expression	Scaling factor
	Gate delay, t_d	$C_p R_p n^2$	r_g
Panel	Offset voltage, ΔV	$V_a C_{gs}/C_{px}$	k'/k^2
charac-	Time constant (charge), τ	$C_{px}/[\beta_0(V_g - V_t - V_d)]$	k/k'
teristic	Time constant (discharge), τ_d	$R_{lc} R_{off} C_{px}/(R_{lc} + R_{off})$	$2(k/k')/(1+kr_p)$

To improve this situation, it is effective to reduce the time constant τ by decreasing C_{px}. We already studied the case of scaling C_{px} to $1/k$. However, the storage capacitance can be decreased further until none remains in the pixels. This unloaded pixel scheme [2.19] is evaluated in this section. We assume the same panel as described above. The only difference is in the C_{st} design. Pixel capacitance C_{px}'' is given by

$$C_{px}'' = C_{gs}/k^2 + r_p C_{lc}, \tag{2.67}$$

and reduction factor k' of the C_{px} is defined as

$$C_{px}''/C_{px} = \left(C_{gs}/k^2 + r_p C_{lc}\right) / \left(C_{gs} + C_{lc} + C_{st}\right) = 1/k'. \tag{2.68}$$

The panel characteristics for the unloaded pixel scheme are summarized in Table 2.3 with this factor of k'. The decay constant τ_d''' is reduced to

$$\frac{\tau_d'''}{\tau_d} = \frac{k}{k'} \frac{R_{lc} + R_{off}}{R_{lc} + kr_p R_{off}}. \tag{2.69}$$

Further reduction yields

$$\tau_d'''/\tau_d = k/k' \quad \text{(for } R_{lc} \gg R_{off}\text{)}, \tag{2.70}$$
$$\tau_d'''/\tau_d = 1/k' r_p \quad \text{(for } R_{off} \gg R_{lc}\text{)}. \tag{2.71}$$

We consider next the offset voltage. The scaling factor for ΔV is k'/k^2. Substitution of $k = 2$ and $k' = 4.6$ as before, yields $k'/k^2 = 1.15$. The offset voltage thus increases 15% compared to the unscaled case, which is within a quite acceptable level. Last feature to be noted is a large aperture ratio of this scheme. Since there is no storage capacitance in the pixel, the improvement in the aperture ratio is substantial. Summarizing the discussion, the unloaded pixel scheme provides a useful tool to design large and high-resolution panels.

The main object of this unloaded pixel scheme is the reduction of charging time constant. This can be estimated from k/k'. Numerical evaluation yields $k' = 4.6$ for $C_{gs} = C_{lc}/3$, $C_{st} = 2C_{lc}$, $k = 2$, and $r_p = 0.639$. Then, the scaling factor of $\tau (= k/k')$ is calculated to be 0.43. This is a considerable and necessary improvement because the scan time decreases inversely proportional to m. Number of scan lines can be increased more than twice without losing the charging characteristics.

The discharging time constant, however, decreases resulting in a lower retention ratio. Numerical evaluation of the worst case of (2.71) yields $\tau_d'''/\tau_d = 0.34$. Though this value seems to be rather low, the decrease in the retention ratio is calculated to be only a few percent. Therefore, this disadvantage is almost completely covered up by an improved charging characteristics.

2.2.5 Fabrication of TFT Panels

The fabrication of a-Si:H TFTs is based on thin-film technology, i.e., thin-film deposition and thin-film pattern etching. This technology is an extended version of crystalline silicon LSI (large-scale integration) technology, although there are some differences between the two. In LSIs, the substrate is a silicon wafer on which a variety of processing steps are performed, such as surface finishing, rinsing, thin-film coating, thin-film deposition, thin-film growth, oxidation, etching, lithography, impurity diffusion, heat treatment, and ion implantation. Silicon is quite stable under extreme conditions which means that processes like oxidation, diffusion, and heat treatment can be carried out at temperatures higher than 1000°C ($T_m = 1420$°C).

However, the TFTs have to be processed on transparent glass substrates, so that the process temperature is limited to below, say, 400°C. Amorphous silicon based technology has an advantage over other technologies in that the processing can all be carried out below 350°C. The process steps include thin-film deposition by plasma CVD (chemical vapor deposition) and metal sputtering, thin-film coating, lithography, oxidation, etching, and rinsing.

The substrate size in TFT fabrication also differs from that of LSIs. LSI wafers are circular in shape and the most popular sizes are below ten in. in diameter. TFTs substrates, on the other hand, are rectangular in shape, and their diagonal sizes range from a few inches to a few feet. Substrates with diagonals of over 40 in. are now used in the production lines of TFT LCDs.

Such large substrates require large optical lithography systems. Both one-to-one projection and step-and-repeat aligners are used in the photolithographic systems for TFT fabrication. The step-and-repeat aligner features accurate alignment and fine patterning, while one-to-one projection aligner features high throughput. The typical design rule for 10-in. class TFT panels is 10 μm, which corresponds to the minimum width of the gate or data buslines. The design rule is now going to shrink from 10 μm to smaller sizes such as 7 μm and further down to 5 μm. The aligner technology has to be compatible with these changes. Generally speaking, the reduction of design rule improves the panel characteristics of TFT LCDs as discussed in the previous section. To achieve fine patterning, a positive photoresist is generally used, where the exposed part of the photoresist is etched away.

Since any dust in the clean room has the potential of making a panel defective, great care has to be taken to get rid of particles. The rinse process is important to produce panels without any point (line) defects. Almost one third of the total processing is devoted to rinsing the panel. The TFT panel

Table 2.4. Composition of 7059 glass (%)

SiO_2	Al_2O_3	Alkaline earth oxide	B_2O_3	As_2O_3
49	10	25	15	1

process is carried out on a glass substrate. Glass has suitable attributes for flat-panel displays, i.e., transparency, rigidity, and thermal stability. In addition to this, the TFT LCD substrates must also be flat, free from surface and internal defects and scratches, resistant to temperature cycles and a variety of chemical etchants, and must have low alkali content.

The most widely used and accepted substrate glass is Corning 7059 glass which is made by the fusion down draw machine [2.21]. The composition of 7059 glass is shown in Table 2.4. A stream of homogeneous molten glass is delivered into a tapered trough at the top of a refractory form called a fusion pipe. When the trough overflows, the glass flows down in sheets on both sides of the fusion pipe. Two glass sheets are manufactured at the same time, and the surfaces of these sheets do not contact any other surfaces during the forming process. In this configuration, glass with a width of approximately 40 in. can be manufactured.

The thermal stability of this glass, which is essentially a function of glass viscosity, is very good. The strain point of 7059 glass, defined as the temperature at which the viscosity is approximately $10^{14.5}$ P, is $\approx 593°C$, while its softening point (viscosity = $10^{7.6}$ P) is $\approx 844°C$. The maximum temperature used in the fabrication process is the strain point minus $25°C$, which leaves a considerable margin for the amorphous silicon, which is processed below $350°C$. In this temperature range, the thermal shrinkage and warp are small since they are related to glass viscosity. The chemical durability of 7059 glass can be estimated from the weight loss under severe etching conditions: when it is dipped in a 10% HF solution for 20 min at room temperature, it loses approximately $10\,mg/cm^2$. This is quite acceptable for the amorphous silicon based panel fabrication process.

The TFT process has to satisfy various requirements since it is a thin-film, low-temperature, large-area, and fine-pattern process. The cell gap (the thickness of the liquid crystal cells) is around $5\,\mu m$. A small gap like this is not compatible with a thick-film process and the TFT process has to be a thin-film process. The requirements for low-temperature result from the substrate being a transparent glass, and fine patterns are required for high resolution. An outline of the process flow is shown in Fig. 2.13, which corresponds to the TFT shown in Fig. 2.6a. The process is essentially the same as TFT fabrication. However, in addition to this, the following parts must be fabricated: gate and data buslines, the storage capacitor, and transparent pixel electrodes.

The transparent pixel electrodes are made by sputtering a target of indium-tin-oxide (ITO) and then patterning it. In a vacuum chamber, the target is sputtered in an atmosphere of argon under a pressure of mTorr. The ITO film is then patterned in an aqueous solution of nitric acid and chloric acid. The ITO film is compatible with thin-film processing and its transmittance is higher than 90% for the light in the visible spectrum and its resistivity is about $10^{-4}\,\Omega\,\mathrm{cm}$.

In Fig. 2.13, four mask steps are shown:

1. The gate electrode of the TFT and the gate busline,
2. the a-Si:H island,
3. the pixel electrode, and
4. the source and drain electrodes and the data busline.

In addition to this, the final passivation layer is deposited and pattern etched. Therefore, five is the standard number of mask steps in the panel process corresponding to the back-channel etched TFT. The minimum pattern size in these steps is determined from such factors as the panel size, the resolution, the aperture, and the productivity. The other design parameters to consider related to the process are the line width, the overlapping width, and separation between different patterns in different masks. These parameters are summed up as a design rule in a particular process line and are observed throughout the whole panel process.

2.3 Thin-Film Transistors

In 1961, Weimer [2.22, 23] proposed thin-film transistors (TFTs). The semiconductor layer and gate insulator of the original TFTs were made of cadmium sulfide and silicon monoxide, respectively. These thin-film layers were deposited in vacuum by evaporation. In spite of the complexity of fabricating TFTs in a vacuum chamber, they obtained both depletion-type and enhancement-type TFTs with good saturation-current characteristics. At first, they applied these TFTs to thin-film logic circuitry and built circuits such as flip-flops, AND gates, and NOR gates for computer applications. Soon after this, the application of TFTs to liquid crystal displays was also proposed and various materials have been studied for this purpose. Among them, CdSe TFTs have been intensively studied [2.24, 25], and TFT-addressed liquid-crystal panels of up to 6 inch × 6 inch in size have been fabricated. Such panels are typically made with Al or Au as metallization, CdSe as the semiconductor, and SiO_2 or Al_2O_3 as the insulator. Their applicability to alphanumeric and video displays has been demonstrated on a 180 × 180 pixel panel.

Hydrogenated-amorphous-silicon thin-film transistors (a-Si:H TFT) were first reported in 1979 [2.26]. The transfer characteristics of an a-Si:H TFT with a gate insulator of SiN are shown in Fig. 2.14 [2.27]. The transfer characteristics correspond to the behavior of the drain current, I_d, as a function

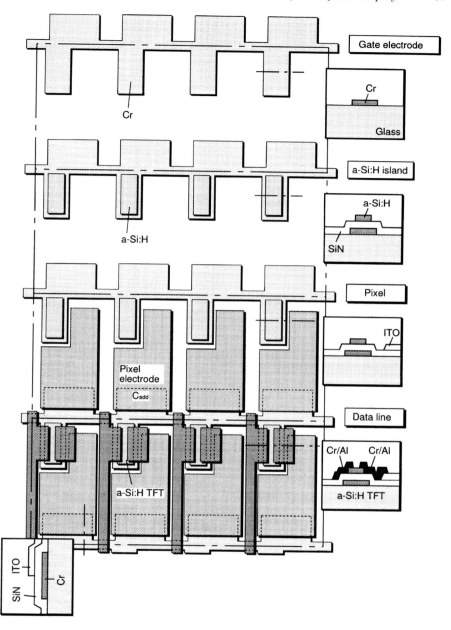

Fig. 2.13. Outline of the process steps involved in fabricating TFT panels. Each step corresponds to a photolithographic mask step. Since the final passivation process is omitted, five mask steps are necessary

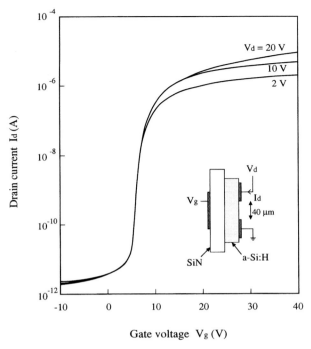

Fig. 2.14. Transfer characteristics of an a-Si:H TFT

of the gate voltage, V_g. As can be seen, the drain current has a wide dynamic range with an on-current in excess of 1 µA and an off-current below 10 pA. The output characteristics of this TFT (i.e., the relation between the drain current, I_d, and the drain voltage, V_d are shown in Fig. 2.15. Saturation of the drain current is clearly demonstrated. The threshold voltage, V_t, of the device was $\approx$ 5 V and the transconductance, g_m, was measured as $g_m = 0.3\,\mu\Omega^{-1}$ at $V_d = 10$ V and $V_g = 20$ V. The field effect mobility was estimated to be 0.4 cm^2/Vs from the transconductance. As a matter of fact, the transfer characteristics of a-Si:H TFTs was implied back in 1972 [2.28] by Spear and LeComber, who used field effect techniques to determine the distribution function of localized states in a-Si:H. Figure 3 of their paper clearly demonstrates the TFT transfer characteristics although the potential was on the order of 1000 V. The gate insulator was a glass substrate with a thickness of 250 µm.

Although liquid-crystal displays were the original application target of a-Si:H TFTs, it was quite a while before the development of an a-Si:H TFT liquid-crystal display was first reported [2.8]. This was a 96 × 96 mm^2 display containing 240 × 240 dots. Since then, a-Si:H TFTs have been widely used to drive active-matrix LCDs.

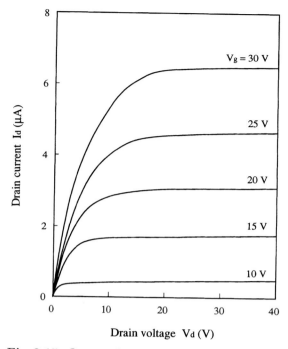

Fig. 2.15. Output characteristics of an a-Si:H TFT. The TFT channel is 500 μm wide and 40 μm long. The a-Si:H and SiN layers are both 0.5 μm thick

2.3.1 Hydrogenated Amorphous Silicon Thin-Film Transistors

The a-Si:H TFT is now widely recognized to be the most important and successful active device for use in active matrix liquid-crystal displays. Three-terminal devices (like TFTs) are more flexible in their operation and have fewer limitations than two-terminal devices (i.e., diodes). However, this is not the only reason for using a-Si:H TFTs as the active devices to drive liquid crystal cells. Hydrogenated amorphous silicon is a material with well-balanced features for electronic applications.

Some of the advantages of a-Si:H are as follows:

1. It can be deposited over a large area: a-Si:H is deposited by plasma chemical vapor deposition (p-CVD) in which silane gas (SiH_4) is decomposed in a plasma excited by rf power. A typical rf frequency is 13.56 MHz, although much lower frequencies (e.g., 100 Hz) can also be used. The deposition system is shown schematically in Fig. 2.16. The diode type reactor contains two electrodes, on one of which the substrates are placed. After the chamber is evacuated, silane gas, hydrogen diluting gas and a doping gas such as phosphine or diborane are introduced into the chamber. The rf power is then applied at a gas pressure of 0.1–1 Torr. The plasma is confined to the space between the two parallel electrodes. The rf power is distributed uniformly

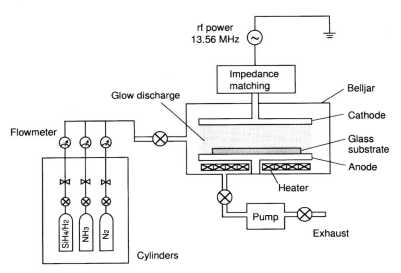

Fig. 2.16. A plasma CVD system for depositing thin films of hydrogenated amorphous silicon and silicon nitride. The silane (SiH$_4$) gas is dissociated in an rf chamber to deposit the a-Si:H film and the ammonia gas and nitrogen gas are added to deposit the SiN layer. Phosphine (PH$_3$) and diborane (B$_2$H$_5$) gas are introduced into the chamber to make n-type and p-type a-Si:H, respectively

over the surface of the electrodes, whose diameter can exceed one meter. Using this technique, films can be deposited with thickness variations of only a few percents or less over the entire surface area. This p-CVD process is also called plasma enhanced CVD (PECVD).

2. It can be deposited at low temperatures: Low temperature deposition is essential in the fabrication of active matrix liquid crystal displays because glass is used as the transparent substrate material. In the p-CVD system, the silane molecules are dissociated by an rf plasma. Consequently, the substrate temperature can be kept low in contrast to thermal CVD systems in which the silane gas is thermally dissociated. The deposition temperature must be low because good-quality a-Si:H films can only be formed if hydrogen is incorporated into the amorphous network; hydrogen will not be retained in the film if the substrate temperature is higher than 450°C.

3. A-Si:H is an amorphous material, so hetero-interfaces can be easily formed while maintaining good interface properties. It is possible to deposit amorphous silicon on various substrates such as insulators (including glass, oxides, and nitrides), metals, and semiconductors. Especially important is the interface between silicon nitride and a-Si:H. The interface properties between the a-Si:H and the SiN gate insulator play a critical role in TFT characteristics. Metals like Cr, Al, Ta, and Mo are used as the metallization material and ITO (indium tin-oxide) is also used as the conductive film. The a-Si:H forms

good interfaces with all these materials, making the processing of TFTs very flexible and allowing a variety of processing schemes to be devised and used in the process line.

4. Amorphous silicon is a hard scratch-resistant material (Vickers hardness HV = 1500–2000 kg/mm^2) and can be finely patterned with photolithographic technology. A minimum dimension for 10-in. diagonal panels of about 10 μm can be achieved without difficulty. The limit of the fine pattern lithographic process of a-Si:H will be well below the submicron level.

5. Hydrogenated amorphous silicon has high electrical resistivity in its undoped state and is highly conductive when doped. The high resistivity of a-Si:H matches the high resistivity of the liquid crystal, i. e., the off-state resistance of a-Si:H TFT is comparable with the resistance of the liquid crystal cell. On the other hand, the on-resistance of a-Si:H TFT is low enough to charge the liquid crystal cell capacitance.

6. Non-toxicity is another favorable feature of a-Si:H.

The inverted staggered-electrode structure is most widely used in TFT LCD fabrication. Figure 2.17a shows a cross-section of the back-channel-etched (BCE) TFT, and Fig. 2.17b shows that of a channel passivated (CHP) a-Si:H TFT: both of these devices have inverted staggered-electrode structure. As shown by the arrows in Fig. 2.17, electrons injected from the source electrode cross the intrinsic a-Si:H layer (i-layer), travel the channel formed at the interface between the gate insulator and the a-Si:H, cross the i-layer again, and reach the drain electrode. Since the channel thickness at the interface is estimated to be several tens of nanometers, the interface properties play a critical role in determining the TFT characteristics. In the back-channel-etched TFT, the channel length (L) is determined by the design rule or the minimum size (S) of fabrication ($L = S$). In the channel-passivated TFT, however, the channel length becomes large: $L = S + 2\Delta L$ where ΔL is the process margin. A process margin of 2 μm for a design rule of 10 μm yields $L = 14$ μm. When the drain voltage becomes high ($V_\mathrm{d} > V_\mathrm{g}$) in this TFT, electrons flow in two ways at the drain: the bottom and the top channel. The voltage applied to the drain electrode forms the top channel as shown in Fig. 2.17b. This effectively reduces the channel length ($S + \Delta L < L < S + 2\Delta L$).

2.3.2 TFT Characteristics

Current–Voltage Characteristics. The current-voltage characteristics of a TFT can be analyzed in essentially the same way as those of crystalline silicon MOSFETs [2.30]. The coplanar structure shown in Fig. 2.18 is used for the analysis. The TFT electrode configuration described in Sect. 2.3.1 differs from this structure in that the carriers cross the a-Si:H thin film before

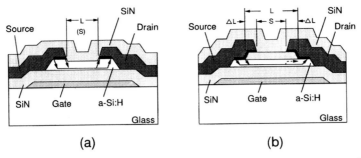

Fig. 2.17a,b. Cross sections of a-Si:H TFTs used in TFT LCD. Both of them are inverted staggered-electrode TFTs with (**a**) back-channel-etched and (**b**) channel-passivated configurations. The arrows correspond to electron channels

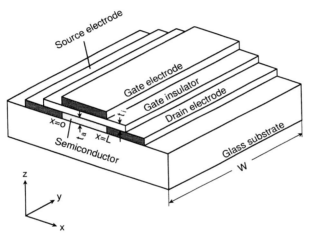

Fig. 2.18. The coplanar TFT structure used in the analysis. In a-Si:H TFTs, the staggered electrode configuration is generally adopted. Therefore, this analytical model is different from the actual device. However the current-voltage characteristics of such devices can be described well by the analytical results obtained using this model

reaching the channel. However, if the a-Si:H film is thin enough, the staggered-electrode configuration can be analyzed in terms of the coplanar TFT structure. The effects of a thicker film will be discussed later in Sect. 2.3.4. The coplanar structure is not common in a-Si:H TFTs, because the technologies essential for fabricating coplanar TFTs (such as impurity diffusion or ion implantation) are not well established in a-Si:H.

In order to formulate the V–I characteristics, the following assumptions are made: (1) the carrier mobility in the channel is constant, (2) the gate capacitance is constant and independent of the gate voltage, (3) the source and drain electrodes are ohmic contacts to the semiconductor, (4) the initial charge density in the semiconductor is n_0, and (5) the gradual channel approximation can be applied. The last condition means that the longitudinal field (E_z) is greater than the transverse field (E_x) in the channel.

The application of a gate voltage V_g induces charge density $\Delta n(x)$ in the channel region. This is given by

$$e\Delta n(x) = (C_i/t_a)[V_g - V(x)], \qquad (2.72)$$

where C_i is the gate capacitance per unit area ($= \varepsilon_i/t_i$), t_a is the a-Si:H thickness, t_i is the gate insulator thickness, and $V(x)$ is the drain voltage at distance x from the source. If the thickness, t_a, is assumed to be sufficiently small, the drain current I_d is given by

$$\begin{aligned} I_d &= t_a W [\sigma_0 + \Delta\sigma(x)] E_x \\ &= t_a W e\mu_n [n_0 + \Delta n(x)] E_x, \end{aligned} \qquad (2.73)$$

where σ_0 and $\Delta\sigma(x)$ are the initial conductivity and the incremental conductivity due to n_0 and $\Delta n(x)$, respectively. From (2.72) and (2.73), I_d is given by

$$I_d = W\mu_n C_i [et_a n_0/C_i + V_g - V(x)] \, dV(x)/dx, \qquad (2.74)$$

which reduces to

$$I_d \int_0^L dx = W\mu_n C_i \int_0^{V_d} [et_a n_0/C_i + V_g - V(x)] \, dV(x). \qquad (2.75)$$

Then, the drain current is given by

$$I_d = \mu_n C_i (W/L) \left[(V_g - V_t) V_d - (V_d^2/2) \right], \qquad (2.76)$$

where $V_t \equiv -et_a n_0/C_i$. The threshold voltage V_t depends on the initial charge density, n_0.

Equation (2.76) is valid for a voltage range of $0 < V_d < V_g - V_t$. Beyond this range, the current is assumed to be constant as in insulated-gate field-effect-transistors. Low V_d values correspond to the region of linear output characteristics where the drain conductance, g_d, and the transconductance, g_m, are given by

$$g_d = \partial I_d / \partial V_d \big|_{V_d \to 0} = \mu_n C_i (W/L)(V_g - V_t), \qquad (2.77)$$
$$g_m = \partial I_d / \partial V_g = \mu_n C_i (W/L) V_d. \qquad (2.78)$$

The drain conductance is a linear function of V_g, and the transconductance is proportional to V_d. Saturation of the drain current occurs when $\partial I_d / \partial V_d = 0$, due to pinch-off of the conducting channel in the neighborhood of the drain. In this case, the saturation current, I_{dsat}, is given by (square-law relationship)

$$I_{dsat} = (1/2)\mu_n C_i (W/L)(V_g - V_t)^2 \qquad (2.79)$$

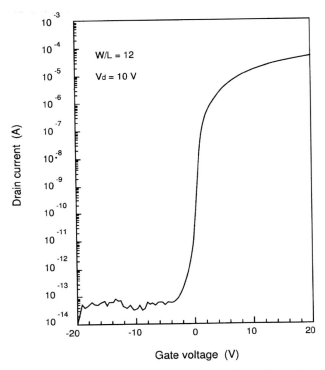

Fig. 2.19. Transfer characteristics of a channel-passivated a-Si:H TFT. The estimated mobility is $1.0\,\text{cm}^2/\text{Vs}$ at a drain voltage of 10 V. The threshold voltage is 1 V

for $V_d < V_g - V_t$. The transconductance in the saturation region is given by

$$g_m = \mu_n C_i (W/L)(V_g - V_t) = 2\mu_n C_i (W/L) I_{dsat}. \tag{2.80}$$

The high-frequency performance of a-Si:H TFTs can be estimated from the gain-bandwidth product, which is equivalent to the maximum operating frequency defined for crystalline silicon devices

$$f_m = g_m/(2\pi C_i WL). \tag{2.81}$$

Substitution of (2.78) into (2.81) yields

$$f_m = \mu_n V_d/(2\pi L^2) \tag{2.82}$$

for $V_d < V_g - V_t$, and

$$f_m = \mu_n (V_g - V_t)/2\pi L^2 \tag{2.83}$$

at the saturation region ($V_d > V_g - V_t$).

The experimental transfer and output characteristics of a-Si:H TFTs were shown in Figs. 2.14, 15. The mobility μ_n of the a-Si:H TFT shown in Fig. 2.14

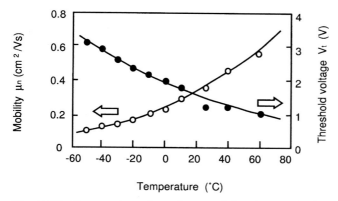

Fig. 2.20. Temperature dependences of the electron drift mobility and the threshold voltage of the a-Si:H TFT

was $\mu_n = 0.4\,\text{cm}^2/\text{Vs}$, but a higher μ_n of about $1.0\,\text{cm}^2/\text{Vs}$ is obtained in TFTs as shown in Fig. 2.19. The transfer characteristics of Fig. 2.19 feature low off-current, low threshold voltage, steep subthreshold slope, and wide dynamic range, i.e., a high on-current and low off-current. The threshold voltage of the TFT shown here is about 1 V, and there is virtually no p-type conduction at negative gate bias voltage, because of the blocking contact at the source and the low hole mobility of a-Si:H. The subthreshold slope is 0.3 V/decade, and the on/off current ratio of I_d is larger than 10^8.

Two important parameters of TFTs are the carrier mobility μ_n and the threshold voltage V_t. In a-Si:H TFTs, the main carriers are electrons whose room-temperature mobility is typically 0.3–0.6 cm^2/Vs. Although higher mobility values approaching $\mu_n = 1\,\text{cm}^2/\text{Vs}$ have been reported, these have only been achieved in carefully prepared samples. The temperature dependence of the mobility and the threshold voltage is shown in Fig. 2.20. The dependence of the mobility is given by

$$\mu_n \simeq \mu_0 (N_C kT/N_T)\exp(-E_a/kT), \tag{2.84}$$

where μ_0 is the extended state electron mobility, N_C is the density of states at the mobility edge, N_T is the effective density of band tail states, and E_a is the activation energy, which reflects the tail state distribution of the a-Si:H. The activation energy in Fig. 2.20 was calculated to be 0.13 eV, and the values of E_a from 0.07 eV [2.31] to 0.16 eV [2.32] have been reported.

Capacitance and Photoconductive Effects. As described in Sect. 2.3.2, the stray capacitance of the TFT induces a dc component in the ac voltage applied to the node (pixel electrode) of a TFT LCD. Since the node potential is alternating, the stray capacitance between the source and the gate, C_{gs}, and that between the drain and the gate, C_{gd}, in the static TFT operation both become important.

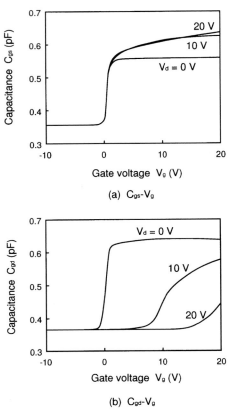

Fig. 2.21a,b. Dependence of stray capacitances, C_{gs} (**a**) and C_{gd} (**b**), on the gate voltage and the drain voltage. The sample is a back-channel-etched a-Si:H TFT

In the simple model, the stray capacitance is determined by the overlap of the gate electrode and the source (or drain) electrode. Therefore, the area used to calculate the capacitance is given by the channel width multiplied by the overlap length. However, the effective overlap length is longer than the geometrical length due to the fringe effect which has been shown to correspond to an extra length of about 1 µm by extrapolating the linear relation between C_{gs} and the overlap length. Moreover, the stray capacitance depends on the gate voltage. When the gate voltage is low, the contribution from the intrinsic amorphous silicon layer must be added to that of the gate insulator. When the gate voltage is turned on, however, an accumulation layer is formed at the interface between a-Si:H and SiN. This reduces the effective thickness of the stray capacitance to the thickness of the SiN layer in its limit, resulting in the increase of capacitance. Figure 2.21 shows how the stray capacitance depends on the gate and drain voltages [2.33]. The dependence of C_{gs} (Fig. 2.21a) shows a rather simple behavior, while that of

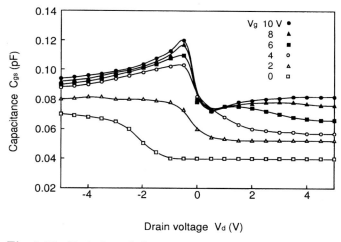

Fig. 2.22. Variation of the parasitic capacitance, C_{gs}, as a function of the drain voltage. The sample is a back-channel-etched TFT with $W/L = 72\,\mu\text{m}/8\,\mu\text{m}$. The overlap length between the gate and the source is $5\,\mu\text{m}$, the a-Si:H thickness is 300 nm, and the threshold voltage is 4.5 V

C_{gd} depends on both V_g and V_d, as shown in Fig. 2.21b. In a TFT LCD panel, we have to consider both of these cases. When the data voltage is positive and the potential of the node to be charged is negative, the stray capacitance behaves as shown in Fig. 2.21a. In the reverse case where the data voltage is negative and the node potential is positive, the stray capacitance behaves as shown in Fig. 2.21b.

The variation of capacitance as a function of the drain voltage is shown in Fig. 2.22 [2.34], and the frequency dependence of C_{gs} is shown in Fig. 2.23 [2.34]. Since the voltage offset is caused by a transient phenomenon, its frequency dependence is important and adds further complexity to this effect.

The photoconductivity of a-Si:H has been widely studied for applications such as solar cells [2.35], photosensors [2.36], and electronic copying machines [2.37]. Figure 2.24 shows the spectral response of a-Si:H pin diodes [2.36]. The spectral response of a-Si:H covers the entire visible spectrum. In TFT LCD applications, however, the photoconductivity of a-Si:H is not at all desirable. Since TFT LCDs are usually operated under strong illumination, this effect must be minimized for maintaining normal TFT operation. In LCD modules, illumination comes both from the bottom and the top. The back light illumination is more intense than the ambient illumination which comes from the top. Therefore, the a-Si:H island is designed to be smaller than the gate electrode as shown in Fig. 2.17. Since the gate electrode blocks an illumination from the backlight, the TFTs do not suffer from leaky photoconductive current.

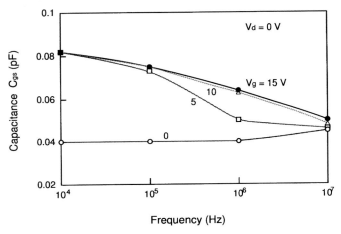

Fig. 2.23. Dependence of C_{gs} on the frequency of ac excitation. The TFT is the same as that of Fig. 2.22

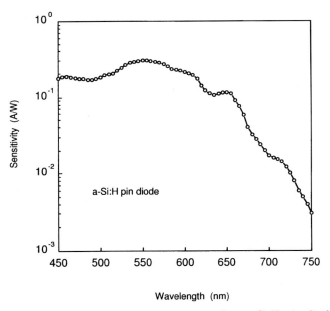

Fig. 2.24. Photoconductive response of an a-Si:H pin diode illuminated from the p$^+$ a-Si:H layer covered with a transparent electrode of ITO (indium tin oxide). The thickness of the p, i, and n layers are 20, 550, and 30 nm, respectively

The photocurrent $I_{\rm ph}$ is given by

$$I_{\rm ph} \propto 1 - \exp(-\alpha_a t_a), \tag{2.85}$$

where α_a is the absorption coefficient of a-Si:H and t_a is the thickness of the film. Naturally, a thinner a-Si:H film is preferable in order to reduce the photocurrent in the reverse biased TFT. The relation between the a-Si:H thickness and the photocurrent or the drain current of the TFT due to photocarriers was measured for the case where the a-Si:H layer is illuminated directly from the top of the TFT. A nonlinear dependence on film thickness is observed, and this dependence is given by

$$I_{\mathrm{ph}} \propto t_a^{3.3}. \tag{2.86}$$

Since a linear dependence is expected from (2.85), other factors such as the hole blocking contact or n$^+$ a-Si:H layer must be taken into account for explaining this discrepancy.

2.3.3 Threshold Voltage Shift

When the voltage is applied to the gate electrode of a-Si:H TFTs, instability is observed in their transfer characteristics. As shown in Fig. 2.25, this instability manifests itself as a parallel shift of the I–V characteristics to the right when the gate bias voltage is positive and to the left when the gate bias is negative. This is the threshold voltage (V_t) shift. The V_t shift dependence on the gate voltage can be represented by $\pm|V_g|^\beta$, where β varies from 1 to 4 [2.38]. ΔV_t depends strongly on the gate bias, while it varies only slightly with the drain voltage. Therefore, the V_t shift data is usually obtained under stress conditions in which the source and drain electrodes were grounded. In obtaining these experimental data, the same TFT can be used repeatedly. The annealing process is effective in recovering the initial I–V characteristics. After heat treatment at 200°C for 30–60 min, the TFT transfer characteristics resume their initial performances.

The experimental formulation of ΔV_t as a function of temperature T, time t, and gate voltage V_g, is given by:

$$\Delta V_t = A\,|V_g|^\beta\,t^\gamma \exp(-E_a/kT), \tag{2.87}$$

where E_a is the activation energy, and A, β, and γ are constants. Figure 2.26 shows the dependence of ΔV_t on the stress time. The constant β, depends more on the bias voltage polarity than the other constants. For a positive bias, β is in the range of 1–2, and for a negative bias, β becomes rather high ranging from 3 to 4. The constant γ varies from 0.3 to 0.4 depending on the device parameters and the bias polarity. The activation energy E_a is in the range of 0.2–0.3 eV [2.31, 38].

The constants in (2.87) depend on the deposition conditions of the a-Si:H and gate insulator. The gas mixing ratio, the rf power, the gas pressure, and the substrate temperature are some of the conditions to be taken into account. The effects of these conditions on ΔV_t are mixed together and there are no universally applicable guidelines for determining the deposition conditions.

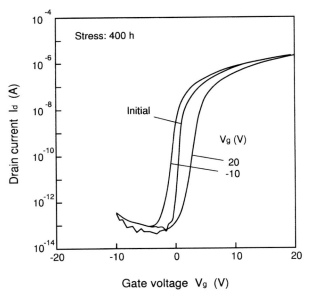

Fig. 2.25. Under the stress voltage application to the gate electrode of an a-Si:H TFT, the transfer characteristics move towards positive and negative directions depending on the polarity of the stress voltage

The relation between the SiN deposition conditions and ΔV_t has been studied since the internal stress of SiN film is high ($\approx 10^9$ dyn/cm^2) and the stress can be both tensile and compressive. The stress-free condition does not necessarily correspond to the lowest V_t shift. The stoichiometric composition (Si$_3$N$_4$) and a substrate temperature of about 300°C are preferred.

The V_t shift mechanism has been discussed by Powell [2.31], who implied from a simple analogy to conventional MOS FETs that the charge trapping in the SiN gate insulator causes this shift. This is supported by the fact that TFTs with silicon-rich SiN are less stable than those with stoichiometric SiN. Another proposed mechanism is that ΔV_t arises due to the formation of bias-induced deep trap levels in the amorphous silicon film. Since the channel is formed in the vicinity of the a-Si:H/SiN interface, these trap levels may also be associated with the interface. In any case, both mechanisms are believed to play roles in the appearance of ΔV_t. At higher gate bias, the ΔV_t of a-Si:H TFTs is due to charge trapping in SiN, mainly through tunneling. The dominant mechanism at lower gate bias, on the other hand, is the creation of states or the breaking of bonds in the a-Si:H film at or near the interface.

The V_t shift has been studied by illuminating a sample with an external light [2.39]. The ΔV_t increases with increasing illumination and this reminds

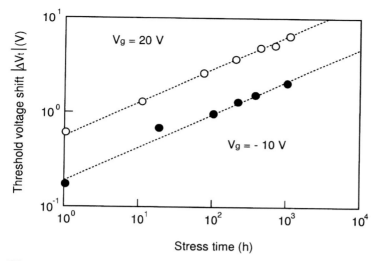

Fig. 2.26. Threshold voltage shift as a function of the stress time. Apart from their absolute values, the dependence is almost the same for both positive and negative voltage stress: $\Delta V_t \propto t^{0.33}$ for $V_g = 20\,\text{V}$ and $\Delta V_t \propto t^{0.34}$ for $V_g = -10\,\text{V}$

us of the defect creation kinetics in bulk a-Si:H or the Staebler-Wronski effect [2.40]. The dependence of the defect creation on the light flux is given by [2.41]

$$N_D \propto G^{0.6} t^{0.33}, \tag{2.88}$$

where N_D is the defect density, G is the illumination intensity, and t is time. The dependence of defect creation on time in (2.88) is similar to the time dependence of ΔV_t shown in Fig. 2.26. However, the carriers generated by the illumination are not directly reflected in ΔV_t, and only a small V_t shift is observed when there is no applied voltage. The photoexcited carriers are distributed in the whole a-Si:H layer and charge accumulation at the a-Si:H/SiN interface does not occur without a bias voltage. When the gate is biased, the photocarriers accumulate at the interface resulting in a larger ΔV_t in the illuminated TFT than in a TFT that is not illuminated.

A large V_t shift observed in discrete a-Si:H TFTs could cause an abnormal operation of TFT panels. However, the TFT LCD panels work normally even after an elongated period of years. This apparent discrepancy is not fully understood, but some explanatory remarks are possible. The TFTs in a LCD panel are operated under stress conditions of both positive and negative gate biases. Positive voltages are applied to the gate electrode in a pulsed mode. During the time from one pulse to another, the gate is biased to negative voltages. Therefore, it is expected that these positive and negative shifts tend to cancel each other out, partially at least. Another factor to be considered is the dependence of V_t shift on the frequency of the voltage pulse. Figure

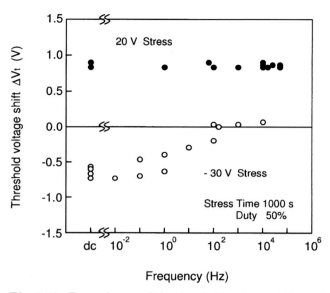

Fig. 2.27. Dependences of the threshold voltage shift on the frequency of the voltage pulse

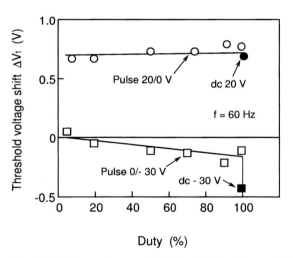

Fig. 2.28. Dependences of the threshold voltage shift on the duty ratio of the pulsed voltage. The total stress times were kept constant for all the data points

2.27 shows this dependence. The V_t shift changes little with frequency when the TFT is operated with positive bias pulses.

However, when the bias voltage is negative, ΔV_t decreases as the frequency of the voltage pulse increases. Moreover, ΔV_t has been found to depend on the duty ratio of the voltage pulse as shown in Fig. 2.28. Although the total

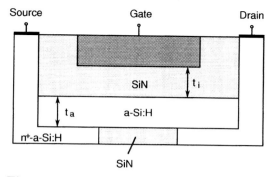

Fig. 2.29. Device structure used in the computer simulation

stress time is kept constant in the experiment, ΔV_t is smaller in pulsed mode than in the dc case [2.42]. It is worth noting that there is a jump in ΔV_t at negative bias as the duty ratio changes from 99 to 100%. These results can be explained by an equivalent circuit model. When the gate voltage is applied to the TFT in a pulsed mode or ac mode, the voltage is divided by the RC circuit of the SiN layer and that of the a-Si:H layer. Consequently, the effective voltage applied to the gate insulator or the electric field at the SiN/a-Si:H interface is dependent on the frequency of the ac pulse.

As described above, TFTs behave in a complex manner when they are operated in pulsed mode. Moreover, instability in TFT operation has been studied over a time span of from 10^3 s to one month (2.6×10^6 s). Then, it is difficult to exactly estimate the V_t shift over a period of years by extrapolating the data of one month. However, it is possible to simulate the TFTs operating on a panel. In the experiment simulating a VGA panel operating at a frame rate of 60 Hz, the V_t shift was shown to saturate after a period of 10^4 s. This indicates that the real lifetime of the TFTs operated in the panel is longer than that predicted from the dc data.

2.3.4 Simulation of TFT Behavior

The electrical characteristics of a-Si:H TFTs can be well represented by the analytical model described in Sect. 2.3.2. However, in order to simulate TFT behaviors more precisely, it is necessary to include the effect of electrons trapped in localized states in the energy gap of a-Si:H and the effect of carriers crossing the intrinsic amorphous layer of a-Si:H in the staggered electrode TFT structure. These effects were simulated in a two-dimensional a-Si:H TFT simulator [2.43] based on a general-purpose three-dimensional crystalline silicon device simulator.

The device structure used in the simulation is shown in Fig. 2.29. The total electron density, n, in the a-Si:H region consists of the electron density n_c in the conduction band and the trapped electron density n_T ($n = n_c +$

n_T). The effect of holes is neglected because the n$^+$ a-Si:H region blocks hole injection and the hole mobility is low. The energy distribution of the electrons is assumed to be given by the quasi-thermal equilibrium condition caused by frequent electron transitions between the extended states and the localized states in the gap. The quasi-Fermi energy is given by the Boltzman distribution function as

$$n_c = N_c kT \exp\left[(-E_c + E_F)/kT\right], \tag{2.89}$$

where N_c is the effective state density at the conduction band edge (mobility edge), k is the Boltzman constant, T is the absolute temperature, E_c is the conduction band edge energy, and E_F is the Fermi energy.

The trapped electron density n_T is given by the Fermi-Dirac distribution as

$$n_T = \int_{E_V}^{E_C} N(E) dE / \{1 + \exp\left[(E - E_F)/kT\right]\}, \tag{2.90}$$

where E_V is the valence band edge energy, and $N(E)$ is the localized state density in the energy gap. From (2.89) and (2.90), n_C and n_T are functions of E_F, and therefore n_T can be regarded as a function of n_C,

$$n_T = f(n_c). \tag{2.91}$$

In the a-Si:H region, the Poisson equation is given by

$$\varepsilon_s \Delta \Psi = e(n_c + n_T - N_D), \tag{2.92}$$

where Ψ is the electrostatic potential, ε_S is the permittivity of a-Si:H, e is the electronic charge, and N_D is the effective donor density. The current continuity equation is given by

$$\nabla J_n = 0 \tag{2.93}$$

$$J_n = \mu_n kT \exp\left(e\Psi/kT\right) \nabla \left[n_c \exp(-e\Psi/kT)\right], \tag{2.94}$$

where J_n is the electron current density and μ_n is the electron mobility.

The potential in the insulator is described by the Laplace equation, and Gauss's law holds for the internal interface boundary. The Dirichlet condition is applicable to the surface electrodes assuming thermal equilibrium and charge neutrality, and the Neumann condition is applicable to the other outer surfaces. Equations (2.92–94) are the basic equations for the steady-state analysis of a-Si:H TFTs. The finite difference method was used to convert these equations to discrete form, and Newton's iterative method was used to solve them.

The sample was assumed to have an a-Si:H thickness t_a, of 0.3 µm, gate insulator (SiN) thickness t_i, of 0.3 µm, gate length of 20 µm and channel length (source-drain spacing) of 13 µm. The overlap between the gate and the source electrode (between the gate and the drain electrode) is 3.5 µm. The effective donor density N_d of the n$^+$ layer was assumed to be 10^{21} cm^{-3}, which

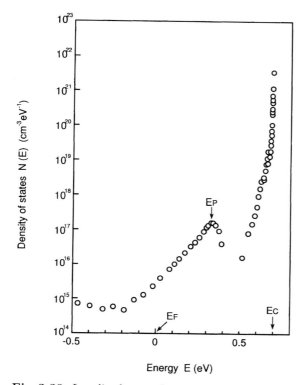

Fig. 2.30. Localized state density distribution of a-Si:H obtained from the quasi-static C–V measurements (voltage scan rate of 0.02 V/s) of an a-Si:H/SiN/metal capacitor. E_c and E_p denote the conduction band edge and the subpeak energy of the gap states, with the origin of the energy at the equilibrium Fermi energy of a-Si:H

is high enough to avoid the effect of series resistance. The specific dielectric constants of a-Si:H and SiN were assumed to be 11.7 and 7.0, respectively.

The localized state density distribution $N(E)$ obtained from C–V measurement of an a-Si:H/SiN/metal capacitor is shown in Fig. 2.30 (flat-band voltage assumed to be $V_{\text{FB}} = -2$ V). This function is expressed as a linear combination of exponential functions as follows:

$$N(E) = N_c \exp\left[(E - E_C)/D\right] + N_P \exp\left[-(E - E_P)/D_1\right] \qquad (2.95)$$

for $E > E_p$, and

$$N(E) = N_P \exp\left[-(E - E_P)/D_2\right] \qquad (2.96)$$

for $E < E_p$, where E_C is the conduction band edge energy, E_P is the subpeak energy shown in Fig. 2.31, D, D_1, and D_2 are the characteristic energies, and N_c and N_P are the characteristic densities of states. Data matching gives us

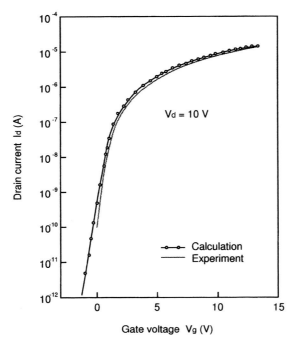

Fig. 2.31. Comparison between the calculated and experimental transfer characteristics of an a-Si:H TFT. The channel length and width are 13 μm and 455 μm, respectively

the values of these parameters: $D = 24.7\,\text{meV}$, $D_1 = 50\,\text{meV}$, $D_2 = 56\,\text{meV}$, $N_c = 1.7 \times 10^{21}\,\text{cm}^{-3}\text{eV}^{-1}$, $N_P = 1.9 \times 10^{17}\,\text{cm}^{-3}\,\text{eV}^{-1}$, and $E_C - E_P = 0.46\,\text{eV}$. The detailed structure of the state density at energies far below E_P peak is disregarded because its effects on the device characteristics are negligible.

Calculated transfer characteristics for the values listed above are compared with the experimental results in Fig. 2.31. The agreement is satisfactory over five orders of magnitude. The electron mobility above the mobility edge was assumed to be $7\,\text{cm}^2/\text{Vs}$. The calculated and experimental output characteristics are shown in Fig. 2.32. Again, the agreement is good enough to simulate even the slight crowding characteristics observed at low drain voltages. The crowding characteristics become more prominent as the a-Si:H layer gets thicker due to the increase in series resistance. The simulated dependence of the threshold voltage V_t, on temperature also agrees with the experimental results as shown in Fig. 2.33. The threshold voltage is defined as the extrapolation of the $I_d^{1/2}$ vs. V_g characteristics.

This simulator provides further information on the a-Si:H TFT characteristics. One example is the C–V characteristics. The capacitive behavior

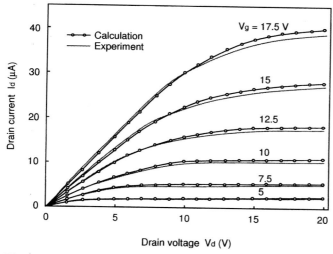

Fig. 2.32. Comparison between the calculated and experimental output characteristics. The effects of a thick (0.3 μm) a-Si:H layer appear as the slight crowding characteristics at low drain voltages. When the output characteristics are calculated with an a-Si:H thickness of 0.4 μm, the crowding characteristics become more prominent representing well the experimental results

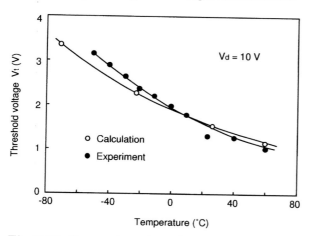

Fig. 2.33. Temperature dependence of the threshold voltage of an a-Si:H TFT. Agreement between the calculated and experimental results is satisfactory

between the gate and source is different from that between the gate and drain as shown in Fig. 2.21. In order to simulate the gate capacitance, the gate charges Q_g are calculated as follows (Gauss's law):

$$Q_g = \varepsilon_i \iint E_n d_s, \tag{2.97}$$

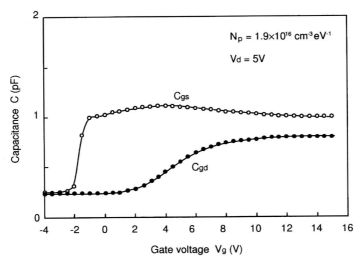

Fig. 2.34. Calculated results of gate-source and gate-drain capacitances as a function of the gate voltage

where E_n is the normal component of the electric field on the gate electrode surface and integration is carried out over the gate electrode surface. An increment of Q_g (i.e., ΔQ_g) divided by an increment of ΔV_d (ΔV_S) gives the drain-gate (source-gate) capacitance. The results are shown in Fig. 2.34. In this calculation, a subpeak density of one tenth the measured value (2.9×10^{16} cm^{-3} eV^{-1}) was substituted, and the drain voltage used in the calculation is 5 V. The experimental behaviors of Fig. 2.21 are seen to be well represented.

The difference between a-Si:H TFTs and MOS FETs is that the carriers (electrons) cross the semiconductor layer at the source and drain electrodes. The simulation results show that the current flow is well confined within the "source path" and the "drain path" (Fig. 2.35) which are the paths of carriers crossing the a-Si:H layer, and that the potential distribution in these paths closely resembles that of the space-charge-limited currents observed in vacuum tubes or in high-resistance semiconductors.

The space-charge-limited current is derived as follows. We suppose that a voltage V, is applied across the thickness t_a, of an a-Si:H layer. Then, the electric field E, across the a-Si:H layer is described by Poisson's equation:

$$\varepsilon_s dE/dy = -en_T = -er_1 n_c, \tag{2.98}$$

where the space charge was approximated by trapped electrons and r_1 is the proportionality factor of n_T/n_C. The current density J is given as:

$$J = -en_c \mu_n E, \tag{2.99}$$

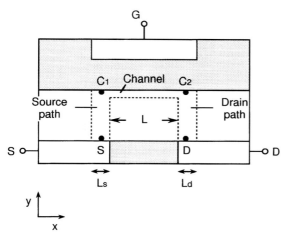

Fig. 2.35. Cross section of an a-Si:H TFT used in the calculation of the analytical model in which the source path and drain pth are defined as indicated

neglecting the diffusion current. Considering the current density to be constant with respect to y, we obtain

$$dE/dy = r_1 J/(\varepsilon_s \mu_n E). \tag{2.100}$$

Integrating (2.100), we get

$$J = 9\varepsilon_s \mu_n/(8r_1 t_a^3) V^2. \tag{2.101}$$

The currents in the source path and in the drain path are then given by

$$I = 9\varepsilon_s \mu_n W L_s/(8r_1 t_a^3) V_{C1}^2, \tag{2.102}$$

$$I = 9\varepsilon_s \mu_n W L_d/(8r_1 t_a^3)(V_d - V_{C2})^2, \tag{2.103}$$

where W is the channel width, and L_s and L_d are the effective overlap lengths in the source path and the drain path, respectively. V_{C1} and V_{C2} are the channel potentials at the opposite ends of the a-Si:H layer with respect to the source and the drain electrode, respectively.

By using the gradual channel approximation, the channel current can be expressed as

$$I = \mu_n \varepsilon_i/(r_0 t_i L) [V_g - V_t - (V_{C1} + V_{C2})/2] (V_{C2} - V_{C1}), \tag{2.104}$$

where ε_i is the permittivity of SiN, V_t is the threshold voltage, L is the channel length, and r_0 is the approximate ratio of n_T/n_c in the channel, which is different from that in the source and drain paths, that is, r_1. From (2.102–104), and assuming $L_s = L_d \, (= L_{ov})$, the following equation for the drain current is obtained:

$$I_d = \mu_n \varepsilon_i W/(r_0 t_i L) [V_g - V_t - (V_d/2)] \left(\sqrt{V_d + V_0} - \sqrt{V_0}\right)^2, \tag{2.105}$$

where

$$V_0 = \frac{8}{9}\frac{r_1}{r_0}\frac{\varepsilon_i}{\varepsilon_s}\frac{t_a^3}{t_i L L_{ov}}(V_g - V_t - V_d/2).\qquad(2.106)$$

This analytical model is a good approximation to the rigorous calculations performed by a computer. The effect of the a-Si:H thickness is condensed in the expression of V_0 in (2.106). It modifies the TFT characteristics, causing them to depart from the ideal co-planar TFT model. One example is the crowding characteristics shown in Fig. 2.32.

2.3.5 Two-Terminal Devices

A nonlinear device has to be associated with each pixel in order to address an active-matrix liquid-crystal display. Two-terminal devices have been studied due to the simpler manufacturing process compared to that of three-terminal devices. Metal-insulator-metal (MIM) proposed in 1983 is a two-terminal device which utilizes tantalum. The side wall of the etched thin tantalum film is anodized to form a 30 to 50 nm-thick tantalum pentoxide [2.44]. This fabrication scheme making use of lateral submicron dimension is effective in reducing the leakage current usually associated with tantalum pentoxide film. The LCD panel of 250 (V)×240 (H) pixels and 100 × 96 mm² area was successfully operated at a duty ratio of 1/256 using this MIM scheme.

Amorphous silicon based diodes also show nonlinear I–V characteristics in both pin [2.45] and Schottky [2.46] devices. The back-to-back diode switch of a-Si:H was discussed for applying them to liquid-crystal displays [2.47]. The features of this type of devices are small number of process steps and small size on a level of 1 μm. As a result of these features, high yield of fabrication

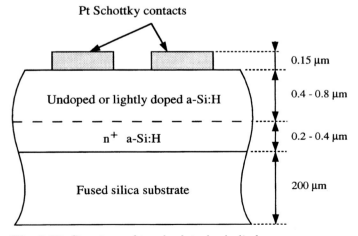

Fig. 2.36. Structure of two back-to-back diodes

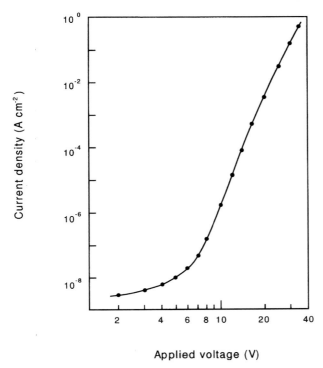

Fig. 2.37. I–V characteristics of an a-Si:H back-to-back diode. The thickness of the undoped a-Si:H layer is 800 nm

is expected. Integration of back-to-back diodes into a matrix of 5×7 dots was demonstrated in 1984 [2.48].

Figure 2.36 shows the cross section of a back-to-back Schottky diode. Evaporation of platinum on top of undoped amorphous silicon layer forms a Schottky junction and n^+-doped amorphous silicon layer beneath the undoped a-Si:H connects two Schottky diodes back to back. The reverse current in Schottky junctions is given by

$$I \propto V^n, \tag{2.107}$$

where n is the voltage power factor. A power factor up to 15 was obtained in this device as shown in Fig. 2.37 and from a highly nonlinear behavior like this, display scan lines more than 1000 have been estimated to be possible.

2.4 Liquid Crystal

The liquid crystal is a medium widely used in display devices for watches, calculators, TVs and PCs. The most prominent feature of this material is that it is operated on a very low power level. The working level can be as low as $1\,\mu W/cm^2$. Since its operating voltage is also low, the liquid crystal fits the display devices to be operated on a battery base. Thus, the application started from the small devices like watches and calculators. However, it now expands from the small monochrome devices to large, high-quality color TVs and computer monitors.

The liquid crystal is an intermediate state of matter between a solid crystal and an isotropic liquid. The thermotropic liquid crystalline phase appears in the temperature range between the melting point, T_m, and the clearing point, T_c. If the material is cooled below T_m, it forms a three-dimensional lattice having long-range order. At T_m, the solid does not immediately turn into an isotropic liquid, but changes into a transitional phase, a liquid-crystalline phase. Material showing this phase is composed of rod-like molecules. Due to this feature, there remains some degree of orientational order of a crystalline phase which is lost in a normal isotropic liquid.

This type of liquid crystal, in which the transitional phase is defined in the temperature range between T_m and T_c, is called thermotropic. In another type of liquid crystal, the amount of solvent defines this phase and it is called lyotropic. Most liquid crystals applied for display devices are thermotropic. Liquid crystal can also be defined from a molecular arrangement or alignment. It can be classified into three types, that is, nematic, cholesteric, and smectic liquid crystals.

Among these three types, the nematic phase is least ordered. The molecules in this phase are aligned with their long axes nearly parallel to each other. Though they have no long-range correlation between their centers of mass, they have an orientational order or a local preferred direction (director). The cholesteric phase of liquid crystals is similar to the nematic phase. In this phase, however, there is spatial variation in the director along the helical axis. In each plane of this structure, the orientational order is the same as in the nematic liquid crystal. The last of the three phases of liquid crystals is the smectic phase. In this phase, the molecules are arranged in a layered structure. In each layer, the liquid crystal molecules are aligned perpendicular to the layer (smectic A). The orientational order in each layer is similar to that in the nematic phase and the molecular centers of gravity have no long-range order. This smectic-A phase is the simplest case of the smectic phase. In the smectic-C phase, the molecules are not normal to the layer, but are somewhat inclined. Due to this layered structure, the smectic phase exhibits a more viscous nature than the nematic phase.

For the display applications of liquid crystal, the nematic phase is most widely used. Microscopic observation of nematic liquid crystal shows the texture patterns which correspond to discontinuities in the local preferred direction of molecules. One classical example of nematic crystal is p-azoxyanisole (PAA):

$$CH_3 - O -⟨⟩- N_2O -⟨⟩- O - CH_3 \ . \tag{a}$$

The two benzene rings are nearly coplanar, and this is a rigid rod about 20 Å long and 5 Å wide. The temperature range showing a nematic phase is between 117–136°C (C117N136I). Another classical example is N-(p-methoxybenzylidene)-p-butylaniline (MBMA):

$$CH_3 - O -⟨⟩- CH = N -⟨⟩- C_4H_9 \ . \tag{b}$$

This is the first material reported to exhibit a nematic phase at room temperature (C22N47I).

In the early stage of applications for multiplexed displays, the cyanobiphenyl and related compounds as shown below have been widely used.

$$R -⟨⟩-⟨⟩- CN, \tag{c}$$

$$RO -⟨⟩-⟨⟩- CN, \tag{d}$$

$$R -⟨⟩- CH = N -⟨⟩- CN, \tag{e}$$

$$R -⟨⟩- COO -⟨⟩- CN. \tag{f}$$

For active matrix liquid crystal displays, however, a new class of materials have been developed. For TFT LCDs, such characteristics as high voltage-retention ratio, low viscosity, large $\Delta\varepsilon$, and small Δn are required. Most widely used materials to match these requirements are 3,4-difluorophenyl compounds [2.49] as shown below.

$$C_3H_7 -⟨⟩-⟨⟩-⟨⟩\substack{F\\F}, \tag{g}$$

$$C_5H_{11} -⟨⟩-⟨⟩- CH_2 \ CH_2 -⟨⟩\substack{F\\F}, \tag{h}$$

$$C_3H_7 -⟨⟩-⟨⟩-⟨⟩\substack{F\ F\\F}. \tag{i}$$

These materials described above have been developed for twisted-nematic (TN) cell configuration. Generally speaking, their anisotropic dielectric constants ($\Delta\varepsilon$) are positive in which the director of molecules tends to align in parallel to the direction of the applied electric field. In the IPS (in-plane switching) mode operation, however, the materials having negative $\Delta\varepsilon$ and high retention ratios are required. For this purpose, 2,3-fluorophenyl compounds [2.50] have been developed as shown below:

$$C_3H_7-\bigcirc-\bigcirc-\underset{F\ F}{\bigcirc}-OC_2H_5, \qquad (j)$$

$$C_5H_{11}-\bigcirc-CO_2-\underset{F\ F}{\bigcirc}-OC_2H_5, \qquad (k)$$

$$C_3H_7-\bigcirc-\bigcirc-CH_2O-\underset{F\ F}{\bigcirc}-OC_2H_5. \qquad (l)$$

The $\Delta\varepsilon$ ranges from -4.4 (j) to -6.0 (l) in these compounds, and the material of (j) exhibits the smectic B phase before the nematic phase appears (C76SB79N186I).

For the super-twisted-nematic (STN) cell use, liquid crystal material should have a large value of the viscosity ratio (k_{33}/k_{11}). Schadt et al. [2.51] reported that this ratio is large in alkenyl compounds. Since then many compounds as shown below have been investigated

$$NC-\bigcirc-\bigcirc-\sim\sim, \qquad (m)$$

$$NC-\bigcirc-\bigcirc-\sim\sim. \qquad (n)$$

The ratio k_{33}/k_{11} of the liquid crystal shown in (m) is 2.13 ($k_{11} = 6.22$, $k_{22} = 3.90$, and $k_{33} = 13.4$) and its $\Delta\varepsilon$ is also large ($\Delta\varepsilon = 10.94$).

2.4.1 Physical Constants of Liquid Crystal

Due to the anisotropic nature of constituent molecules, the physical constants of liquid crystal behave in a complex manner. The optical and electrical characteristics of liquid crystalline cells depend on the behaviors of such constants as the dielectric constants, refractive indices, elastic constants, and coefficients of viscosity. The basic aspects of these constants are discussed here.

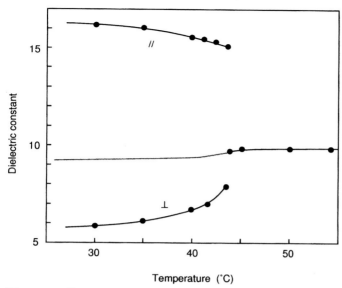

Fig. 2.38. Temperature dependence of the dielectric constant of cyano-biphenyl: C_7H_{15}-◯◯-CN at a frequency of 100 kHz. This material exhibits a large positive anisotropy ($\varepsilon_\parallel > \varepsilon_\perp$). The average of the dielectric constant $\varepsilon_{av} = (\varepsilon_\parallel + 2\varepsilon_\perp)/3$, denoted by *the middle line*, is nearly equal to the isotropic dielectric constant above the clearing temperature

Dielectric Constant and Refractive Index. Assuming that the z-axis is parallel to the director (average long axes direction of liquid crystal molecules), we obtain the diagonalized expression for dielectric constant. The displacement vector $\boldsymbol{D}$ is given by

$$\boldsymbol{D} = \begin{bmatrix} \varepsilon_\perp & 0 & 0 \\ 0 & \varepsilon_\perp & 0 \\ 0 & 0 & \varepsilon_\parallel \end{bmatrix} \boldsymbol{E} \tag{2.108}$$

where $\boldsymbol{E}$ is the electric field in a vector form, $\varepsilon_\perp$ and $\varepsilon_\parallel$ are the dielectric constants for the directions perpendicular and parallel to the director, respectively. Since $\varepsilon_\parallel$ is different from $\varepsilon_\perp$ in a nematic liquid crystal, the displacement vector is not parallel to the electric field. The displacement vector becomes parallel to $\boldsymbol{E}$ only when the electric field lies on the x–y plane or is parallel to the z-axis.

For display application, a large absolute value of $\Delta\varepsilon(=\varepsilon_\parallel - \varepsilon_\perp)$ is preferable since it is directly related to the threshold voltage of liquid crystal cell (2.131, 164). The dielectric constants ($\varepsilon_\parallel, \varepsilon_\perp$) are dependent on temperature, as shown in Fig. 2.38 [2.52]. It also depends on the frequency of electric excitation. In measuring the capacitance of the plane capacitor filled with nematic liquid crystal, the ac frequency is set low enough (1 kHz–10 kHz) to yield static permittivities. In the optical frequency range, the dielectric permittivities are obtained from the refractive index as shown below:

$$\varepsilon_\parallel/\varepsilon_0 = n_\parallel^2, \tag{2.109}$$

$$\varepsilon_\perp/\varepsilon_0 = n_\perp^2. \tag{2.110}$$

The liquid crystal exhibits anisotropy of refractive indices or birefringence, as do optically uniaxial crystals. A uniaxial crystal has two principal refractive indices n_0 and n_e which correspond to the refractive indices for an "ordinary" ray and for an "extraordinary" ray, respectively. In the nematic phase, the optical axis of the uniaxial crystal coincides with the director of the liquid crystal. Thus, we have the same expression for the birefringence as in the uniaxial crystal

$$\Delta n = n_\parallel - n_\perp. \tag{2.111}$$

The temperature dependence of the refractive indices shows almost the same behavior as the dielectric constant. The average value of the refractive indices in the nematic phase is given by the relationship

$$n_{\mathrm{av}}^2 = (n_\parallel^2 + 2n_\perp^2)/3. \tag{2.112}$$

This average value n_{av} is different from the refractive index in the isotopic phase, but the difference is small. This situation also holds for the average value of the dielectric constant $\varepsilon_{\mathrm{av}}$ calculated from the experimental values of $\varepsilon_\parallel$ and $\varepsilon_\perp$ as

$$\varepsilon_{\mathrm{av}} = (\varepsilon_\parallel + 2\varepsilon_\perp)/3. \tag{2.113}$$

In a number of strongly positive materials ($\Delta\varepsilon > 0$), the calculated mean dielectric constant $\varepsilon_{\mathrm{av}}$ is a few percentage points smaller than the isotropic value. This relation can be observed in Fig. 2.38 and has been confirmed in the theoretical studies.

Elastic Constants. In uniaxial liquid crystals, the preferred orientation of constituent molecules is given by the director vector $\boldsymbol{n}$. The rod-like molecules of the nematic liquid crystal are aligned nearly parallel to the long axes of neighboring molecules. This orientation of alignment at each point $\boldsymbol{r}$ is given by the vector $\boldsymbol{n}(\boldsymbol{r})$. The orientation varies gradually and continuously from point to point within the medium. This transition of the director from one direction to the other induces a curvature strain in the medium.

By introducing a local right-handed system of Cartesian coordinates x, y, z with $\boldsymbol{n}$ parallel to the z-axis at the origin, we can define the curvature strain tensors $n_{i,j}$ ($n_{i,j} = \partial n_i/\partial x_j$), which correspond to the changes in the orientation of the director.

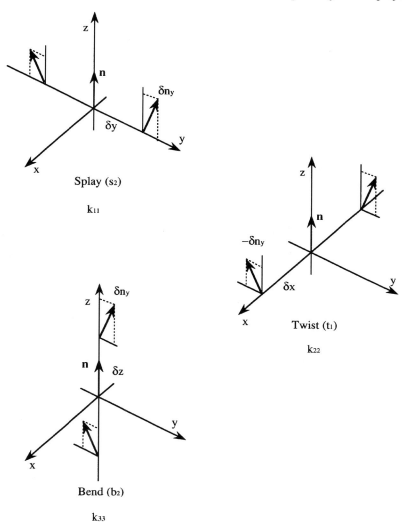

Fig. 2.39. Deformation of nematic-liquid-crystal molecules. Three types of deformation: splay, twist, and bend, are shown for variations in the director orientaion along the z-axis

The strain tensor $n_{i,j}$ has three types of components: splay, twist, and bend deformation.

$$\begin{aligned} \text{Splay}: & \quad s_1 = \partial n_x/\partial x; \quad s_2 = \partial n_y/\partial y; \\ \text{Twist}: & \quad t_1 = -\partial n_y/\partial x; \quad t_2 = \partial n_x/\partial y; \\ \text{Bend}: & \quad b_1 = \partial n_x/\partial z; \quad b_2 = \partial n_y/\partial z. \end{aligned} \qquad (2.114)$$

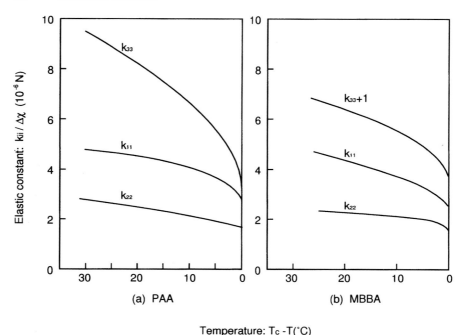

Fig. 2.40a,b. Temperature dependence of the elastic constants of nematic liquid crystals: (a) PAA, (b) MBMA. $\Delta\chi$ is the anisotropy of the magnetic susceptibility ($\Delta\chi = \chi_\parallel - \chi_\perp$:dimensionless quantity)

These deformations are shown in Fig. 2.39. For these deformations in an incompressible fluid and for isothermal deformation, the free energy per unit volume F is given as

$$F = F(n_i, n_{i,j}). \tag{2.115}$$

For infinitesimal deformation, this F value of a deformed liquid crystal relative to that in the state of uniform orientation, ΔF, can be expanded as

$$\Delta F = \frac{1}{2}\left[k_{11}(\nabla\bm{n})^2 + k_{22}(\bm{n}\nabla\times\bm{n})^2 + k_{33}(\bm{n}\times\nabla\times\bm{n})^2\right]. \tag{2.116}$$

The three elastic constants k_{11}, k_{22}, and k_{33} are called the splay, twist, and bend constants, respectively. These constants correspond to the three types of deformation shown in Fig. 2.39, and their temperature dependencies are shown in Fig. 2.40 [2.53].

Viscosity. Nematic liquid crystal flows like a conventional organic liquid having similar molecules. However, its behavior is rather complex and diffi-

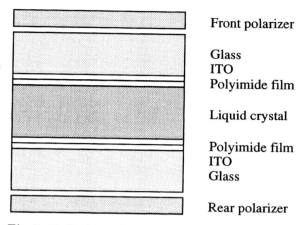

Fig. 2.41. Basic configuration of a TN cell

cult to study due to the anisotropy of nematics. The translational motions of molecules are coupled to inner, orientational motions and the flow is dependent on the angles the director forms with the flow direction and with the velocity gradient. Thus, in most cases, the flow disturbs the alignment and causes the director to rotate. Conversely, rotation of the director induces back flow in the surrounding nematics. As a result, the theoretical and experimental studies are more complex than those for isotropic liquids.

The viscous-stress tensor $\boldsymbol{\sigma}$ has the elements σ'_{ij} and these elements are given as a function of the velocity-gradient tensor and the angular velocity of the director $\boldsymbol{n}$ relative to its surroundings. Six coefficients used to describe the tensor elements σ'_{ij} have the dimensions of viscosity and they, from α_1 to α_6, are called Leslie coefficients. Only five coefficients out of six are independent since there is a relation derived by Parodi. The rotational viscosity coefficient γ_1 used to define the response time of the liquid crystal cell is related to the Leslie coefficients as follows.

$$\gamma_1 = \alpha_3 - \alpha_2. \tag{2.117}$$

2.4.2 Twisted-Nematic Cell

The electro-optical effect in a twisted nematic (TN) liquid crystal cell described in 1971 [2.6] is now widely used in active-matrix liquid-crystal displays. As shown in Fig. 2.41, a thin (5 μm) layer of nematic liquid crystal is placed between two glass plates provided with a transparent and conductive coating. On both plates the orientation of liquid crystal molecules or *director*, is aligned to be nearly parallel to the surface of the glass plate (planar or homogeneous alignment), and as shown schematically in Fig. 2.42, on each plate this orientation is twisted 90° with respect to that of the other plate, so that the orientation is continuously twisted from the bottom to the top

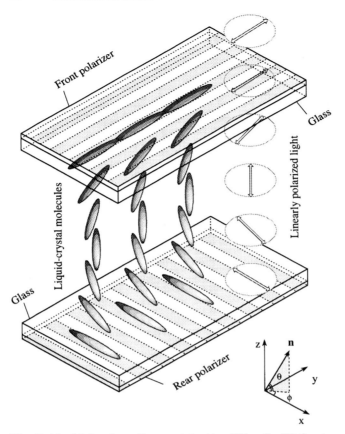

Fig. 2.42. Molecular alignment in the TN cell. Without any applied field, the orientation of molecules rotates 90° from the bottom glass plate to the top glass plate. Linearly polarized light is also rotated by 90° as it travels through the cell

glass plate by 90°. The homogeneous alignment of liquid-crystal molecules on each plate is produced by rubbing the surface of a polyimide thin film in the proper direction with a fabric. Since a cell thickness of 5 μm is much greater than the wavelength of light, the polarization plane of linearly polarized light travelling normal to the glass plates rotates with the liquid crystal axis. The 90° twist of the director should then lead to a 90° rotation of the linearly polarized light. Therefore, in this normally white (NW) mode of operation where the two polarizers are set perpendicular to each other, the light is transmitted through the cell when there is no applied voltage.

A voltage applied to the cell modifies the director orientation. When the applied voltage is below a threshold level there is no change of the orientation, but at the threshold the molecular orientation begins to be aligned to the electric field and tends to be perpendicular to the glass plate. At volt-

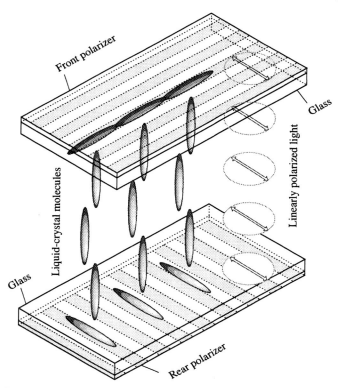

Fig. 2.43. When a sufficiently high field is applied to the cell, the molecules are aligned perpendicular to the glass (homeotropic alignment) and the rotation of the polarized light does not take place. Light is thus blocked by the top polarizer and does not come out of the module

ages well above the threshold, the alignment of the molecules is completely parallel to the field except the regions adjacent to the surface of the glass (homeotropic alignment). In this state, shown in Fig. 2.43, the TN cell becomes optically inactive and linearly polarized light travels through the cell without any rotation of the polarization. Then, the light polarization becomes perpendicular to the output polarizer resulting in no transmission of light.

The originally proposed TN cell used a parallel scheme: the two polarizers were parallel to each other. In this NB (normally black) configuration, the light is not transmitted in a field-free state and is transmitted when the applied voltage is above the threshold. When there is no bias voltage, however, transmission is suppressed to zero only for monochromatic light of a wavelength λ of $2d\ \Delta n/\sqrt{3}$. Therefore, in a practical display in which a broad-band illumination is used, there is a small amount of light leakage as shown in Fig. 2.44. In the NB mode this leakage reduces a contrast ratio, which is the ratio between the "on" and "off" transmissions.

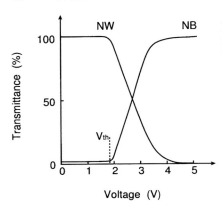

Fig. 2.44. Normally black (NB) and normally white (NW) modes of operation. These modes correspond to the parallel and the perpendicular polarizer schemes, respectively

In the normally white mode of operation, however, there is little reduction in the contrast ratio since this ratio is governed by an "off" transmission corresponding to a high bias voltage. When a sufficiently high voltage is applied the liquid crystal molecules are aligned parallel to the electric field or perpendicular to the glass substrate, and there is no rotation of electric vector of the polarized light. Transmission is therefore suppressed completely regardless of the wavelength of the light. The transmission in the "on" state is wavelength dependent, but this transmission does not have a critical effect on the contrast ratio. A contrast ratio higher than 100 can be readily obtained in the normally white mode of operation.

The threshold voltage (V_{th}) of the TN cell is defined as the exprapolated voltage of the transition region from the homeotropic to the homogeneous alignment as shown in Fig. 2.44. The V_{th} of the cell is designed to be 2–3 V, and is determined by the choice of the liquid-crystal materials to be mixed. The maximum operating voltage, or the peak amplitude of the data signal of the display, is about 7–8 V. This low voltage and the low current due to the high resistance of the liquid crystal contribute very much to the low power consumption of the display. The current density of the working display is less than $0.1\,\mu\text{A}/\text{cm}^2$ and the power level is therefore on the order of $1\,\mu\text{W}/\text{cm}^2$. The power required for a 10 in. display is less than $300\,\mu\text{W}$ as far as the liquid crystal is concerned.

Electrical Properties of TN Cell. First, we consider the threshold voltage of a TN cell. The liquid crystal in a TN cell is sandwiched between the two glass plates. The molecules at the boundaries are strongly coupled to the surfaces of these plates. The director orientations at the boundaries differ from each other by an angle of ϕ_0, which commonly takes a value of $\pi/2$. The director orientation is a function of z (in the coordinate system shown in Fig. 2.42) and is parallel to the x axis at $z = 0$. The director at $z = d$ is twisted by ϕ_0 from the x-axis. When there is no external field the director is written as $\boldsymbol{n} = [\cos\phi(z), \sin\phi(z), 0]$, and the boundary condition is $\phi(0) = 0$, and $\phi(d) = \phi_0$. When an external electric field is applied parallel to the z-axis, the director is affected by this field and tends to be aligned to the field. Writing the deviation from the x–y plane as $\theta(z)$, we obtain

$$\boldsymbol{n} = [\cos\theta(z)\cos\phi(z), \cos\theta(z)\sin\phi(z), \sin\theta(z)]. \tag{2.118}$$

The boundary conditions of $\theta(z)$ are again those of a strong anchoring: $\theta(0) = \theta(d) = 0$. The free energy F of this system per unit area is then given by

$$F = (1/2)\int_0^d [(k_{11}\cos^2\theta + k_{33}\sin^2\theta)(d\theta/dz)^2 \\ + \cos^2\theta(k_{22}\cos^2\theta + k_{33}\sin^2\theta)(d\phi/dz)^2 - \Delta\varepsilon E^2\sin^2\theta]dz, \tag{2.119}$$

where k_{11}, k_{22}, and k_{33} are respectively the constants of splay, twist, and bend stresses. From the condition that the free energy takes a minimum value, we obtain the following equations (Euler's equation):

$$\frac{d}{dz}\left[\cos^2\theta(k_{22}\cos^2\theta + k_{33}\sin^2\theta)\frac{d\phi}{dz}\right] = 0 \tag{2.120}$$

and

$$(k_{11}\cos^2\theta + k_{33}\sin^2\theta)d^2\theta/dz^2 + (k_{33} - k_{11})\sin\theta\cos\theta(d\theta/dz)^2 \\ + \sin\theta\cos\theta[2k_{22}\cos^2\theta + k_{33}(\sin^2\theta - \cos^2\theta)](d\phi/dz)^2 \\ + \Delta\varepsilon E^2\sin\theta\cos\theta = 0. \tag{2.121}$$

For small $\theta(z)$, (2.120) and (2.121) are linearized to

$$d^2\phi/dz^2 = 0, \tag{2.122}$$

and

$$k_{11}d^2\theta/dz^2 + [(2k_{22} - k_{33})(d\phi/dz)^2 + \Delta\varepsilon E^2]\theta = 0. \tag{2.123}$$

From (2.122) and the boundary condition, we obtain

$$\phi(z) = \phi_0 z/d, \tag{2.124}$$

and (2.123) is thus reduced to

$$k_{11}d^2\theta/dz^2 + C_0\theta = 0, \tag{2.125}$$

where

$$C_0 = \Delta\varepsilon E^2 + (2k_{22} - k_{33})(\phi_0/d)^2. \tag{2.126}$$

Then, for $C_0 > 0$, the solution of (2.126) is written as

$$\theta(z) = C_1\cos\left[(C_0/k_{11})^{1/2}z\right] + C_2\sin\left[(C_0/k_{11})^{1/2}z\right], \tag{2.127}$$

where C_1 and C_2 are constants. The boundary conditions $\theta(0) = \theta(d) = 0$ yield,

$$C_1 = 0, \quad \sin[(C_0/k_{11})^{1/2}d] = 0. \tag{2.128}$$

From (2.128) we obtain

$$[C_0/k_{11}]^{1/2} d = m\pi \quad m = 0, 1, 2 \ldots. \tag{2.129}$$

The case of $m = 0$ corresponds to that in which there is no external field. The angle θ starts to tilt at the threshold field E_{th}, and this corresponds to the case of $m = 1$

$$\left(\frac{\Delta\varepsilon E_{\text{th}}^2 + (2k_{22} - k_{33})(\phi_0/d)^2}{k_{11}} \right)^{1/2} d = \pi. \tag{2.130}$$

From (2.130) for $\phi_0 = \pi/2$, we obtain the threshold voltage $V_{\text{th}}(= E_{\text{th}}d)$ of the TN cell as

$$V_{\text{th}} = \pi \left\{ [k_{11} + (k_{33} - 2k_{22})/4]/\Delta\varepsilon \right\}^{1/2}. \tag{2.131}$$

The threshold voltage of TN cell is seen to be independent of the cell gap.

The response time of the TN cell is derived from the equation of motion of the director, which describes the balance between the torques due to the elastic and viscous forces and the external field. Equating the left-hand side of (2.123) to $\gamma_1(\partial\theta/\partial t)$, we obtain

$$k_{11}\partial^2\theta/\partial z^2 + [\Delta\varepsilon E^2 + (2k_{22} - k_{33})(\phi_0/d)^2]\theta = \gamma_1 \partial\theta/\partial t, \tag{2.132}$$

where t is the time and γ_1 is the coefficient of the rotational viscosity (2.117). Equation (2.132) holds only for small θ but is adequate for the present purpose. If we assume

$$\theta = \theta(z)[1 - \exp(t/\tau_R)], \tag{2.133}$$

where τ_R is the rise-time constant, (2.132) is reduced to

$$k_{11} \frac{d^2\theta}{dz^2} + [(2k_{22} - k_{33})(\phi_0/d)^2 + \Delta\varepsilon E^2 - \gamma_1/\tau_R]\theta(z) = 0, \tag{2.134}$$

where we further assumed $\theta = \theta(z)t/\tau_R$ for small t. The same procedure as in (2.123) leads us to the next equation

$$\left\{ [(2k_{22} - k_{33})(\pi/2d)^2 + \Delta\varepsilon E^2 - \gamma_1/\tau_R]/k_{11} \right\}^{1/2} d = \pi. \tag{2.135}$$

From (2.135), the time constant is given as

$$\tau_R = \gamma_1 d^2/[\Delta\varepsilon(V^2 - V_{\text{th}}^2)], \tag{2.136}$$

where V_{th} is given by (2.131).

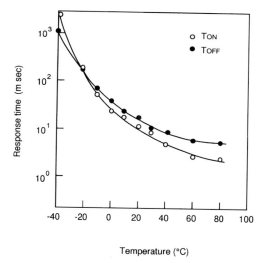

Fig. 2.45. Temperature dependence of the response time of the TN cell with a 4.8 μm cell gap

The decay-time τ_D is obtained following the same procedure:

$$\tau_D = \gamma_1 d^2 / \left(\Delta\varepsilon V_{\text{th}}^2\right). \tag{2.137}$$

We can see from (2.136) and (2.137) that the time constant is proportional to the square of the cell gap. Reducing the cell gap is therefore an effective way to shorten the cell's response time [2.55].

For a display like an LCD as well as a CRT, the rise and fall times are defined for convenience as the times when the brightness reaches, after the turn-on and turn-off of the drive voltage, 90% and 10%, respectively of full brightness. These turn-on and turn-off times (t_{ON} and t_{OFF}) are given by

$$t_{\text{ON}} = 2.3\tau_R, \quad \text{and} \quad t_{\text{OFF}} = 2.3\tau_D. \tag{2.138}$$

An example of the temperature dependence of t_{ON} and t_{OFF} is shown in Fig. 2.45 [2.56]. The turn-on time is usually shorter than the turn-off time, and at room temperature these times are on the order of 10 ms. Therefore, the response time of a TN cell is well acceptable for such application as TVs and computer terminals.

As described in Sect. 2.3.2, the dc voltage offset appears on the pixel electrode because of the parasitic capacitance of the a-Si TFTs. This offset ΔV is given by (2.30) and will be easier to cancel out if the capacitance of liquid crystal cell, C_{lc}, is constant. Unfortunately, C_{lc} depends on the applied voltage and the dc voltage offset changes from pixel to pixel and from time to time.

The variation of C_{lc} with the bias voltage as shown in Fig. 2.46 is due to an anisotropic dielectric constant of liquid crystal. When the bias voltage

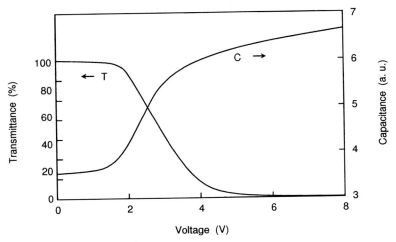

Fig. 2.46. Dependence of the TN cell capacitance on the applied voltage. Also shown is the dependence of the NW TN cell transmittance

is lower than the threshold voltage, the capacitance of the twisted-nematic liquid-crystal cell is given by

$$C_{lc} = \varepsilon_\perp A/d, \tag{2.139}$$

where d is the cell gap, A is the effective area of the pixel electrode, and $\varepsilon_\perp$ is the dielectric permittivity for the direction perpendicular to the director. Above threshold, the capacitance begins to increase from the value of (2.139). When the bias voltage is sufficiently high, C_{lc} is given by

$$C_{lc} = \varepsilon_\parallel A/d, \tag{2.140}$$

where $\varepsilon_\parallel$ is the dielectric permittivity for the direction parallel to the director. It is worth noting that the dielectric permittivity used here is for the low frequency electric field. Because the electric field is applied perpendicular to the glass substrate, it is perpendicular to the director or the optical axis of the liquid crystal below threshold. On the other hand, the electric vector of the linearly polarized light is parallel to the glass and parallel to the director. Therefore, below threshold the dielectric permittivity is $\varepsilon = \varepsilon_\parallel$ for the optical wavelength, whereas it is $\varepsilon = \varepsilon_\perp$ for the static field. The situation is reversed, however, well above the threshold. As shown in Fig. 2.43, the director of the liquid crystal becomes vertically aligned and the dielectric permittivity becomes $\varepsilon = \varepsilon_\perp$ for the optical wavelength and $\varepsilon = \varepsilon_\parallel$ for the static field.

Optical Properties of TN Cell. The molecules in the TN cell undergo a 90° rotation from the top to the bottom of the liquid crystal cell. The material used in these devices is birefringent and has a positive dielectric constant; that is,

$$n_e - n_o = \Delta n > 0, \tag{2.141}$$
$$\varepsilon_\| - \varepsilon_\perp = \Delta\varepsilon > 0, \tag{2.142}$$

where n_o and n_e are the ordinary and extraordinary refractive indices. In a liquid crystal this birefringence is given by

$$\Delta n = n_\| - n_\perp. \tag{2.143}$$

The 90° rotation of molecules in this system is associated with the rotation of the plane of polarization of light transmitted through the cell. If polarized light is incident on the cell in such a way that the polarization is parallel to the director at the entrance, its polarization is rotated 90° as it exits the liquid crystal (Fig. 2.42).

When the electric field is applied along the z-axis, the tilt angles of the molecules are modified to become realigned parallel to the field. If the field becomes high enough, the molecular realignment is completely parallel to the field and the TN cell has no optical activity. Under crossed polarizers as shown in Fig. 2.42, this TN device exhibits a transmitting behavior (normally white) when there is no applied field (Fig. 2.42) and is switched to an extinction state when an electric field is applied (Fig. 2.43).

The propagation of the polarized light in an anisotropic medium like a liquid crystal can be analyzed by solving the equation that governs the evolution of the ellipse of polarization of a light wave ψ [2.57, 58]. The equation is a first-order ordinary differential equation for $\psi = \psi(z, \psi_0)$, where ψ_0 is the value at $z = 0$; that is, in the incident state of polarization. ψ is a single complex variable which describes the ellipse of polarization of the incident light wave. Both the azimuth ϕ and ellipticity e can be calculated from ψ as follows [2.59]

$$e = (|\Psi| - 1)/(|\Psi| + 1), \quad \phi = (1/2)\arg\Psi. \tag{2.144}$$

In a TN cell where the director is parallel to the x-axis at $z = 0$ and it is aligned to the y-axis at $z = d$, the polarization is described by the following equation

$$\frac{d\Psi}{dz} = \left[\frac{i\pi\Delta n}{\lambda}\exp\left(\frac{-i\pi z}{d}\right)\right]\Psi^2 - \left[\frac{i\pi\Delta n}{\lambda}\exp\left(\frac{i\pi z}{d}\right)\right]. \tag{2.145}$$

The general solution of this equation is obtained by putting it in the form

$$\Psi = f(z)\exp(i\pi z/d). \tag{2.146}$$

Then the polarization at the exit plane ($z = d$) is expressed as

$$\Psi = -\frac{\sqrt{1+u^2} - i(1+u)\tan\left[(\pi/2)\sqrt{1+u^2}\right]}{\sqrt{1+u^2} + i(1-u)\tan\left[(\pi/2)\sqrt{1+u^2}\right]}, \tag{2.147}$$

where $u = 2d\Delta n/\lambda$ for the total twist angle of $\pi/2$.

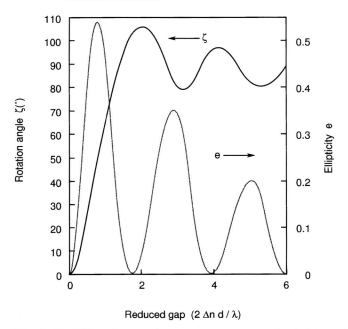

Fig. 2.47. Ellipticity e and the rotation angle ζ $[=(\pi/2)+\phi]$ at the exit plane of a TN cell. The most common choice of $u(=2d\Delta n/\lambda)$ for a display cell is $u=\sqrt{3}$

From (2.144) and (2.147), we obtain

$$e = \tan\left((1/2)\sin^{-1}\left\{\frac{2u}{1+u^2}\sin^2\left[(\pi/2)\sqrt{1+u^2}\right]\right\}\right), \quad (2.148)$$

$$\tan(2\phi) = \frac{2\sqrt{1+u^2}\tan\left[(\pi/2)\sqrt{1+u^2}\right]}{1+u^2-(1-u^2)\tan^2\left[(\pi/2)\sqrt{1+u^2}\right]}. \quad (2.149)$$

These equations define the polarization state of the radiation emerging from the TN cell. Figure 2.47 shows the ellipticity e of (2.148) and the rotation angle of ζ $[=(\pi/2)+\phi]$ of (2.149) as a function of the reduced cell gap u $(=2\Delta n\,d/\lambda)$.

Both the ellipticity and the angle of rotation oscillate with u, but tend to be independent of u when it is large. In the limit of large u,

$$u(=2\Delta n\,d/\lambda) \gg 1 \quad (2.150)$$

the plarization of the incident light is rotated exactly 90° as the light traverses the cell and, regardless of its wavelength, is linearly polarized to the y-axis when it exits the cell. This condition is called Mauguin's limit. To reach this limit, however, an impractically large value of d is necessary; and in practice a

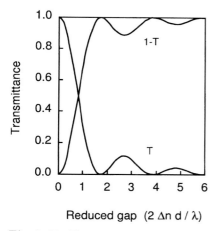

Fig. 2.48. Transmission of the TN cell as a function of $u = 2\Delta n\,d/\lambda$. The curves T and $1-T$ correspond to the transmittances of the normally black cell and the normally white cell, respectively

smaller value of u is chosen. In this case, the emitted light becomes elliptically polarized as it exits the cell. The exiting light is linearly polarized only when the following equation is satisfied:

$$2\Delta n\,d/\lambda = \sqrt{m^2 - 1}, \tag{2.151}$$

where m is an integer. The smallest and most commonly used cell gap is for $m = 2$ and is given by

$$d = (\sqrt{3}/2)\lambda/\Delta n. \tag{2.152}$$

The transmittance of radiation, T, is obtained from (2.148) and is given for a system with parallel polarizers (NB cell) as

$$T = \frac{\sin^2\left[(\pi/2)\sqrt{1+u^2}\right]}{1+u^2}, \tag{2.153}$$

and for a system with crossed polarizers (NW cell) as

$$T' = 1 - T. \tag{2.154}$$

Equations (2.153) and (2.154) are plotted in Fig. 2.48.

As before, the transmission oscillates with u; but for a u value corresponding to (2.152) there is no transmission of light for parallel polarizers and 100% transmission of light for crossed polarizers when no field is applied. For a given cell gap of d, however, this condition is achieved only for the monochromatic light with a wavelength given by

$$\lambda = 2\Delta n\,d/\sqrt{3}. \tag{2.155}$$

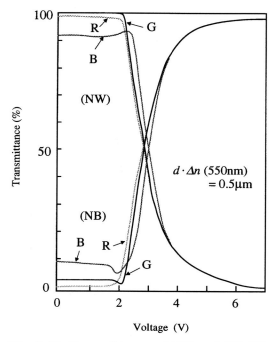

Fig. 2.49. Transmittances of NW and NB TN cells as a function of the applied voltage. A $\Delta n\,d$ value of 0.5 µm is most commonly used in the practical display. The wavelength dependence of the trasmission below threshold is critical in determining the contrast ratio of the NB cell

In TFT LCDs, three colors (R, G, and B) are used and it is not possible for one cell to satisfy (2.155) for all these colors. And because each color has a bandwidth of about 100 nm, it is not possible to satisfy (2.155) even for one color. For a wavelength other than that given by (2.155) the emitted light is elliptically polarized and the optic axis does not coincide with the y-axis. Figure 2.49 shows the transmission of the display panel vs. drive voltage for the NB and NW cells [2.60]. In the NB or parallel polarizer system, there is a substantial leakage of light when no field is applied and a scatter in the transmission is observed for the R, G, and B colors. Since the contrast ratio CR of the display is defined as

$$\mathrm{CR} = T_{\mathrm{on}}/T_{\mathrm{off}}, \tag{2.156}$$

where T_{on} and T_{off} are the full transmittance and the off transmission, the leakage and scatter in the off-state transmission greatly lowers the contrast ratio.

Even in the NW-mode operation shown in Fig. 2.49, the scatter in the transmission when no field is applied is obvious. In this operational mode, however, the contrast ratio can be made high by increasing the maximum operating voltage. When the voltage is increased, the transmission is reduced and its wavelength dependence disappears. The influence of the scattering of the field-free transmission on the contrast ratio is not large and a contrast ratio higher than 100 can be obtained relatively easily over a full wavelength range. Because of this effect, the normally white mode is preferred and is used most widely.

The viewing-angle characteristics of the display are also important. Liquid crystal displays have an inherent drawback in that their viewing-angle characteristics are not as good as those of CRTs. The transmission of LCDs shows complex behavior as the viewing angle is changed, since the optical path length depends on the angle of incidence. Berreman calculated [2.61] the viewing-angle characteristics of the cell for the coordinate system as shown in Fig. 2.42. The tilt angle θ and the azimuthal or turn angle ϕ are defined there and the director at the bottom surface is oriented parallel to the x-axis and the director at the top surface lies parallel to the y-axis. Light rays travel along the z-axis. The electric field is applied in the z-direction.

The tilt and turn angles θ and ϕ are shown in Fig. 2.50 as functions of z. The assumptions made when calculating these functions were that the cell gap is 10 µm, the liquid crystal is MBBA -doped to give positive dielectric anisotropy, the ordinary and extraordinary refractive indices for He-Ne laser light (633 nm) are

$$n_o = 1.54 \quad (\varepsilon_\perp/\varepsilon_0 = 2.37), \tag{2.157}$$
$$n_e = 1.75 \quad (\varepsilon_\parallel/\varepsilon_0 = 3.06), \tag{2.158}$$

and the glass and the ordinary axis of the liquid crystal have matching refractive indices. It was also assumed that $k_{11}/k_{33} = 0.79$ and $k_{22}/k_{33} = 0.48$.

Figure 2.51 shows the computed transmittance as a function of angle of incidence for various field strengths shown in Fig. 2.50. In this example, the polarizer transmits optic electric vectors along the y-axis (the director at the bottom is along the x-axis) and the analyzer is parallel to the polarizer. This configuration differs from the commonly used one in which the polarizer transmits optic electric vectors along the x-axis. But because curves for the case when the polarizer is oriented parallel rather than perpendicular to the x-axis differ only about one percent from the curves shown in Fig. 2.51, these results can be extended to the commonly used TN cell configuration.

To improve the viewing-angle characteristics, such schemes have been proposed as a two-domain TN cell [2.62, 63], in-plane-switching (IPS) mode [2.64], and vertical alignment (VA) mode [2.65]. The IPS mode will be described in detail in the next section.

2.4.3 In-Plane-Switching Cell

Birefringence plays a key role for applying liquid crystal as a display medium. However, this birefringence effect causes the viewing-angle characteristics of

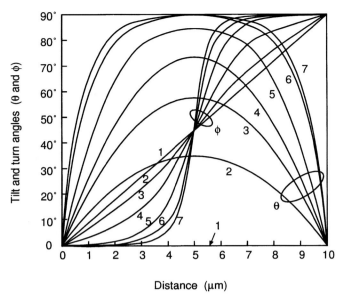

Fig. 2.50. Tilt (θ) and turn (ϕ) angles of the TN cell as functions of z. Numbers in the figure correspond to voltages normalized to V_{th}: $V/V_{\text{th}} =$ (1) ≤ 1, (2) 1.083, (3) 1.295, (4) 1.69, (5) 2.56, (6) 3.42, (7) 4.12

LCDs rather inferior to other displays such as CRTs and PDPs (plasma display panels). The multidomain scheme mentioned in the previous section presents one of the solutions to overcome this viewing-angle problem. This multidomain mode utilizes the TN mode where voltages are applied vertically to the glass substrate. The in-plane-switching (IPS) mode utilizes transverse field effects by an in-plane electrode structure [2.64]. The in-plane interdigital electrodes were used to demonstrate electro-optic effects to be used for display devices. Experiments were performed on three kinds of molecular orientations: 1) homeotropic, 2) homogeneous, and 3) 90°-twisted mode [2.67]. The molecular orientations of liquid crystals are normal and parallel to plates in homeotropic and homogeneous mode, respectively. Modulation of optical transmission were observed in all modes including the schemes of crossed and parallel polarizers.

In 1992, the positive influence of the IPS mode on the viewing-angle characteristics was mentioned [2.64]. Both computer simulation and experimental results showed improved viewing-angle dependence of optical transmission. In 1995, the TFT LCDs with in-plane-switching mode were reported to exhibit wide viewing-angle characteristics [2.68]. The normalized transmission (T/T_0) of the IPS mode under crossed polarizers is given by [2.69]

$$T/T_0 = \sin^2(2\alpha_\ell) \sin^2(\pi d \Delta n/\lambda), \tag{2.159}$$

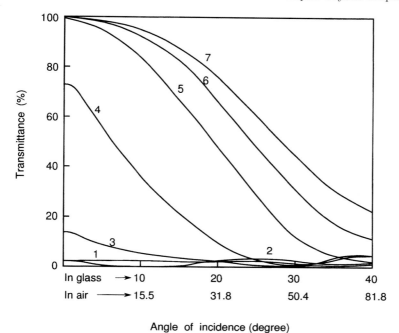

Fig. 2.51. Transmittance as a function of the angle of incidence in the x–y plane of Fig. 2.42. Curves are for the seven values of V/V_{th} given in Fig. 2.50

where α_ℓ is the angle between the liquid crystal director and the polarizer axis, d is the liquid crystal cell gap, Δn is the anisotropy of refractive indices, and λ is the wavelength. The maximum transmission is obtained when $\alpha_\ell = 45°$ and this theoretical prediction is in accordance with the experimental voltage-dependent transmission profile as shown in Fig. 2.52. This situation is schematically depicted in Fig. 2.53.

Switching behavior of the IPS mode can be analyzed from the free energy of the system. Figure 2.54 shows the definition of twist angle ϕ of deformation by an in-plane field. The free energy F per unit volume is given by as a function of the twist angle ϕ

$$F = \frac{1}{2}\int_0^d \left[k_{22}\left(\frac{d\phi}{dz}\right)^2 - |\Delta\varepsilon|\, E^2 \sin^2\phi \right] dz, \tag{2.160}$$

where k_{22} is the elastic constant of twist deformation and $\Delta\varepsilon$ is the dielectric anisotropy. Euler's equation defines the condition that (2.160) takes a minimum value. From this, we obtain the following equation

$$k_{22}(d^2\phi/dz^2) + |\Delta\varepsilon|\, E^2 \sin\phi \cos\phi = 0. \tag{2.161}$$

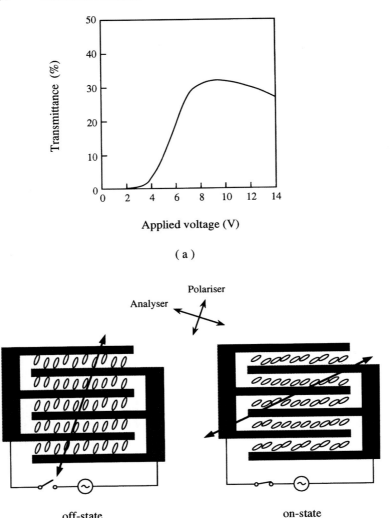

Fig. 2.52. (a) Transmittance dependence on the applied voltage of the IPS cell with a negative dielectric anisotropy; (b) Molecular alignments for the on- and off-state of the IPS mode are shown schematically

Equation (2.161) is reduced to

$$\sqrt{|\Delta\varepsilon|/k_{22}}Ed = m\pi , \qquad (2.162)$$

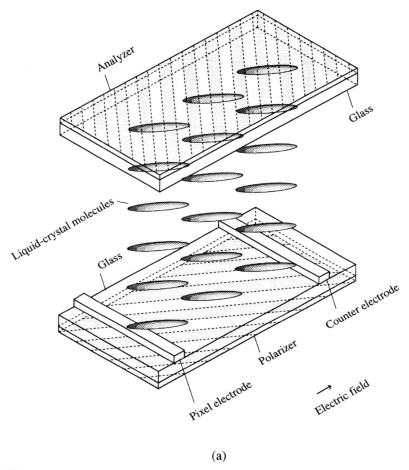

(a)

Fig. 2.53a,b. Molecular orientations of the crossed polarizer IPS system: (**a**) off-state without any applied voltage and (**b**) on-state with a voltage applied between planar electrodes

if we assume small twist angles and the boundary conditions of $\phi(0) = \phi(d) = 0$. The threshold field E_{th} where the angle ϕ begins to twist, corresponds to $m = 1$, and is given by

$$E_{\text{th}} = (\pi/d)\,(k_{22}/|\Delta\varepsilon|)^{1/2}\,. \tag{2.163}$$

Then, the threshold voltage V_{th} of the IPS mode is given by

$$V_{\text{th}} = (\pi s/d)\,(k_{22}/|\Delta\varepsilon|)^{1/2}\,, \tag{2.164}$$

where s is the distance between in-plane electrodes.

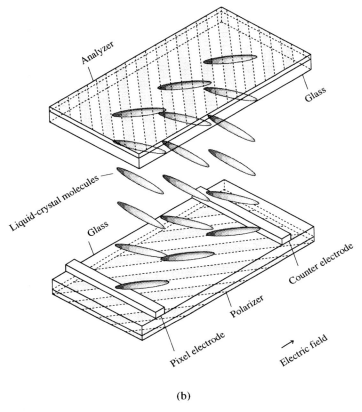

Fig. 2.53a,b (continued).

The dependence on (s/d) gives us some options in determining the threshold voltage of the IPS mode in contrast to the TN mode operation where V_{th} has no dependence on physical parameters. The cell gap d is determined from (2.159), and is given by

$$\Delta n\, d = \lambda/2 \tag{2.165}$$

to obtain a maximum transmittance. The IPS mode can be realized by using both positive and negative dielectric anisotropy, $\Delta \varepsilon$. The pretilt angle of liquid crystal molecules is homogeneous with respect to the glass substrate. With respect to the in-plane electrodes, molecules are aligned with an angle of 2–20 degrees for positive dielectric anisotropy and 70–88 degrees for negative anisotropy material [2.68]. Figure 2.52 shows an example of transmittance vs. applied voltage for the IPS cell with interdigital electrodes. The liquid crystal has a negative dielectric anisotropy and the normally-black mode operation is obtained by a crossed polarizer system. The feature of the IPS mode is a realization of pure black in an off-state, and the color fidelity is

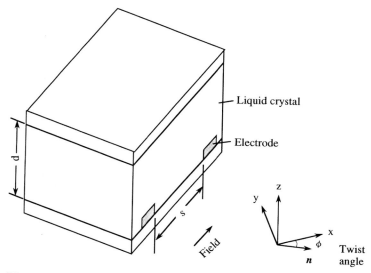

Fig. 2.54. Twist angle of deformation of the IPS liquid-crystal cell

also improved over TN mode displays. The viewing-angle characteristics have been improved to a level of ±70 degrees (CR > 10) in both horizontal and vertical directions.

2.4.4 Super-Twisted Nematic (STN) Cell

The transmission characteristics of the twisted-nematic cell shows a moderate transition from on- (off-) to off- (on-) state above the threshold voltage. This transition has a linear feature and the gray-scale representation is produced relatively easily, especially in TFT LCDs. In the passive-matrix liquid crystal display, however, this slow transition becomes a liability rather than an asset. The multiplexing in the passive-matrix LCDs with large information capacities requires a fast transition from the off-state to the on-state above the threshold. Otherwise, the contrast ratio of these displays becomes low and the viewing angle becomes narrow when the rows and columns of passive-matrix liquid-crystal displays are highly multiplexed. According to Alt and Pleshko [2.70], a matrix with m rows is driven optimally when the applied rms voltages are chosen as follows:

$$V_s = \left[(m^{1/2} + 1)/(m^{1/2} - 1)\right]^{1/2}, \tag{2.166}$$

and

$$V_{ns} = 1, \tag{2.167}$$

where V_s and V_{ns} are respectively the rms voltages to the select pixel and to the non-select pixel [2.71]. As m, the number of scan lines, is increased the

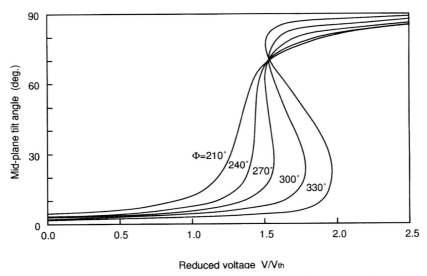

Fig. 2.55. Theoretical curves of the tilt angle of local directors in the midplane of an STN cell as a function of the reduced voltage V/V_{th} where V_{th} is the Freedericksz threshold voltage of a non-twisted layer with zero pretilt [2.72]

select voltage approaches the non-select voltage of (2.167). If $m = 100$, the select voltage becomes $V_s = 1.11$. If $m = 400$, $V_s = 1.05$. Therefore, the select voltage is higher than the non-select voltage by only several percent.

Super-twisted nematic (STN) liquid crystal displays were developed to satisfy the requirements described by (2.166) and (2.167). The STN is a cell with a twist angle of about 270° and with a relatively high pretilt angle. The phenomenon produced with this kind of cell was originally referred to as the supertwisted birefringent effect (SBE) [2.72]. The original SBE technology has been modified (lower pretilt angles, etc.) and the term STN (super-twisted nematic) [2.73, 74] is now used most commonly. The basic principle of an STN cell, however, is the same as that of an SBE cell. Figure 2.55 shows the voltage dependence of the director in the midplane of a chiral nematic layer with a 28° pretilt angle at both boundaries. The bistable range appears as the total twist angle ϕ is increased above 245°, and the technologically convenient twist angle of 270° is generally used.

A high pretilt angle, on the order of 5° and 30°, is required at both interfaces to ensure that only deformations of a twist angle of about 270° occur in the display. With low pretilt angles, a distortion with 180° or less twist becomes more stable. The polarizer setting of STN cells also differs from that of TN cells. For a nematic layer with a 270° left-handed twist, the optimum state is obtained when the front polarizer is oriented so that (1) the plane of vibration of the E vector makes a 30° angle with the projection of the layer and (2) the rear polarizer is at an angle of about 60° with respect to the projection of the director at the rear boundary. This orientation is required because of the residual twist and retardation of the select state. As a result of the interference of the optical normal modes propagating in the layer, the display has a yellow birefringence color in the non-select state. Rotation of one of the polarizers by an angle of 90° results in an image, complementary to the previous "yellow mode," in which the select state is colorless and bright and the non-select state is dark blue ("blue mode"). Recent development of the retardation film has made the "white mode" STN display possible.

The threshold voltage of the STN display is given by [2.75]

$$V_{th} = \sqrt{\{4\pi(d/p)k_{22} - [k_{33} - 2(k_{33} - k_{22})\cos^2\theta_0]\phi_0\}\phi_0/\Delta\varepsilon}, \qquad (2.168)$$

where ϕ_0 is the total twist angle, θ_0 is the pretilt angle, p is the chiral pitch, and d is the cell spacing. The threshold voltage is 2–3 V, and the response time of the STN cell is on the order of a few hundred ms at 20°C. A contrast ratio of 10:1 or more can be obtained and the viewing cone makes an angle of 30° with the vertical.

References

2.1 P. K. Weimer: Proc. IRE-AIEE Solid State Device Res. Conf. (Stanford, California), 1961.
2.2 P. K. Weimer: Thin film transistors. In Field Effect Transistors, edited by J. T. Wallmark and H. Johnson. Prentice Hall (New Jersey), 1966.
2.3 B. J. Lechner, F. J. Marlowe, E. O. Nester and J. Tults: Liquid crystal matrix displays. Proc. IEEE, **59**, 1566–1579, 1971.
2.4 G. H. Heilmeier, L. A. Zanoni and L. A. Barton: Dynamic scattering: a new electrooptic effect in certain classes of nematic liquid crystals. Proc. IEEE, **56**, 1162–1171, 1968.
2.5 T. P. Brody, J. A. Asars and G. D. Dixon: A 6×6 inch 20 lines-per-inch liquid-crystal display panel. IEEE Trans. Electron Devices, **ED-20**, 995–1001, 1973.
2.6 M. Schadt and W. Helfrich: Voltage-dependent optical activity of a twisted nematic liquid crystal. Applied Physics Letters, **18**, 127–128, 1971.
2.7 P. G. LeComber, W. E. Spear and A. Gaith: Amorphous-silicon field-effect device and possible application. Electronics Letters, **15**, 179–181, 1979.
2.8 T. Tsukada: State-of-the-art of a-Si TFT/LCD. Transaction of the Institute of Electronics, Information and Communication Engineers (Japan), **J76-C-II**, 177–183, 1993.
2.9 M. Ohta, M. Oh-e, K. Kondo: Development of Super-TFT-LCDs with in-plane-switching display mode, Proc. Int. Display Res. Conf., S30-2, 707–710, 1995 (Hamamatsu).
2.10 Y. Kaneko, A. Sasano, and T. Tsukada: Analysis and design of a-Si TFT/LCD panels with a pixel model. IEEE Transaction Electron Devices, **ED-36**, 2953–2958, 1989.
2.11 T. Tsukada (1993). Development of aluminum gate thin-film transistors based on aluminum oxide insulators. In Amorphous Insulating Thin Films. Material Research Society Symposium Proceedings (Boston, 1992), edited by J. Kanicki et al., 371–382. Pittsburgh: Material Research Society.

2.12 F. Morin: Electrooptical performance of a TFT-addressed TNLC panel. Proc. 3rd International Display Research Conference (Kobe, 1983). 412–414, California: SID, 1983.

2.13 Y. Kaneko, Y. Tanaka, N. Kabuto, and T. Tsukada: A new address scheme to improve the display quality of a-Si TFT/LCD panel. IEEE Transaction Electron Devices, **ED-36**, 2949–2952, 1989.

2.14 Y. Nanno, Y. Mino, E. Takeda, and S. Nagata: Characterization of sticking effects of TFT-LCD. In Digest of Technical Papers of the Society for Information Display International Symposium (Las Vegas, 1990), 404–407, California: SID, 1990.

2.15 Y. Kanemori, M. Katayama, K. Nakazawa, H. Kato, K. Yano, Y. Fukuoka, et al.: 10.4-in-diagonal color TFT-LCDs without residual images. In Digest of Technical Papers of the Society for Information Display International Symposium (Las Vegas, 1990), 408–411, California: SID, 1990.

2.16 Y. Nasu, S. Kawai, S. Kisumi, K. Oki, and K. Hori: Color LCD for character and TV display addressed by self-aligned a-Si:H TFT. In Digest of Technical Papers of the Society for Information Display International Symposium (San Diego, 1986), 289–292, California: SID, 1986.

2.17 K. Asama, T. Kodama, S. Kawai, Y. Nasu, and S. Yanagisawa: A self-alignment processed a-Si TFT matrix circuit for LCD SID panels. In Digest of Technical Papers of the Society for Information Display International Symposium (Philadelphia, 1983), 144–145, California: SID, 1983.

2.18 H. Yamamoto, H. Matsumaru, K. Tsutsui, N. Konishi, M. Nakatani, K. Shirahashi, A. Sasano, and T. Tsukada: A new a-Si TFT with Al_2O_3/SiN double-layered gate insulator for 10.4-inch diagonal multicolor display. In Technical Digest of the International Electron Devices Meeting (San Francisco, 1990), 851–854, New York: IEEE, 1990.

2.19 T. Tsukada: Scaling theory of liquid-crystal displays addressed by thin-film transistors, IEEE Tr. Electron Devices, **45**, 387–393, 1998.

2.20 R. H. Dennard, F. H. Gensslen, H-N. Yu, V. L. Rideout, E. Bassous, and A. R. LeBlanc: Design of ion-implanted MOSFET's with very small physical dimensions, IEEE J. Solid-State Circuits, **SC-9**, 256–268, 1974.

2.21 W. H. Dumbaugh, P. L. Bocko: Substrate glasses for flat-panel displays. In Digest of Technical Papers of the Society for Information Display International Symposium (Las Vegas, 1990), 70–72, California: SID, 1990.

2.22 P. K. Weimer: An evaporated thin-film triode, In Proceedings of the IRE-AIEE Solid State Device Research Conference (Stanford, California, 1961), 1961.

2.23 P. K. Weimer: The TFT - a new thin-film transistor, Proc. IRE, **50**, 1462–1469, 1962.

2.24 T. P. Brody, J. A. Asars, and G. D. Dixon: A 6×6 inch 20 lines-per-inch liquid-crystal display panel, IEEE Trans. Electron Devices, **ED-20**, 995–1001, 1973.

2.25 F-C. Luo, W. A. Hester, and T. P. Brody: Alphanumeric and video performance of a 6" × 6" 30 lines-per-inch thin-film transistor-liquid crystal display panel, In Digest of Technical Papers of the Society for Information Display International Symposium (San Francisco, 1978). California: SID, 1978.

2.26 P. G. LeComber, W. E. Spear, and A. Gaith: Amorphous-silicon field-effect device and possible application, Electronics Letters, **15**, 179–181, 1979.

2.27 A. J. Snell, K. D. Mackenzie, W. E. Spear, and P. G. LeComber: Application of amorphous silicon field effect transistors in addressable liquid crystal display panels, Applied Physics, **24**, 357–362, 1981.

2.28 W. E. Spear and P. G. LeComber: Investigation of the localized state distribution in amorphous Si films, Journal of Non-Crystalline Solids, **8-10**, 727–738, 1972.

2.29 Y. Okubo, T. Nakagiri, Y. Osada, M. Sugata, N. Kitahara, and K. Hatanaka: Large-scale LCDs addressed by a-Si TFT array, In Digest of Technical Papers of the Society for Information Display International Symposium (San Diego, 1982), California: SID, 1982.

2.30 S. M. Sze: Physics of Semiconductor Devices, pp. 567–573. New York: Wiley-Interscience, 1969.

2.31 M. J. Powell: Charge trapping instabilities in amorphous silicon-silicon nitride thin-film transistors, Applied Physics Letters, **43**, 597–599, 1983.

2.32 N. Lustig and J. Kanicki: Gate dielectric and contact effects in hydrogenated amorphous silicon-silicon nitride thin-film transistors, Journal of Applied Physics, **65**, 3951–3955, 1989.

2.33 M. Ohta, M. Tsumura, J. Ohida, J. Ohwada, and K. Suzuki: Active matrix network simulator considering non-linear C-V characteristics of TFTs intrinsic capacitances, In Proceedings of the 12th International Display Research Conference (Hiroshima, 1992), 431–434. California: SID, 1992.

2.34 M. Nakazato and T. Higuchi: Capacitance-voltage characteristics of a-Si TFTs, In Proceedings of the 12th International Display Research Conference (Hiroshima, 1992), 439–442, California: SID, 1992.

2.35 D. E. Carlson and C. R. Wronski: Amorphous silicon solar cell, Applied Physics Letters, **28**, 671–673, 1976.

2.36 H. Yamamoto, T. Baji, H. Matsumaru, Y. Tanaka, K. Seki, T. Tsukada, et al.: High speed contact type linear sensor array using a-Si pin diodes, In Extended Abstracts of the 15th Conference on Solid State Devices and Materials (Tokyo, 1983), 205–208, The Japan Society of Applied Physics, 1983.

2.37 I. Shimizu, T. Komatsu, K. Saito, and E. Inoue: A-Si thin film as a photoreceptor for electrophotography, Journal of Non-Crystalline Solids, **35 & 36**, 773–778, 1980.

2.38 Y. Kaneko, A. Sasano, T. Tsukada, R. Oritsuki, and K. Suzuki: Improved reliability in amorphous silicon thin-film transistors, In Extended Abstracts of the 18th International Conference on Solid State Devices and Materials (Tokyo, 1986). 669–702. The Japan Society of Applied Physics, 1986.

2.39 M. Katayama, H. Morimoto, S. Yasuda, T. Takamatu, H. Tanaka, M. Hijikigawa: High-resolution full-color LCDs addressed by double-layered gate-insulator a-Si TFTs, In Digest of Technical Papers of the Society for Information Display International Symposium (Anaheim, 1988), 310-313, California: SID, 1988.

2.40 D. L. Staebler and C. R. Wronski: Reversible conductivity changes in discharge-produced amorphous Si, Applied Physics Letters, **31**, 292–294, 1977.

2.41 R. A. Street: Hydrogenated Amorphous Silicon, p. 216. Cambridge: Cambridge University Press, 1991.

2.42 R. Oritsuki, T. Horii, A. Sasano, K. Tsutsui, T. Koizumi, Y. Kaneko, and T. Tsukada: Threshold voltage shift of a-Si TFTs during pulse operation, In Extended Abstracts of the International Conference on Solid State Devices and

Materials (Yokohama, 1991), 635–637. The Japan Society of Applied Physics, 1991.

2.43 T. Toyabe, H. Masuda, Y. Kaneko, A. Sasano, H. Fukushima, and T. Tsukada: A two-dimensional numerical model of amorphous silicon thin-film transistors, In Technical Digest of the International Electron Devices Meeting (Washington, D. C., 1986), 575–578, IEEE, 1986.

2.44 S. Morozumi, T. Ohta, R. Araki, T. Sonehara, K. Kubota, Y. Ono, T. Nakazawa, H. Ohara: A 250 × 240 element LCD addressed by lateral MIM, Proc. Intn'l Display Res. Conf., **10-3**, 404–407, 1983 (Kobe).

2.45 R. A. Gibson, P. G. LeComber, and W. E. Spear: The characteristics of high current amorphous silicon diodes, Appl. Phys., **21**, 307–311, 1980.

2.46 N. Szydlo, E. Chartier, N. Proust, J. Magarino, and D. Kaplan: High current post-hydrogenated chemical vapor deposited amorphous silicon p-i-n diodes, Appl. Phys. Lett., **40**, 988–990, 1982.

2.47 D. G. Ast: Materials Limitations of amorphous-Si:H transistors, IEEE Tr. Electron Devices, **ED-30**, 532–539, 1983.

2.48 N. Szydlo, E. Chartier, J. N. Perbet, N. Proust, J. Magarino, and M. Hareng: Integrated matrix-addressed LCD using amorphous-silicon back-to-back diodes, Proc. SID (Society for Information Display), 265–268, 1984.

2.49 Y. Goto, T. Ogawa, S. Sawada and S. Sugimori: Fluorinated liquid crystals for active matrix displays, Mol. Cryst. Liq. Cryst., **209**, 1–7, 1991.

2.50 V. Reiffenrath, J. Krause, H. J. Plach and G. Weber: New liquid-crystalline compounds with negative dielectric anisotropy, Liq. Cryst., **5**, 159–170, 1989.

2.51 M. Schadt, M. Petrzilka, P. R. Gerber and A. Villiger: Polar alkenyls: physical properties and correlations with molecular structure of new nematic liquid crystals, Mol. Cryst. Liq. Cryst., **122**, 241–260, 1985.

2.52 D. Lippens, J. P. Parneix and A. Chapoton, Journal de Physique, **38**, 1465, 1977.

2.53 W. H. de Jeu, W. A. P. Claassen and A. M. J. Spruijt: The determination of the elastic constants of nematic liquid crystals. Mol.Cryst. and Liq. Cryst., **37**, 269–280, 1976.

2.54 M. Schadt and W. Helfrich: Voltage-dependent optical activity of a twisted nematic liquid crystal, Applied Physics Letters, **18**, 127–128, 1971.

2.55 E. Jakeman and E. P. Raynes: Electro-optic response times in liquid crystals, Physics Letters, **39A**, 69–70, 1972.

2.56 K. Katoh, S. Imagi and N. Kobayashi: Active-matrix-addressed color LCDs for avionic application. In Digest of Technical Papers of the Society for Information Display International Symposium (Anaheim, 1988), 238–241. California: SID, 1988.

2.57 C. H. Gooch and H. A. Tarry: The optical properties of twisted nematic liquid crystal structures with twist angles $\leq 90°$, Journal of Physics D: Applied Physics, **8**, 1575–1584, 1975.

2.58 C. H. Gooch and H. A. Tarry: Optical characteristics of twisted nematic liquid-crystal films, Electronics Letters, **10**, 2–4, 1974.

2.59 R. M. A. Azzam and N. M. Bashara: Simplified approach to the propagation of polarized light in anisotropic media-application to liquid crystals, Journal of the Optical Society of America, **62**, 1252–1257, 1972.

2.60 F. Funada, M. Okada, N. Kimura and K. Awane: Selection and optimizing of liquid crystal display modes for the full color active-matrix LCDs, Journal of the Institute of Television Engineers, **42**, 1029–1034, 1988 (In Japanese).

2.61 D. W. Berreman: Optics in smoothly varying anisotropic planar structures: application to liquid-crystal twist cells, Journal of the Optical Society of America, **63**, 1374–1380, 1973.

2.62 K. H. Yang: Two-domain twisted nematic and tilted homeotropic liquid crystal displays for active matrix applications. In Proc. International Display Research Conference (San Diego, 1991), 68–72, California: SID, 1991.

2.63 Y. Koike, T. Kamada, K. Okamoto, N. Ohashi, I. Tomita and M. Okabe: A full-color TFT-LCD with a domain-divided twisted-nematic structure. In Digest of Technical Papers of the Society for Information Display International Symposium (Boston, 1992), 798–801, California: SID, 1992.

2.64 R. Kiefer, B. Weber, F. Windscheid, G. Baur: In-plane switching of nematic liquid crystals, Proc. Intn'l Display Res. Conf., P **2-30**, 547–550, 1992 (Hiroshima).

2.65 K. Ohmuro, S. Kataoka, T. Sasaki, Y. Koike: Development of super-high-image-quality vertical-alignment-mode LCD, Digest Tech. Papers Society for Information Display Intn'l Symposium, 845–848, 1997.

2.66 R. A. Soref: Transverse field effects in nematic liquid crystals, Appl. Phys. Lett., **22**, 165–166, 1973.

2.67 R. A. Soref: Field effects in nematic liquid crystals obtained with interdigital electrodes, J. Appl. Phys., **45**, 5466–5468, 1974.

2.68 M. Ohta, M. Oh-e, K. Kondo: Development of Super-TFT-LCDs with in-plane switching display mode, Proc. Intn'l Display Res. Conf., S **30-2**, 707–710, 1995 (Hamamatsu).

2.69 M. Oh-e and K. Kondo: Electro-optical characteristics and switching behavior of the in-plane switching mode, Appl. Phys. Lett., **67**, 3895–3897, 1995.

2.70 P. M. Alt and P. Pleshko: Scanning limitations of liquid-crystal displays, IEEE Transactions on Electron Devices, **ED-21**, 146–155, 1979.

2.71 J. Nehring, A. R. Kmetz: Ultimate limits for matrix addressing of rms-responding liquid-crystal displays, IEEE Transactions on Electron Devices, **ED-26**, 795–802, 1979.

2.72 T. J. Scheffer and J. Nehring: A new, highly multiplexable liquid crystal display, Applied Physics Letters, **45**, 1021–1023, 1984.

2.73 F. Leenhouts and M. Schadt: Electro-optics of supertwist displays; dependence on liquid crystal material parameters, In Proc. 6th International Display Research Conference (Tokyo, 1986), 388–391. California: SID, Tokyo: ITE, 1986.

2.74 K. Kinugawa, Y. Kondo, M. Kanasaki, H. Kawakami and E. Kaneko: 640×480 pixel LCD using highly twisted birefringence effect with low pretilt angle, In Digest of Technical Papers of the Society for Informatio Display International Symposium (San Diego, 1986), pp. 122–125. California: SID, 1986.

2.75 P. A. Breddels and H. A. van Sprang: An analytical expression for the optical threshold in highly twisted nematic systems with nonzero tilt angles at the boundaries, Journal of Applied Physics, **58**, 2162–2166, 1985.

3 Laser Crystallization for Polycrystalline Silicon Device Applications

James B. Boyce and Ping Mei

Xerox Palo Alto Research Center, Palo Alto, CA 94304, USA
E-mail: boyce@parc.xerox.com; pmei@parc.xerox.com

Abstract. Pulsed excimer-laser processing of amorphous silicon on non-crystalline substrates is an important processing technology for large-area polysilicon electronics, such as flat-panel displays and two-dimensional imaging arrays. The technique allows for the creation of CMOS polysilicon on glass substrates, the integration of polysilicon silicon and amorphous devices on the same glass substrate, and the fabrication of self-aligned amorphous silicon thin-film transistors via laser doping. Materials studies show that laser-crystallized polysilicon contains larger grains with fewer defects than polysilicon prepared by other techniques and exhibits large lateral grain growth in a narrow range of excimer laser energy density, with a corresponding peak in the electron mobility. This interesting materials phenomenon provides the opportunity to create excellent polysilicon for device applications, using an appropriate region of parameter space to avoid a non-uniform grain-size distribution and large surface roughness. Also, laser-processing enhancements, such as laser doping and fabrication of self-aligned transistors, provide additional tools to fabricate unique devices. These materials and device processing issues are described, and the device results are presented. For the devices, emphasis is placed on homogeneity and stability as well as on important performance parameters, such as thin-film transistor mobility and leakage currents.

3.1 Introduction

Over the last decade, interest in polycrystalline-silicon (poly-Si) devices has been driven largely by the rapid growth of large area electronic systems, specifically, flat-panel active-matrix displays and two-dimensional imagers [3.1–21]. Currently, amorphous silicon (a-Si) is the semiconductor used in the electronics of these large area systems, since it can readily be deposited on large glass substrates and processed into thin film transistors at low temperatures. By low temperature is meant below about 600°C which is a typical softening point of the glass substrates used for the large-area systems. This can be increased somewhat for the newer glasses, such as Corning 1737, which has a softening point of 660°C [3.22]. In any case, the separation point between high and low temperature processing is in the vicinity of 600°C, and we will use it to specify the dividing point between the two regions with the understanding that this value is approximate and may be increased somewhat as the substrates improve.

Processing of a-Si is compatible with this low-temperature glass but its low speed and low current carrying capability limit it to pixel switch applications. The driver circuits such as, shift registers, multiplexers, and amplifiers, require the faster performance of crystalline silicon. These circuits are currently external to the flat panel array and require costly packaging to assemble. Polycrystalline silicon has an electron mobility that is several orders of magnitude larger than that of amorphous silicon, typically $\approx 100\,\mathrm{cm^2/Vs}$ versus $< 1\,\mathrm{cm^2/Vs}$. This allows poly-Si to be applicable to the peripheral drive circuitry which requires higher speeds and/or higher current carrying capability than amorphous silicon can provide. In addition, poly-Si provides CMOS capability that is missing for a-Si due to its poor p-type device performance. Poly-Si CMOS allows the fabrication of low power-consumption circuits. As a result, poly-Si can provide the functionality and performance necessary for low-power driver circuits that can be fabricated on the glass panel. This architecture reduces the number of external connections, as illustrated in Fig. 3.1, and thereby the packaging costs.

To create the polycrystalline silicon, a variety of techniques have been employed, ranging from direct deposition of poly-Si to laser crystallization of deposited a-Si precursors. The former typically requires temperatures in excess of 600°C to obtain good electronic grade material. As a result, this technique is limited to wafer-sized fused-silica substrates for applications in smaller electronic systems such as projection displays. To be compatible with the glass substrate, the two crystallization processes are thermal crystallization in the solid phase or laser crystallization from the melt [3.17]. The latter is the focus here since excimer-laser crystallization of amorphous silicon is a preferable processing technique, as described in Sect. 3.2.

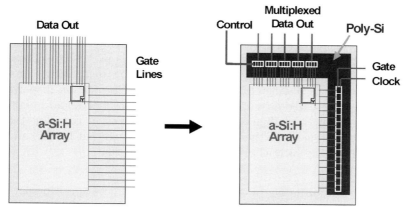

Fig. 3.1. An illustration of the simplified packaging allowed by poly-Si peripheral circuits on an amorphous Si large-area array [3.15]

For a viable laser crystallization process, high quality poly-Si has to be formed over the entire area of the glass plate or edge regions for peripheral circuits. Poly-Si is rich in defects, primarily associated with the grain boundaries. These defects create charge-trapping centers that limit the electrical transport through the material and degrade other device parameters as well [3.23]. Nonetheless, device-grade material can and has been made using laser crystallization and devices with excellent characteristics have been fabricated. The homogeneity of this material and the devices, however, has been problematic [3.25]. The grain size and trap-state density depend strongly on the crystallization and other processing parameters. Any variations in grain size, surface topography, interface structure, and the like translate into variations in the device parameters. Obtaining device-grade material and achieving homogeneity for large-area poly-Si electronic circuits is a major focus of this chapter.

Section 3.2 describes the crystallization process and the properties of the resulting poly-Si material. Emphasis here is placed on excimer-laser crystallized poly-Si. Laser doping of the poly-Si is also discussed. Section 3.3 describes the fabrication and the characteristics of poly-Si thin film transistors (TFTs). Section 3.4 presents a hybrid amorphous and poly-Si processing technique and describes its advantages and the circuit results. Section 3.5 concludes this chapter with a brief discussion of applications of poly-Si.

3.2 Laser Processing of Polysilicon

3.2.1 Polysilicon

Until recently, the primary application of polysilicon was in gate electrodes for metal-oxide-semiconductor (MOS) transistors in integrated circuits [3.23, 24]. Now with the rapid growth of large-area electronics and its need for higher performance devices than provided by a-Si, poly-Si is finding another major application. Currently, several of the major display manufacturers have prototype poly-Si display fabrication facilities and have plans to scale these up to mass production capacities of order 100,000 panels per month. These are large-area displays, e.g., 12.1 in, on glass, requiring low temperature processing. This activity is distinct from the smaller-sized projection displays that are currently fabricated on 4 in fused silica substrates by standard high-temperature processing techniques. There is also interest in reducing the processing temperature even further to well below 600°C so that flexible and less expensive substrates such as plastics can be used [3.26].

Growth of Polysilicon. Poly-Si can be prepared by direct deposition onto a substrate or by crystallization of an amorphous Si precursor. The poly-Si prepared by direct deposition exhibits a columnar structure and small grains. As a result, its electrical properties are poor, with low mobilities, typically about $5\,\mathrm{cm^2/Vs}$ [3.13]. Crystallization of a-Si precursors, either by solid phase crystallization or by laser crystallization, yields better poly-Si with higher mobilities.

The deposition of the precursor Si films is most often done by chemical vapor deposition (CVD) from silane gas (SiH_4), although sputtered a-Si has also been studied. Two CVD techniques are used: low-pressure chemical vapor deposition (LPCVD) at temperatures near 600°C (550–700°C) and plasma enhanced chemical vapor deposition (PECVD) at temperatures near 250°C. For LPCVD, the Si deposited in the upper part of the temperature range is polycrystalline, whereas that in the lower part of the range is amorphous. The temperature at which the transition from amorphous to crystalline occurs can be lowered by reducing the silane partial pressure and also by using disilane rather than silane [3.27]. A fully amorphous film is preferred for the precursors since any poly-Si grains present in the starting film provide a large number of nucleation sites for subsequent grain growth during the crystallization of the remainder of the film. This yields poly-Si with smaller grains and a corresponding lower mobility than poly-Si crystallized from fully amorphous precursors.

Films prepared by standard PECVD deposition are amorphous although microcrystalline silicon can be obtained by reducing the silane partial pressure and increasing that of hydrogen gas. PECVD a-Si:H can be used as a poly-Si precursor, but it contains large amounts of hydrogen, about 10 at. % for electronic grade a-Si:H. Since the hydrogen evolves during the crystallization process, it complicates the crystallization but is not a serious barrier. Typically an extra dehydrogenation step is required before PECVD a-Si:H is crystallized. For solid-phase crystallization the dehydrogenation and crystallization anneals can be combined. For laser crystallization, a dehydrogenation anneal or a laser heating at low laser fluences is used to remove the bulk of the H before a high-laser-fluence crystallization step. This avoids film ablation and cracking caused by the rapid evolution of the H during laser crystallization.

Crystallization of LPCVD or PECVD a-Si films on glass substrates is achieved either by solid-phase crystallization (SPC) at temperatures near 600°C or by laser heating. The former technique (SPC) became of interest for TFT work when it was determined that the field effect mobility was an order of magnitude larger than for as-deposited poly-Si, i.e., about 40 rather than $5\,\mathrm{cm^2/Vs}$ [3.28]. But due to the low temperatures (below $\approx$ 600°C) required by the glass substrates, the thermal-anneal times are too long for a viable manufacturing process. In addition, extra steps of high-dose Si implants [3.29] are often required for the LPCVD-deposited precursors to ensure a fully amorphous starting material for the best grain growth. In addition, the grain structure for thermally crystallized poly-Si is inferior to that of laser crystallized films, as seen in Fig. 3.2 [3.30, 31]. The laser-processed film has large, columnar grains, which have few intra-granular defects. The thermally

Fig. 3.2a,b. TEM cross section of two low-temperature processed poly-Si films on glass, providing a comparison of typical grains from (**a**) an excimer laser-crystallized film and (**b**) a thermally-crystallized film [3.31]

crystallized poly-Si, on the other hand, has smaller grains with a variety of intra-granular defects, mainly threading dislocations and stacking faults. These comparisons with laser processing have led to the decline in interest in SPC for poly-Si device fabrication. The current procedure of choice is laser crystallization and so is the process discussed in detail below.

Electrical Properties of Polysilicon. Before discussing laser crystallization, we briefly describe the electrical properties and defects in poly-Si. Although its electrical properties are superior to those of a-Si, grain boundary effects diminish its performance compared with that of single-crystal silicon. The grain boundaries disrupt the periodic structure of the crystallites, presenting discontinuities that restrict the current flow. In addition, they contain broken and strained Si-Si bonds that create trapping states that lie within the Si bandgap. These traps diminish all the important TFT parameters, namely, the mobility and subthreshold slope are lowered, the threshold voltage and leakage current are increased, and the stability is diminished.

The effects of these traps and barriers at the grain boundaries on the conductivity and mobility have been thoroughly studied for doped poly-Si [3.23]. The variation in the mobility of the carriers at room temperature as a function of doping level is shown in Fig. 3.3 [3.32]. The remarkable feature in these data is the minimum in the mobility that is not seen in single-crystal Si. This minimum is due to the effects of the poly-Si grain boundaries. As carriers are introduced into the poly-Si through doping, the free carriers become trapped at defect sites at the grain boundaries and are depleted from the grains. This causes the height of the electronic barrier at the grain boundaries, V_B, to increase with doping level as more charge is localized there. In addition, these carriers are immobilized by the traps at the grain boundaries and are lost to the conduction process. The barrier height increases with doping concen-

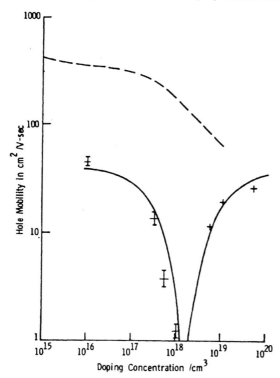

Fig. 3.3. The variation in the room-temperature Hall mobility of the carriers as a function of B doping level in polysilicon (data points) compared with that in single-crystal silicon (*dashed line*) from Seto [3.32]. The *solid line* is the theoretical calculation for poly-Si described by the model of Fig. 3.4

tration, N_d, until all the traps, of density N_T, are filled. For $N_d > N_T$, the additional carriers added by doping screen the trapped charge with the result that the barrier height drops with increasing doping concentration. This is shown schematically in Fig. 3.4 for a one-dimensional model with grains of length L [3.23].

In this model V_B is maximum at a carrier concentration of $N_d = Q_T/L$, where Q_T is the number of traps per unit area of grain boundary. The grain boundary is assumed to be of negligible thickness compared with L. When the carrier concentration is sufficient to fill all the traps, i.e., $N_d > Q_T/L$, then V_B decreases with increasing carrier concentration according to

$$V_B = qQ_T^2/8\varepsilon N_d , \qquad (3.1)$$

where q is the electronic charge and ε is the dielectric constant of poly-Si. From Fig. 3.4 and (3.1), it is evident that one requires large grains (large L) with few defects (small Q_T) for optimal electrical properties of the poly-Si. As a result, the efforts at producing good quality poly-Si have concentrated

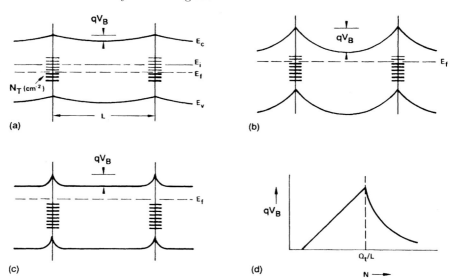

Fig. 3.4a–d. Schematic representation of the height of the electronic barrier, qV_B, at the grain boundaries. As shown in (**d**), it increases with doping level as more charge is localized there, reaches a maximum at the critical concentration, $N_d = Q_T/L$, and then decreases. Shown also are the energy diagrams for dopant concentrations (**a**) below, (**b**) at, and (**c**) above the critical concentration [3.23]

on these two aspects of the material: (1) creating large-grain poly-Si through growth and re-growth techniques, such as, laser crystallization, and (2) defect reduction through growth and passivation techniques.

Passivation of Defects in Polysilicon. Defect reduction is an important step in the production of device quality poly-Si for TFT applications [3.33]. Creating large-grain material is helpful in this regard, but passivation has the largest effect. Passivating the trap states reduces the deleterious effects of the grain boundaries since, once passivated, they no longer provide active levels within the bandgap to act as traps, thereby reducing Q_T. Since the grain boundary traps are associated with dangling or strained bonds, terminating or modifying them with hydrogen as in the case of a-Si:H and the Si-SiO$_2$ interface can passivate them. H passivation reduces the number of trap states, as shown in Fig. 3.5 [3.34] for laser-crystallized poly-Si. The dangling-bond spin density varies from 1 to $6.7 \times 10^{18}/\text{cm}^3$ in as-crystallized poly-Si for the different processing conditions used. The minimum in each of the curves corresponds to a maximum in the grain size as a function of laser fluence. This defect concentration is reduced by at least an order of magnitude for each of these poly-Si films by hydrogen passivation. The as-passivated spin density is about $1 \times 10^{17}/\text{cm}^3$ for all the films, independent of the starting defect concentration. This is a saturation value and is not lowered further by

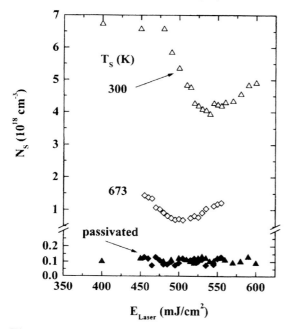

Fig. 3.5. Defect concentration from ESR measurements for laser crystallized poly-Si as a function of laser fluence before (*open symbols*) and after passivation (*solid symbols*). The triangles and diamonds represent data from samples crystallized at two different substrate temperatures (300 K and 673 K, respectively). The minimum in each of these curves corresponds to a maximum in the grain size. Hydrogenation was achieved by exposing the samples to monatomic hydrogen at 350 K for 2 h [3.34]

additional hydrogenation. A similar saturation level of defects of $10^{17}/\text{cm}^3$ is obtained for as-deposited poly-Si when optimized hydrogenation conditions are used [3.35]. This hydrogen passivation process reduces the number of traps, thereby reducing the number of trapped carriers. This causes a lowering of the grain boundary barrier height, V_B, and a corresponding improvement in conductivity and mobility. As a result, hydrogen passivation is a vital step in the production of poly-Si devices with good electrical properties as discussed in Sect. 3.3.

3.2.2 Laser Crystallization

Crystallization of a-Si using continuous wave lasers has been studied extensively in the past [3.36], but is incompatible with low-temperature substrates, since the film and substrate temperatures are heated to well above 600°C. A pulsed laser, on the other hand, provides the means of melting the Si film rapidly enough [3.37] that a low-temperature substrate is not damaged. Using UV light is also advantageous since it is highly absorbed in the Si film rather than the substrate. As a result, the most commonly used laser crystallization technique for a-Si on glass is performed with pulsed rare-gas halogen excimer lasers.

Excimer Laser. The short wavelength, high intensity, and short pulse length characteristics of the excimer laser make it a favorable tool for crystallizing amorphous silicon on glass substrates [3.38]. The short wavelength in the ultraviolet (308 nm for a XeCl excimer) ensures that the high laser energy is absorbed in the thin silicon film and not the substrate. For example, the absorption depth in silicon is about 7 nm for a XeCl excimer radiation, compared with a typical film thickness of order 100 nm. Also the reflectivity of Si in the UV varies only slightly with temperature and phase. The reflectivity at 308 nm is $\approx$ 60% for crystalline Si and $\approx$ 70% for liquid Si. As a result, there is no complication due to a large and discontinuous changes in the energy being absorbed by the film while it heats up and melts, as is the case of IR laser crystallization.

The short excimer laser pulse length of 20–200 ns, ensures that the silicon film is heated rapidly above the melting point and solidifies quickly, with the heat flowing to the unheated substrate. Typical solidification times are on the order of 100 ns [3.10] and are sufficiently short that low melting point substrates, such as glass, do not have time to flow. Thus excimer-laser crystallization does not melt the substrate and can therefore be used to produce poly-Si on glass, as well as other temperature sensitive materials such as plastic. In addition, it is well suited, due its spatial selectivity, to producing hybrid amorphous and poly-Si material and/or devices in neighboring regions of the same glass substrate. Due to the rapid heat up and cool down of this process, neighboring material and devices maintain their integrity while selected regions are being crystallized. This provides significant simplifications in circuit design, fabrication and packaging for large area electronic devices. With peripheral poly-Si electronics, such as shift registers and multiplexers, the number of external lines that have to be attached to the plate is substantially reduced, along with the packaging costs.

The raw excimer laser beam is spatially non-uniform and rounded. Since the grain growth depends strongly on the laser fluence, a uniform beam is required [3.39–41]. The sought-after beam profile is a top hat or a mesa with steep slopes. To this end, the beam is homogenized using a homogenizer that creates a more uniform beam with a top-hat beam profile. But for currently available excimer lasers and optical systems, the beam is inhomogeneous despite the homogenizer. The homogenizer averages out some spatial variations and creates a top-hat beam profile, but it also produces many interfering plane waves. The interference results in variations in the beam intensity that, for currently used optics, has a spatial frequency of order 10 μm microns, comparable to device dimensions. So even for the best currently available optical system, non-uniformities exist in the beam profile and these can translate into inhomogeneities in the laser-crystallized silicon.

One method to diminish the inhomogeneity further is to expose each point of the amorphous silicon to multiple shots while the beam is scanned over the film in steps that are smaller than the beam size [3.7, 40, 41]. This multiple-shot crystallization provides a partial averaging of the beam inhomogeneities, but the final pulse at each point is expected to imprint any beam inhomogeneities into the poly-Si.

Heat Flow Analysis. The rapid heating of the film by the laser pulse and its subsequent rapid cool down are described by the heat diffusion equation [3.10, 13]. In one-dimension,

$$\rho C_\mathrm{p}(\delta T/\delta t) = \delta/\delta z \left[\kappa(\delta T/\delta z)\right] + \alpha I(z,t) \,, \tag{3.2}$$

where z is the distance into the film and substrate. $I(z,t)$ is the energy deposited in a unit area of Si per unit time by the excimer laser, given by,

$$I(z,t) = I_0(t)\,(1-R)\,\exp(-\alpha z) \,, \tag{3.3}$$

where ρ, C_p, and κ are the film and substrate parameters, density, specific heat, and thermal conductivity, respectively. They are each functions of temperature, T, and can differ substantially for the film and substrate. R and α are the optical reflectivity and optical absorption coefficient of the Si film and are functions of temperature and laser wavelength.

Numerical solutions of this heat-flow equation for the case of a thin a-Si film on a glass or fused silica substrate with temperature dependent parameters (specific heat, thermal conductivity, density, and reflectivity) provides useful information and insight into the heating and melting during laser crystallization. A function of interest from these numerical calculations is the temperature profile (temperature versus depth and time) for a given laser pulse. If the laser pulse is ideal with a top-hat shaped spatial profile and a square temporal shape, two parameters, the energy density and the pulse duration, provide a complete description. These two parameters serve as approximate descriptors for pulses of an actual laser beam shape that is typically an asymmetric Gaussian in time and a rounded mesa in space. Some programs, such as LIMP [3.42], provide calculation results using the actual beam temporal profile, $I_0(t)$, as an input function. LIMP is a one-dimensional simulation so it does not take into account lateral heat flow through the silicon between the melted and unmelted areas or the spatial non-uniformities of the beam profile. Nonetheless, it and similar routines provide useful information on the heating and melting process.

Figure 3.6 shows the results of such a model calculation for the case of an actual asymmetric XeCl excimer laser pulse whose full width at half maximum is about 20 ns and whose fluence is approximately 550 mJ/cm^2. The film is 100 nm of LPCVD a-Si on a fused silica substrate. The film melts

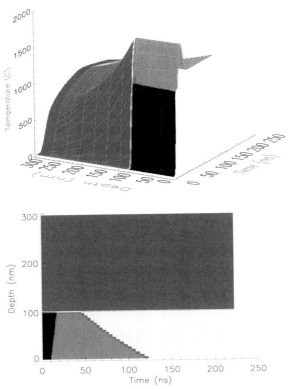

Fig. 3.6. Model calculations using the one-dimensional heat-flow equations for the case of 100 nm of a-Si on a fused silica substrate irradiated by a XeCl excimer laser pulse, 20 ns wide and about 550 mJ/cm^2 in intensity (*black* = a-Si; *gray* = liquid Si; *light gray* = poly-Si; *dark gray* = fused silica substrate)

during the pulse, is superheated well above the melting point of Si, remains molten at the interface for $\approx$ 30 ns, and solidifies completely in an additional $\approx$ 70 ns, with a melt-front velocity of $\approx$ 0.7 m/s. It should be noted that the surface region of the substrate is also heated to above the melting point of Si, thereby exceeding the softening point of glass and even that of quartz. Due, however, to the brevity of this high-temperature state, the substrate material does not have sufficient time to flow and become distorted under normal processing conditions. Extreme superheating can, however, damage the substrate interface, as discussed below.

This value of 550 mJ/cm^2 used in the calculation of Fig. 3.6 is well above the melt-through value for this case of a 20 ns pulse and a 100 nm thick film. The model calculations show that the surface melts at $\approx$ 100 mJ/cm^2 and the film melts all the way through at $\approx$ 300 mJ/cm^2. It should be mentioned that the specific laser fluence at which the various phenomena, such as melt through, occur depends also on the pulse duration and shape. Using (3.2) and (3.3) for the case of a square laser pulse of duration τ, one can derive the threshold energy density to raise the film surface to the melting point to be [3.13]

$$F_\mathrm{T} = (T_\mathrm{m} - T_0)\rho C_\mathrm{P}(\pi D\tau)^{1/2}/2(1-R)\,, \tag{3.4}$$

where T_m is the melting temperature, T_0 is the film temperature before irradiation, and D is the thermal diffusion coefficient in the film, $D = \kappa/\rho C_\mathrm{p}$. This expression shows that the threshold fluence for surface melting varies as $\sqrt{\tau}$ for a square pulse. The expression is different for other pulse shapes, but this result shows the basic variation in F_T with the temporal shape and width of the pulse. In this article we will quote specific values for the fluence at which various phenomena occur. It will be understood that these fluence values are only appropriate for the specific pulse shape and duration used in the measurements, as well as for a given spatial variation.

3.2.3 Grain Growth

Both LPCVD amorphous Si (containing little or no hydrogen) and PECVD a-Si:H (containing about 10 at. % H) have been processed successfully using laser crystallization. In the latter case, the H has to be removed either by a thermal anneal or by a stepped increase in laser exposure [3.39, 46, 47]. In both cases a remarkable phenomenon occurs in the grain growth as a function of laser energy density. During the laser crystallization process, large grains, which can be as large as 100 times the film thickness, can form [3.40–45]. Large lateral grain growth occurs in a very narrow range of laser fluences about a characteristic value, F_p [3.15, 49]. This critical fluence depends on the characteristics of the laser pulse (wavelength, shape, and pulse length) and the film thickness and the thermal properties of the substrate. For example, it is approximately $550\,\mathrm{mJ/cm^2}$ for an asymmetric Gaussian beam with a full-width at half-maximum of 20 ns and for 100 nm of Si on glass or quartz [3.15]. F_p is one of the three characteristic fluence values that separate the grain growth into four regimes. The other two characteristic fluences are the surface-melt threshold fluence and the melt-through fluence. For the asymmetric Gaussian beam above and for 100 nm of Si on glass, these two fluences are $\approx 100\,\mathrm{mJ/cm^2}$ and $\approx 300\,\mathrm{mJ/cm^2}$, respectively, and $F_\mathrm{p} \approx 550\,\mathrm{mJ/cm^2}$.

At laser fluences below $300\,\mathrm{mJ/cm^2}$, partial melting occurs and the average grain size is small and heterogeneous, as seen in Fig. 3.7. Below the surface-melt threshold fluence, about $100\,\mathrm{mJ/cm^2}$, only sample heating occurs and little or no crystallization results (Fig. 3.7a). Above the surface-melt threshold fluence, partial melting of the film occurs. In this case, the grains are larger in the surface region where the primary melt occurs and smaller near the interface where explosive crystallization happens, as seen in Fig. 3.7b–d.

The phenomenon of explosive crystallization has been well studied [3.50–53]. Briefly, explosive crystallization occurs when a rapidly moving ($>10\,\mathrm{m/s}$)

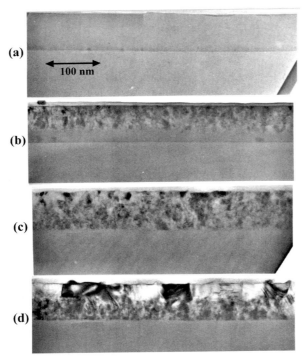

Fig. 3.7a–d. Cross sectional TEM images of 100 nm of laser-crystallized silicon on quartz, showing the heterogeneous structure of the film for laser fluences below the melt-through value. The laser fluences are (**a**) just below the surface-melt threshold of $\approx 100\,\mathrm{mJ/cm^2}$, (**b**) $150\,\mathrm{mJ/cm^2}$, (**c**) $200\,\mathrm{mJ/cm^2}$, and (**d**) $250\,\mathrm{mJ/cm^2}$, below the melt-through value of $\approx 300\,\mathrm{mJ/cm^2}$. The small-grain material near the interface has been created by explosive crystallization

molten layer propagates through the film, resulting in fine-grain poly-Si. Due to the lower melting temperature of a-Si with respect to that of crystalline Si ($\approx 1400\,\mathrm{K}$ versus $1687\,\mathrm{K}$), the primary melt created at the surface of the film by the absorbed laser pulse is supercooled and, therefore, solidifies into poly-Si. The latent heat released during this solidification heats the underlying a-Si at the interface between the solidifying-liquid and the a-Si solid below. This heat is sufficient to melt a layer of the underlying a-Si due to the larger latent heat of fusion for crystalline Si ($1800\,\mathrm{J/g}$ versus $\approx 1300\,\mathrm{J/g}$ for a-Si). This thin molten layer is supercooled and, therefore, highly unstable. It solidifies and the process repeats, creating another thin molten layer below that propagates downward into the a-Si film at high velocity, greater than $10\,\mathrm{m/s}$. The rapid cooling and solidification of the Si causes the fine-grain structure.

Above the melt-through threshold ($\approx 300\,\mathrm{mJ/cm^2}$), complete melting of the film occurs. The grains become homogeneous in size throughout the depth

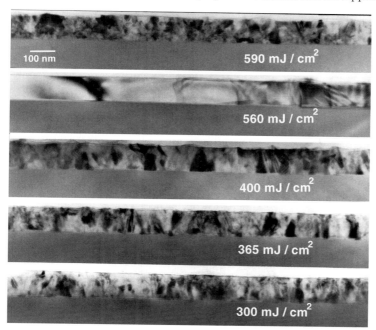

Fig. 3.8. Cross sectional TEM images of 100 nm of silicon on quartz, laser crystallized with 100 laser shots per point for laser fluences above the melt-through value of about $300\,\mathrm{mJ/cm^2}$ [3.15]

of the film; they have a columnar morphology; and their lateral extent is comparable to the film thickness, as seen in Fig. 3.8 [3.15]. The grain size increases only gradually with increasing fluence. In the vicinity of F_p ($\approx 550\,\mathrm{mJ/cm^2}$), a rapid increase in the lateral grain size occurs. Grains as large as several microns, or about 100 times the film thickness, can result. Above this peak fluence, the average grain size drops and a heterogeneous mixture of large and small grains is often observed.

Figures 3.9 and 3.10 further document the development of the large grains for the case of a scanned laser beam that is strongly overlapped so that each point of the film is exposed to multiple-laser shots [3.15]. Figure 3.9 shows the laser fluence dependence of the Si (111) X-ray peak intensity, the average grain size from the X-ray results, and the Hall electron mobility for phosphorous-doped material (about 1 at. % P). The peak in the X-ray intensity, observed in the vicinity of $550\,\mathrm{mJ/cm^2}$, is also evident in the average grain size extracted from the X-ray data and in the Hall mobility measurements. The maximum mobility occurs for the largest grain size, as is to be expected from the discussion above, i.e., due to the fact that the grain boundaries present a dominant scattering mechanism for the carriers in poly-Si.

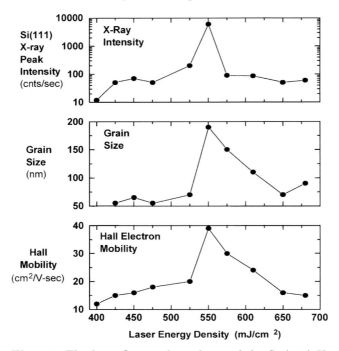

Fig. 3.9. The laser fluence dependence of the Si (111) X-ray peak intensity, the average grain size from the X-ray results, and the Hall electron mobility for phosphorous-doped material, using 100 shots per point for a 100 nm thick a-Si starting film on fused silica substrates [3.15]

The full extent of the grain growth is evident in the TEM cross-sections and planar views of Fig. 3.10 where the peak in the grain size near $F_\mathrm{p} \approx 550\,\mathrm{mJ/cm^2}$ is seen directly. Note the large grains in the vicinity of F_p and the smaller grains above and below this laser energy density. The planar views of Fig. 3.10 show that the grains in material crystallized near F_p are greater than a micron in lateral extent, much larger than the film thickness of 100 nm. Note also the narrowness of the region of large lateral grain growth. As is evident in Fig. 3.10, a change in fluence of only 5% above or below F_p leaves the large grain region and creates smaller grains of order of 100 nm, approximately the sample thickness.

This increase in grain size with fluence can be understood in terms of seeded growth from the liquid phase. In the low fluence regime, the film is only partially melted and the amorphous Si below the melt undergoes explosive crystallization that yields fine grain material [3.50–53], as discussed above. The melted portion of the film re-grows vertically into larger grain material, with nucleation sites presumably provided by the material below the primary melt. The two-layer structure of the larger-grain re-grown melt

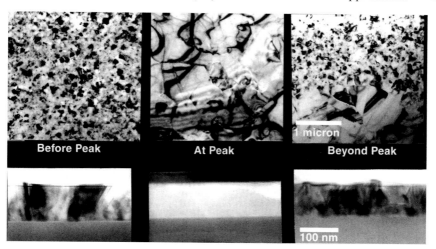

Fig. 3.10. TEM cross sections and planar views of 100 nm of Si laser crystallized below (530 mJ/cm^2), at (550 mJ/cm^2), and above (570 mJ/cm^2) the laser fluence for large lateral grain growth [3.15]

and the finer-grain explosive crystallization layer are readily seen in Fig. 3.7. The fraction of the film that is melted increases with increasing laser fluence. At the melt-through threshold, the entire film is melted and a smaller number of nucleation sites exist. These sites nucleate the grains that grow from the substrate to the surface, yielding poly-Si with a columnar morphology. If the nucleation sites are associated with regions of crystalline order or with actual crystallites at the interface, then the heating provided by the increasing laser energy would be expected to destroy more of these nucleation sites at the film/substrate interface. The decrease in the nucleation site density reduces the number of competing crystals and thereby allows for larger lateral grain growth during solidification and growth from the sparse nucleation sites. At F_p the areal density of nucleation sites has reached a minimum, allowing for a maximum in the grain size [3.41]. The grain growth at this point is still limited since heat loss to the substrate leads to supercooling and the formation of additional nucleation sites [3.44]. The results on 100 nm thick films indicate that the grain size limit several microns.

Beyond F_p, the decrease in grain size at the higher energies can be attributed to a dramatic increase in the number of nucleation sites during the rapid cooling of the molten film. One mechanism is the creation of small grain material by homogeneous nucleation from a supercooled melt [3.43, 44]. It is assumed that all the nuclei at the film-substrate interface have been destroyed at these high laser fluences. For grain growth to begin, nucleation of crystals has to occur. But for these thin films, the cooling rate is high so that supercooling of the melt can result before nucleation. If this supercooling is sufficiently large, i.e., of order 500°C, then copious nucleation can occur, resulting in small-grain material. In fact, if the cooling rate is even more rapid, exceeding about 10^{10} K/s, then crystallization is suppressed and amorphization occurs [3.50, 54, 55]. It turns out, however, that for pulsed laser crystallization of a-Si on fused silica substrates, this high cooling rate can only be achieved for films thinner than about 24 nm [3.56]. Films 24 nm or thinner can be quenched into the amorphous state, whereas thicker films are either a mixture of a-Si and poly-Si or fully polycrystalline.

In addition to homogeneous nucleation from a supercooled melt, another possible nucleation source in this high fluence regime is the creation of nuclei at the substrate due to substrate damage [3.45, 57]. A roughening of the interface is observed in the TEM cross sections of Fig. 3.10 for the above-peak region for a multiple-shot crystallization. That substrate damage can occur is reasonable since one-dimensional heat flow calculations indicate that the silicon/substrate interface can approach 2000°C at these laser energy densities (Fig. 3.6). Both mechanisms can provide numerous nucleation sites, resulting in the frustration of large lateral grain growth and the production of smaller grain material. For device fabrication, this small-grain material above the peak in undesirable so that crystallization below and near the peak is recommended.

3.2.4 Surface Roughening

Surface roughening also occurs in thin silicon films upon crystallization. Surface roughness in laser crystallized poly-Si has been attributed to the differences in the latent heat and thermal conductivity between polycrystalline and amorphous Si [3.9], to hydrogen evolution in PECVD Si [3.58], to optical interference effects, and to the freezing of capillary waves excited in the molten silicon [3.49]. For the latter model, the volume increase of silicon upon freezing (liquid: $2.53\,\text{gm/cm}^3$; solid: $2.30\,\text{g/cm}^3$) can drive the denser liquid silicon toward grain boundaries. This grain boundary region is the last to freeze, and the extra silicon accumulated there is frozen into elevated structures, enhancing the roughness. This effect is most pronounced in the laser fluence regime of large lateral grain growth, i.e., near $550\,\text{mJ/cm}^2$ for 100 nm thick films. The micrographs of Fig. 3.11 show the hillocks formed at the grain boundaries at this fluence. The elevations at the grain boundaries of this large grain material are separated by about 1.5 μm and can be as high as the film thickness.

Atomic force micrographs (AFM) yield results consistent with these TEM data. Fig. 3.12a shows a $5 \times 5\,\text{μm}^2$ area of a film crystallized by a single laser shot at $\approx 550\,\text{mJ/cm}^2$. The image shows the boundary region between large lateral grains (right) and fine grains crystallized from the supercooled melt (left). The fine grain area has a roughness on the order of 6 nm rms (root-mean-square). The large grain region has a much larger roughness of about 40 nm rms. The areas also show ridges at grain boundaries and hillocks up to 140 nm high at points of confluence for several or more grains. Figure 3.12b

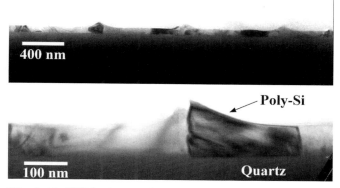

Fig. 3.11. TEM micrographs of 100 nm thick films laser crystallized at 550 mJ/cm², showing the hillocks formed at the grain boundaries [3.49]

shows a 5 × 5 μm² area of a film crystallized with 100 shots at ≈ 510 mJ/cm². The surface is substantially smoother (< 4 nm rms) and the grains have intragranular ridges and hillocks. The morphologies are consistent with the large grains in the single shot sample having grown from a single, small nucleus, and the large grains in the multi-shot sample having developed from a distributed nucleus or multiple coherent nuclei (i.e. a partially melted grain). These AFM results show that the rms roughness increases by about an order of magnitude for material crystallized at the fluence of the peak. This large roughness occurs over a very narrow range of laser energy densities, a full width at half maximum of only 30 mJ/cm² or a mere 5% of the center fluence. This is the very narrow range of large lateral grain growth discussed above and is a problematic region for device fabrication, since roughness is known to be detrimental to carrier transport [3.59]. As discussed below on Sect. 3.3.2, the adjustment of the processing parameters to avoid this regime yields excellent material and devices with good parameters and uniformity.

3.2.5 Laser Doping

Excimer laser processing provides other unique capabilities in the device-fabrication process, one of which is the doping of the TFT source/drain contacts. The traditional doping process for crystalline silicon is ion implantation, which requires a subsequent thermal process to anneal out the ion damage and to activate the carriers. For low-temperature poly-Si, this is the maximum process temperature, approaching 600°C [3.18]. Recently, it has been realized that laser doping has the potential of simplifying this process and avoiding any high-temperature processing steps [3.10, 60–65]. It provides the possibility of creating self-aligned and short-channel TFTs, as well as lowering the processing temperature. For a-Si TFTs, this process has the added advantage that both crystallization and doping of the S/D contacts

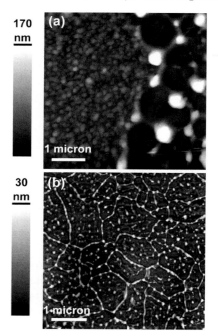

Fig. 3.12. (a) AFM image of the surface of poly-Si crystallized by a single laser shot at ≈ 550 mJ/cm². The left half has supercooled to produce small grains, while the right half has undergone large lateral grain growth, forming ≈ 1 μm grains and large hillocks. (b) AFM image of 100 shot crystallization at ≈ 510 mJ/cm². Ridges and hillocks occur both within and between the grains. In each figure the light regions are the high points [3.49]

can be performed in a single step. Due to the short excimer-laser pulse length, the silicon melts and solidifies in times on the order of 100 ns. These times are, however, sufficiently long to allow for the diffusion of dopant impurities throughout the molten regions of the film. The diffusion length for P in molten Si for this typical laser-processed melt time of 100 ns is about 70 nm, comparable to the film thickness. As a result, the laser crystallization process can be augmented by including a laser doping procedure, thereby reducing the number of process steps in the fabrication of thin-film poly-Si devices. The current laser process for thin-film poly-Si devices involves the laser crystallization of an amorphous silicon film, followed by ion implantation for the source and drain contacts. The implantation step requires a subsequent activation anneal, which, for time considerations, has to be performed as near as possible to the softening temperature of the glass substrate. Laser doping has the potential of simplifying this process and avoiding any high temperature steps. Since the entire process is compatible with low-temperature glass substrates as well as other temperature sensitive materials, the laser crystallization/doping procedure can also be used to produce hybrid amorphous and poly-Si material and/or devices in neighboring regions of the same substrate.

In the laser doping process, a laser pulse briefly melts a surface layer in a doping region, allowing the dopant species to be introduced into the molten material [3.10]. When the molten layer solidifies, the dopant species are distributed and electrically activated in the layer. The source of the dopant atoms can be either a deposited layer, such as a heavily phosphorous-doped or boron-doped amorphous silicon layer [3.65] or a dopant-containing spun-on glass [3.10, 63], or a gas containing the dopant species, such as PF_5 or BF_3 [3.60–62, 64], at the silicon surface. In each case the laser melts the silicon film and the dopant atoms diffuse into the molten region. In the solid-source doping case, the doping source is melted along with the silicon. In the case of gas immersion laser doping (GILD), pyrolosis occurs to disassociate the molecules before dopant diffusion proceeds. This process is illustrated schematically in Fig. 3.13 for the case of a gas-phase dopant source used to dope a silicon wafer [3.66]. An Al mask is patterned and used to shield regions of the Si wafer where doping is not to occur. This is the process appropriate for TFT device fabrication where only the source and drain regions are doped and the channel region remains undoped, being protected by the mask.

Dopant Distribution. Figure 3.14 shows the boron dopant depth profiles measured by secondary ion mass spectrometry (SIMS) for samples doped at three different laser energy densities [3.66, 67]. These data reveal that the doping level and doping depth are both controlled by the laser melting process. Since the boron diffusion coefficient in molten Si is about $\approx 10^{-4}$ cm^2/s, which is very fast compared with the solid phase diffusion rate ($\approx 10^{-11}$ cm^2/s), the dopant diffusion occurs primarily in the liquid phase. A higher laser doping energy results in a longer melt duration and deeper melting depth which leads to a higher doping level and deeper doping depth. During the solidification, the solid/liquid interface is moving toward the surface and the dopants are diffusing in the opposite direction. The overall effect is that the doping is slightly shallower than the melting depth.

The melt time can be measured directly using the transient reflectivity of a HeNe probe laser (wavelength = 633 nm) during the laser doping process. Since molten Si is a metal, silicon's reflectivity increases by about a factor of about 2 upon melting, from 40% to about 70% at 633 nm. This provides a direct measure of the total melt time. The melt times can be related to the melt depth using heat flow calculations of the type discussed above in Sect. 3.2.2. Good agreement is found between the calculated melt depth and the doping depth measured by SIMS, as in Fig. 3.14 [3.66]. It should be mentioned that the actual dopant profiles are sharper than the measured SIMS profiles due to the finite depth resolution of SIMS from ion mixing. So the doping depth corresponds to the point at which the SIMS concentration deviates from a constant slope. Also with increasing laser energy, the melt

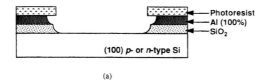
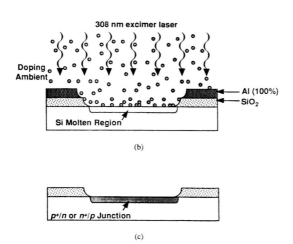

Fig. 3.13a–c. Schematic of the laser doping process for the case of a gas-phase dopant source used to dope a Si wafer. A photomasking step, (**a**), is required to create a mask, an Al reflector in this example. This defines the regions to be doped by the laser (**b**). The mask is removed to yield a doped region adjacent to an undoped region, (**c**), as in a source/drain contact for a TFT [3.66]

duration increases, allowing more dopants to be incorporated in the silicon film.

The results above are for a silicon wafer, but similar results apply for a-Si or poly-Si films on a glass substrate. Figure 3.15 shows the SIMS profiles for 100 nm of poly-Si on quartz, doped with phosphorous by the GILD process using PF_5 gas [64]. This figure shows the P profiles for a laser fluence of $560\,\mathrm{mJ/cm^2}$, above the melt-through threshold, which from heat flow calculations is about $480\,\mathrm{mJ/cm^2}$. There is a flat concentration profile throughout the thickness of the film except for a small drop near the substrate, which is pronounced at a low number of laser shots. On the other hand, at $365\,\mathrm{mJ/cm^2}$, below the melt-through threshold, the phosphorous concentration profiles are not flat throughout the film, as shown in Fig. 3.16. The concentration is seen to fall off at about one decade per 20 nm beyond a depth of 40 nm. This is the approximate instrumental resolution for the SIMS sputtering conditions that were used. So, in fact, the SIMS profile is consistent with a dopant profile that is essentially constant to a depth of about 40 nm (the melt depth at $365\,\mathrm{mJ/cm^2}$), where it falls to the background level of

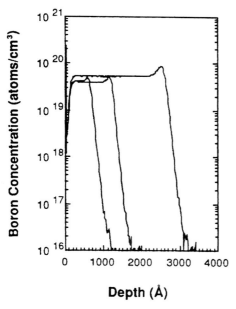

Fig. 3.14. Boron doping profiles in GILD-doped Si wafers from SIMS measurements for increasing laser fluence from left to right of approximately 760, 930, and 1280 mJ/cm² [3.67]

10^{18}/cm³. Heat flow simulations that agree with the total integrated concentrations are shown in Fig. 3.16. The difference in sharpness of the profiles is due to the finite SIMS resolution.

Electrical Properties. The electrical properties of the laser-doped films are determined using conductivity and Hall effect measurements. Typical results on the sheet resistance versus number of laser shots are shown in Fig. 3.17 [64] for several laser energy densities and for the same experimental conditions of Fig. 3.15 and Fig. 3.16. A typical sheet resistance needed for the source and drain contacts in a poly-Si TFT is about $1000\,\Omega/\square$. So these results indicate that approximately 100 shots are required to achieve this value as long as the laser energy density exceeds $400\,\mathrm{mJ/cm^2}$ for these 100 nm thick films on quartz. For 1000 shots, resistances below $30\,\Omega/\square$ are achieved. For lower laser fluences, less efficient doping occurs due to the shorter melt time and a corresponding smaller incorporation time.

The carrier concentration, extracted from Hall measurements, rises almost linearly with shots for the GILD process at a rate of about 2×10^{18}/cm³ per pulse [3.64], equivalent to a phosphorous dose of 2×10^{13}/cm² per pulse. Thus about 100 shots are required to achieve an ion-implantation dose of 2×10^{15}/cm², typical of that used for poly-Si source and drain contacts.

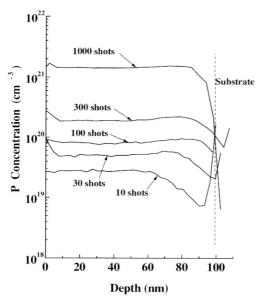

Fig. 3.15. SIMS data showing phosphorous depth profiles in 100 nm thick Si films on quartz substrates, laser doped at a fluence of 560 mJ/cm^2, above the melt-through threshold, with differing numbers of laser shots [3.64]

The variation of carrier concentration with laser fluence is small and only substantial at low shot densities and laser fluences below 400 mJ/cm^2. Similar results have been obtained in the case of B for p-type doping.

For GILD, a large number of pulses are required to achieve high doping levels. In fact, the P dopant rate of 2×10^{13}/cm^2/pulse is two orders of magnitude less than the number of Si surface sites available for a dopant gas molecule to adsorb. As a result, more pulses are required to compensate for the low efficiency per pulse. To dope throughout the entire film thickness, high laser fluences are needed. However, high fluence and large number of shots are the conditions under which film ablation often occurs. To avoid ablation and still heavily dope throughout the film thickness, a two-step GILD doping procedure has been used. In addition, solid source doping yields high dopant levels with only a few laser shots [3.10]. The solid-source doping procedure, however, requires an additional deposition step. This extra step in simplified by using spun-on-glass as the dopant source. Nonetheless, deposited solid sources provide an efficient source of dopants and thereby may be a more appropriate procedure for realizing laser doping for the fabrication of devices with fewer laser shots.

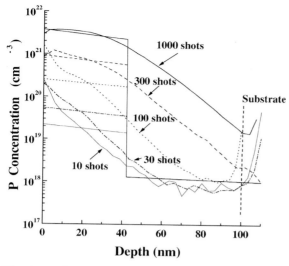

Fig. 3.16. SIMS data showing phosphorous depth profiles in 100 nm thick Si films, laser doped at a fluence of 365 mJ/cm^2, below the melt-through threshold, with differing numbers of laser shots. Heat flow simulations that agree with the total integrated concentrations are also shown. The difference in profiles sharpness is due to the SIMS resolution [3.64]

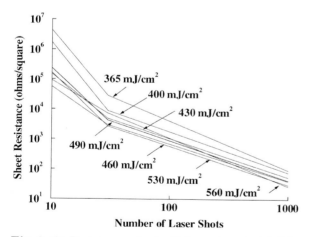

Fig. 3.17. Resistance of poly-Si, laser doped with P by the GILD process at several laser fluences [3.64]

3.3 Low-Temperature Poly-Si Devices

Three classes of laser processed device are discussed. First, self-aligned a-Si TFTs are devices that have low parasitic capacitance between the source/drain and the gate. These devices are fabricated using laser doping of the source and drain contacts and are described in Sect. 3.3.1. Second, CMOS poly-Si TFTs enable low power-consuming circuits. Such devices are required for any high-performance circuits such as the peripheral drivers. CMOS poly-Si is discussed in Sect. 3.3.2. Third are hybrid a-Si/poly-Si devices and circuits, which take advantage of the low leakage currents of a-Si and the faster performance of poly-Si, and are described in Sect. 3.4.

The main challenges in applying the laser crystallization technique in high-volume manufacturing of TFTs are; (1) improving the uniformity of all the device performance parameters over a large area, (2) reducing the TFT leakage currents to satisfy the requirements for pixel switching, and (3) improving the device stability to ensure reliable circuit performance over the period of operation. In this section, low-temperature poly-Si device fabrication and the issues of device performance, including uniformity, leakage current and stability, are discussed.

3.3.1 Device Fabrication

Device Architectures. Common poly-Si TFT architectures include top-gate co-planar, top-gate staggered, and bottom-gate staggered structures, as shown in Fig. 3.18. A major advantage of the co-planar structure, Fig. 3.18a, is that the source/drain contacts are self-aligned with respect to the channel region which minimizes the parasitic capacitance of the TFT. The smaller parasitic capacitance results in a smaller feed-through voltage on the pixel electrode arrays for display applications, reducing image flicker and sticking. For imagers, the smaller capacitance leads to better signal-to-noise ratios. The self-aligned structure also lends itself to scaling down the channel dimensions to improve the pixel aperture ratio for display or imaging applications.

An advantage of the staggered structures (Fig. 3.18b,c) is the reduction in the electric field in the drain region. Since the source/drain contacts are made on opposite sides of the channel layer relative to the gate, the electric field near the drain region is smaller than in the coplanar structure, resulting in reduced field-dependent leakage currents and a reduced kink effect. The field dependence of the leakage current are discussed in Sect. 3.3.3. The bottom-gate staggered structure, which is not self-aligned using conventional processing, can be made self-aligned using laser doping, as discussed below.

Due to the features of the laser-crystallized poly-Si described previously, the performance characteristics of the top- and the bottom-gate TFTs are quite different. The critical channel region in the top-gate TFTs is the surface layer of the Si film. Since the top layer of the poly-Si film exhibits larger grains and a smaller number of intra-granular defects, the top-gate TFTs usually have higher mobility than that of bottom-gate TFTs. However, as described in Sect. 3.2.4, the surface roughness due to the large lateral grain growth can result in diminished and non-uniform device performance for the top-gate TFTs.

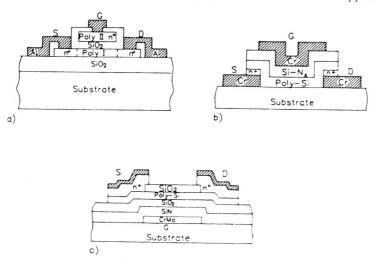

Fig. 3.18a–c. Cross sections of the various poly-Si TFT architectures: (**a**) top-gate coplanar, (**b**) top-gate staggered, and (**c**) bottom-gate staggered structures [3.13]

An advantage of the bottom-gate TFT is that the problem of the surface roughness is avoided. Their main disadvantage, however, is their relatively low mobility, due, in part, to the relatively high defect density near the bottom layer of the Si film, the TFT channel. Also, the laser processing energy density is limited by the thermal sensitivity of the gate insulator. The laser crystallization is performed on an a-Si film that has been deposited on the gate insulator. This structure is sensitive to heating from the laser processing which can damage the gate insulator, as indicated in Fig. 3.10. A lower laser energy density is required to avoid this damage to the gate insulator. As a result, the maximum electron field-effect mobility that has been obtained for bottom-gate TFTs is less than that for top-gate structures [3.86].

Process Flow. A typical process flow for the fabrication of top-gate, coplanar poly-Si TFTs is shown in Fig. 3.19. After the deposition of a buffer oxide on the glass substrate, a thin silicon layer, ≈ 100 nm thick, is deposited either by LPCVD or PECVD. Laser annealing using either a scanning beam or a large-area, non-scanned beam, then crystallizes the film. After crystallization, a CVD oxide, ≈ 100 nm thick, is deposited, followed by a polysilicon gate deposition of thickness ≈ 350 nm. After gate definition, either a phosphorus ion implant or a boron implant is performed to make the source and drain regions of the n-MOS or p-MOS devices, respectively. Typical implant conditions are 2×10^{15} cm^{-2} and 80 keV for phosphorus and 2×10^{15} cm^{-2} and 30 keV for boron. A 600°C furnace anneal is used to activate the source drain implants. This is the highest temperature step in the fabrication pro-

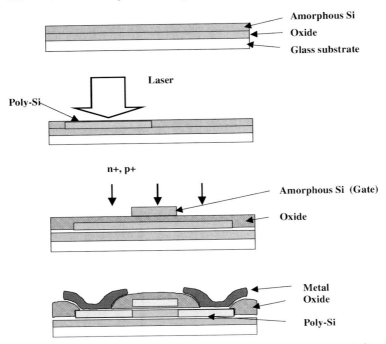

Fig. 3.19. Fabrication steps for top-gate, coplanar poly-Si TFTs [3.18]

cess. The devices then must be hydrogenated. Typically several hours in a plasma hydrogenation system at 350°C is necessary. A crossover isolation oxide can then be deposited and contacts etched to the source, drain and gate of the TFT. Then metal is deposited and patterned to form interconnects for the completed device. Using this fabrication procedure produces, TFTs are made with excellent device characteristics, as presented in Sect. 3.3.2.

As mentioned above, the 600°C furnace anneal for the dopant activation is the highest-temperature step in the poly-Si TFT fabrication process. This activation anneal can be eliminated by using laser doping rather than ion implantation for doping the source/drain contacts. This doping procedure is described in Sect. 3.2.5. It has yielded excellent devices for several different processes and device structures and has been studied by a number of groups.

Self-Aligned a-Si TFTs. Laser doping can be used to fabricate self-aligned, bottom gate a-Si TFTs. As shown in Fig. 3.18, bottom-gate structures, such as the conventional bottom-gate a-Si:H TFT, are not self-aligned. In the a-Si TFT case, depositing phosphorous-doped a-Si:H with subsequent patterning forms the source/drain contacts. Since it is difficult to selectively etch doped a-Si over the intrinsic a-Si, a top passivation island is utilized as the etch-stop to form the source/drain electrode. As a result, there is significant overlap

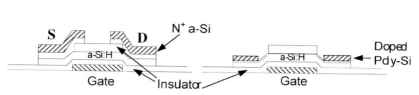

Fig. 3.20a,b. Device structures of bottom-gate a-Si:H TFTs (**a**) with the conventional doped a-Si source/drain contacts and (**b**) with laser-doped source/drain contacts [3.18]

of the source/drain electrodes with the channel region, causing additional parasitic capacitance between the gate and the source/drain. In addition, the non-self-aligned structure limits the scaling down of the channel dimension, which affects the pixel fill factor. Using laser doping, bottom-gate TFTs with a self-aligned geometry can be fabricated, as shown in Fig. 3.20 for the case of the conventional a-Si:H TFT structure. This structure reduces the parasitic capacitance and lowers the contact resistance. A factor of 20 lower contact resistance of the laser-doped source/drain has been achieved. This has also enabled a-Si TFTs with short-channel lengths down to 1 µm and with reasonable performance to be made [3.19].

3.3.2 CMOS Device Performance

The various device parameters of interest are determined from the measured transfer characteristics of the transistor, the source-drain current, I_{sd}, as a function of the gate voltage, V_g, for various source-drain voltages, V_{sd}. The parameters of interest are the on-off ratio and the leakage current, the threshold voltage, the field effect mobility, and the subthreshold slope. Also, the uniformity in these parameters across the wafer or plate is important for any circuit applications. Transfer curves for a top-gate poly-Si TFT fabricated in the manner described in Fig. 3.19 are shown in Fig. 3.21, for an n-MOS TFT with channel width/length ratio (W/L) of 50 µm/15 µm. A laser energy density of 500 mJ/cm^2 and 5 shots was applied to the 50 nm LPCVD silicon channel layer. The device has a high on-off ratio (nearly 10^8), a threshold voltage less than 2 V, a field effect mobility greater than 150 cm^2/Vs, and a leakage current of 0.3 pA/µm at 10 V drain bias. In addition, the transistor turns on abruptly, having a subthreshold slope of better than 0.3 V per decade of source-drain current. These excellent device characteristics indicate that the laser-crystallized material is also of high quality. The device parameters are discussed further below and in subsequent sections.

Subthreshold Slope. The subthreshold slope, S, parameterizes how abruptly the transistor turns on. It is degraded by defects that trap charge in the

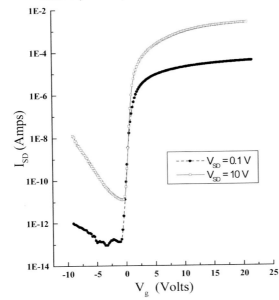

Fig. 3.21. n-MOS TFT device transfer characteristic with channel $W/L = 50/15$ for a laser energy density of $500\,\mathrm{mJ/cm^2}$ and 5 shots applied to a 50 nm LPCVD silicon channel layer. Curves for two different source-drain voltages are shown [3.18]

space charge region of the transistor. For a single-crystal MOSFET, $S \approx (1 + C_\mathrm{D}/C_\mathrm{g})$, where C_g is the gate capacitance and C_D is the space charge capacitance, which varies as the square root of the space charge density, $\sqrt{N_\mathrm{SC}}$. As a result, S increases as the square root of N_SC [3.13]. It is about $0.16\,\mathrm{V/decade}$ for a MOSFET with a substrate doping level of $10^{16}/\mathrm{cm}^3$ and a gate oxide thickness of 150 nm.

The value of $0.3\,\mathrm{V/decade}$ for the poly-Si TFT of Fig. 3.21 corresponds to a larger space charge density of order $10^{17}/\mathrm{cm}^3$, due to the large number of trapping states at the grain boundaries and intra-granular defects, as discussed in Sect. 3.2. It is also consistent with the number of dangling bond defects determined from ESR measurements after hydrogenation, as shown in Fig. 3.15. The value before hydrogenation is even larger by a factor of about 50, corresponding to a poorer $S \approx 1\,\mathrm{V/decade}$. The dramatic effect of hydrogenation on N_SC that is shown in Fig. 3.15 is also evident in the reduction in S shown in Fig. 3.22, from $1.24\,\mathrm{V/decade}$ before hydrogenation to $0.24\,\mathrm{V/decade}$ after hydrogenation. These TFTs were hydrogenated in a hydrogen plasma for 8 h at 350°C. Also, the dramatic improvement in the leakage current by 4 orders of the magnitude and in the threshold voltage is

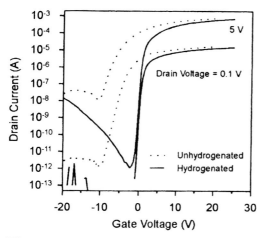

Fig. 3.22. $I_{sd} - V_g$ characteristics for a TFT with $W = 10\,\mu\text{m}$ and $L = 10\,\mu\text{m}$, showing the dramatic performance improvement obtained from hydrogenation in the leakage current, threshold voltage, and subthreshold slope. The subthreshold slope improves from 1.24 V/decade to 0.24 V/decade, the leakage current improves by four orders of magnitude, and the threshold voltage improves from -10.3 V to -0.2 V [3.21]

readily apparent in these data. Hydrogenation is evidently an important step in fabricating TFTs with good parameters, as discussed further in Sect. 3.3.3 below.

Mobility. A striking feature of the laser crystallization process is the formation of large grains in a narrow range of the laser energy density, as described in Sect. 3.2.3. This feature affects the TFT mobility and uniformity. Below the critical value, F_p, the TFT mobility increases gradually with laser energy density. This improvement in the TFT performance is due to the reduction of the intra-grain defects and the small and gradual increase in grain size. Near the region of the critical fluence, the same region for producing the maximum grain size, the TFT mobility increases abruptly and reaches a maximum value due to the large lateral grain growth. When the laser energy density is above the critical value, the TFT mobility decreases and the distribution of the TFT performance becomes non-uniform, as does the grain size.

Figures 3.23 and 3.24 show plots of mobility versus laser energy density. The data are for various numbers of laser shots for n-MOS and p-MOS transistors fabricated as described in Sect. 3.3.1 and Fig. 3.19. For both n-type and p-type transistors, the mobility peaks near $500\,\text{mJ/cm}^2$. This peak in mobility corresponds to the peak in the grain size for the experimental conditions of these measurements, namely 50 nm thick films (both LPCVD and PECVD) and a SOPRA XeCl laser with a 160 ns pulse width. Beyond this peak there

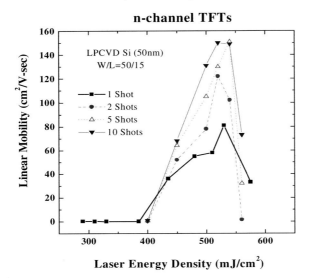

Fig. 3.23. Field effect mobility for top-gate, n-MOS TFTs as a function of laser energy density. Data are presented for various numbers of laser shots for n-MOS transistors fabricated in 50 nm thick LPCVD silicon [3.18]

is a sharp drop in mobility as the grain size drops and the ablation threshold is approached. Multiple shot annealing improves device performance, with most of the improvement seen in the first five laser shots. This corresponds to an increase in grain size and grain quality with the number of pulses.

Uniformity. Although devices with excellent characteristics have been obtained, the uniformity of the devices has been problematic [3.7, 9, 15]. Any variation in grain size, surface topography, interface structure, and the like translates into variations in the device parameters. This nonuniformity can be significant in the overlap regions for scanned laser crystallization [3.15] and is alleviated by either heating the substrate to temperatures of about 400°C during crystallization [3.7] or using a small pitch scan, i.e., one with step size much less than the beam dimension [3.9]. The inhomogeneity is lessened by these approaches but is not eliminated. The root cause is the distribution in grain size and surface roughness created by portions of the laser beam that have fluences near the fluence of the grain size peak, F_p. The sharpness of the large lateral grain growth regime exacerbates this effect since small variations in the beam intensity, either spatially in the beam profile or temporally from shot to shot, can create rough, large-grain regions in an otherwise smooth, uniform film.

In the regions with roughened surfaces, the TFT mobility drops and the threshold voltage increases. Roughness is known to be detrimental to carrier transport [3.59] and therefore creates these undesirable device characteristics.

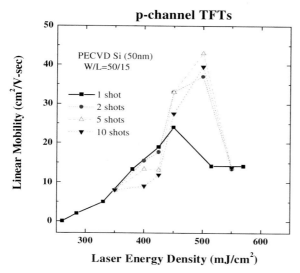

Fig. 3.24. Field effect mobility for top-gate TFTs versus laser energy density for p-MOS devices fabricated in 50 nm PECVD silicon [3.18]

Overall, as much as a 50% variation in mobility can be observed [3.15], far too big for any large area electronic applications. This is a worse case scenario that can be alleviated by crystallizing below the fluence of large grain growth. Good uniformity can be obtained by keeping the beam fluence below F_p [3.17]. The fluence must be sufficiently below F_p so that all artifacts such as interference fringes and beam fluctuations will not carry the fluence above F_p. There is a trade-off, however. As the fluence decreases, so do the grain size and the mobility. Yet the distribution in mobility over the devices on the plate will also decrease. It is a matter of creating a few spectacular devices with a wide variation versus creating all good devices with good parameters in a narrow distribution. The good devices are what are desired in large-area electronics and can readily be achieved by keeping the beam fluence below F_p.

Alternative methods to improve the device uniformity by controlling the grain size distribution have been proposed. These include seeded grain growth [3.68] and a controlled heat flow process [3.69]. In the seeded grain growth method, crystalline seeds are generated in a defined pattern to match the design of the devices. In the controlled heat flow process, the lateral heat flow is manipulated by patterning of materials with different thermal conductivity or by thermal isolation of the device region. Bottom-gate TFTs with maximum mobility of $600\,\text{cm}^2/\text{Vs}$ have been fabricated in laser crystallized poly-Si films over a SiO_2 membrane [3.70]. Using the thermal-isolation, membrane structure, large grains of several tens of microns have been generated in a pre-defined pattern, as shown in Fig. 3.25.

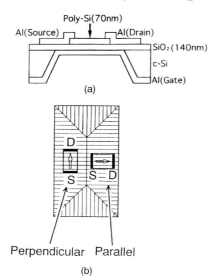

Fig. 3.25a,b. Schematics of bottom gate TFTs fabricated on a membrane, (**a**) cross section, and (**b**) top view [3.70]. Due to the lateral heat flow pattern, there are dominant orientations of the grain boundaries. Two kinds of TFTs were fabricated with the channel length parallel or perpendicular to the grain boundaries respectively. The TFT field effect mobility of the TFT with the channel length parallel to the grain boundaries is twice as that with the channel length perpendicular to the grain boundaries

Various methods for reducing the surface roughness to improve performance and uniformity have been proposed. Multiple laser irradiation is as an efficient way to reduce the surface roughness [3.12]. A laser crystallization process using two different fluence values is also reported to alleviate the surface roughness by re-planarization, yielding a small roughness of about ±5% [3.9]. In addition to modifying the laser process, a chemical-mechanical polishing has been proposed to improve the device performance for the top-gate poly-Si TFTs [3.71].

For the bottom-gate TFTs, the effect of the surface roughness is less detrimental to the device performance compared with that of the top-gate TFTs. The device uniformity may be improved by combining multiple irradiation with a lowering of the laser energy density below the critical value and by choosing the bottom-gate TFT structure. Figure 3.26 shows results from an experiment to obtain uniform TFT characteristics with a process compatible with a-Si TFTs. A 4 mm wide beam with an energy density of $\approx$ 75% of the critical value crystallized a swath of Si film that was fabricated into an array of bottom-gate TFTs. The mobility has an acceptable 10% variation over the 4 mm width of the beam then deviates substantially at the beam edges. By increasing the laser fluence to the critical energy value, the TFT mobility can be increased substantially. However, some of the TFTs exhibit large gate leakage due to damage to the gate insulator.

3.3.3 Device Leakage Currents

The leakage current requirements for the poly-Si TFTs in the peripheral driver circuits of active matrix arrays are not as stringent as that for the

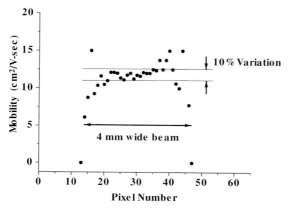

Fig. 3.26. Transistor mobility as a function of position for bottom-gate TFTs fabricated on a glass wafer and crystallized by a 4 mm wide beam [3.15]

pixel switches. A major challenges for poly-Si pixel switch is to achieve a sufficiently low leakage current to prevent the pixel from losing charge during a frame time. The allowed maximum leakage current can be estimated as,

$$I_{\text{Leak}} < \eta Q / \tau_{\text{fr}} \, , \tag{3.5}$$

where τ_{fr} is the frame time (33 ms for a video rate), η is the quality factor for the application (the percentage of charge-loss allowable), and Q is the charge stored on the pixel capacitor. The maximum charge stored in the pixel depends on the storage capacitor and the operation voltage:

$$Q = C_{\text{store}} V_{\text{op}}, \quad C_{\text{store}} = \frac{\varepsilon A}{d} \, . \tag{3.6}$$

For a display array with a pixel size of $100 \times 100\,\mu\text{m}^2$, the pixel capacitance from (3.6) is about 0.25 pF. Under an operating voltage of 10 V and a quality factor of 2%, the maximum leakage current for the pixel switch is about 1.5 pA. The restrictions on the leakage current are more stringent when the pixel size is smaller or the operating voltage is lower. For imaging applications the frame time is often much larger (≈ 1 s) requiring a much smaller value of the leakage current allowable for the pixel switches.

Poly-Si TFTs exhibit higher off-state (leakage) currents compared with a-Si:H TFTs or single crystalline MOSFETs. Under a gate-off condition, the leakage current exhibits an exponential dependence on the bias on the gate and the drain. In addition, the leakage current is asymmetric in most cases when source and drain are reversed, due to an asymmetric spatial distribution of the defects in the channel, as discussed below. Figure 3.27 illustrates the difference in the leakage current when the source and drain are reversed, while the on-state behavior is unchanged [3.72].

The field-dependent leakage current in poly-Si TFTs originates from intragrain and grain boundary defects or traps. Several authors have proposed

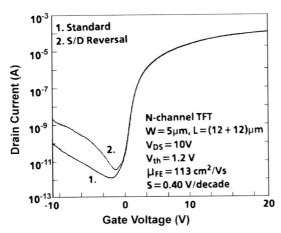

Fig. 3.27. Leakage current characteristics of a dual gate n-channel TFT with reversed source/drain connections [3.72]

models for the leakage current, most of which are based on field-emission or field-enhanced tunneling at trap sites in the high-field region near the drain [3.73–75]. Various experimental results show that the leakage current is a direct consequence of the high drain field in the poly-Si. The leakage current may vary from device to device depending on the spatial location of the defect as well as the energy of the traps in the poly-Si bandgap [3.75]. For example, when the traps are near the edge of the gate, where the electric field is relatively low, the leakage current is low. As the traps move further from the gate edge toward the source, the leakage current at first rises as the traps cross the peak channel electric field, and then decrease as they move nearer to the source.

Currently, the most efficient method to reduce the leakage current is to incorporate hydrogen into the TFT channel layer to passivate the trap states. The hydrogenation methods include exposure to a hydrogen plasma at an elevated temperature [3.73], hydrogen ion implantation followed by thermal activation of hydrogen [3.76], and deposition of a thin hydrogenated silicon nitride film followed by thermal anneal to drive hydrogen from the nitride film into the silicon channel [3.77]. Figure 3.22 shows the effects of hydrogenation on the leakage current reduction. After 8 h of hydrogenation at 350°C, the leakage current is reduced by 4 orders of the magnitude. In addition to hydrogenation, fluorine passivation and ammonium plasma passivation are also reported to effect leakage current reduction [3.78, 79].

Reducing the electric field strength along the channel in the drain depletion region can minimize the TFT leakage current. The field can be reduced by offsetting the source/drain contacts where a space of order a few microns is introduced between the source/drain regions and the channel region defined by the gate. The field can also be reduced by introducing lightly doped

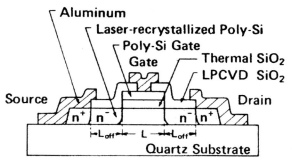

Fig. 3.28. Schematic cross-sectional view of a lightly doped source/drain structure for leakage current reduction [3.80]

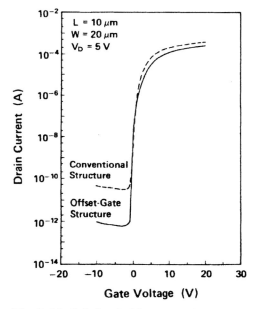

Fig. 3.29. Subthreshold current characteristics in a conventional poly-Si TFT and an offset-gate-structure poly-Si TFT [3.80]

source and drain regions, as shown in Fig. 3.28. In this case, the lightly doped regions were formed by phosphorus implantation with a dose of $10^{12}/\text{cm}^2$, compared with a dose of $10^{15}/\text{cm}^2$ for the n^+ contacts. Figure 3.29 shows a comparison of the leakage current between a conventional top-gate poly-Si TFT and a TFT with the source/drain-offset structure. The offset-structure TFT has a leakage current reduction of one to two orders of magnitude [3.80]. Although the source/drain-offset structure increases the device contact resistance, the on-state current reduction is insignificant as long as the design of the offset region is appropriate.

3.3.4 Device Stability

The stability of poly-Si TFTs is of importance for large area electronic applications, especially for the driver circuitry where the drive currents and frequencies are high. The various types of poly-Si TFT degradation occur under a combination of gate and drain bias stress. The degradation includes any or all of the following: an increase in the leakage current, a shift in the threshold voltage, a decrease in the on-current, and a hysteresis in the on-current. The first two are the most troublesome. Most degradation occurs when a device is stressed in the "on-state". Unlike single-crystal silicon, which degrades due to hot-carrier effects, poly-Si TFTs can degrade significantly when the device is in the on-state with a high channel current flowing, even when the source-drain bias is low [3.81]. In this case, both the threshold voltage and the leakage current increase. In addition, stressing using ac stress conditions leads to a more severe performance loss than comparable steady-state, dc stress conditions.

Several mechanisms have been proposed to account for the instability [3.82–84]. Hack et al. [3.82] proposed that changes in the characteristics of stressed poly-Si TFTs are caused by the generation of defects in the poly-Si in response to changes in the free carrier densities. There are two types of the stress-related defects, each with a different position within the energy bandgap. Stressing in the TFT linear operation region ($V_g > V_{sd}$) produces defects around or below midgap. The defects generated are distributed symmetrically spatially from source to drain, causing a threshold voltage shift. Degradation under saturation conditions ($V_g < V_{sd}$) results in additional defects being created higher up in the bandgap. These defects are in the high field region near the drain, which causes asymmetric device behavior. The origins of both types of defects are related to weak silicon bonds. These stress-induced defects are metastable since annealing of these devices at about 300°C will eliminate the defects and restore the device performance to its original state.

Fonash et al., [3.84] have described a reversible passivation/depassivation stress phenomenon. After an on-state stress with V_{sd} of 20 V, the device exhibits a lower leakage current when measured with the source/drain voltage polarity reversed relative to its polarity used for stressing. The same device was then subjected to a second on-state stress with the source/drain stress voltage reversed. The leakage current after this stress was lower when measured with the source/drain polarity the same as that of the first stress voltage (Fig. 3.30). Since the poly-Si devices, which were not hydrogenated, do not exhibit this behavior, the authors attribute this phenomenon to hydrogen release or capture at defects and positively charged hydrogen motion in the stress electric field.

The high field related instability may be reduced by the same methods used in the leakage reduction described in Sect. 3.3.3. For example, the device degradation for the staggered TFT structure, which has a lower field near the

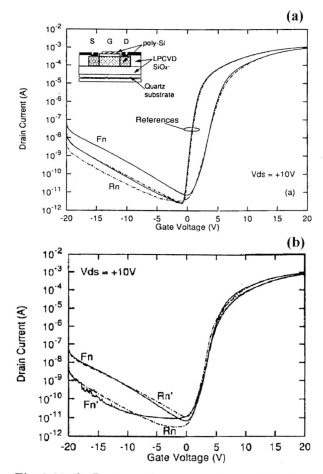

Fig. 3.30a,b. Device performance under on-state stress with (**a**) forward and then (**b**) reverse source/drain bias of 20 V. Fn and Fn' are measured for the first and second stresses with the source/drain connected with the same polarity as that of the stress voltage. Rn and Rn' are measured for the first and second stresses with the source/drain polarity reversed relative to that of the stress [3.84]

drain, is less severe compared with that of the planar structure. Figure 3.31a shows the stress effect of a double, bottom-gate staggered poly-Si TFT made by laser crystallization. The stress was performed under linear conditions for 4 h. There is a small positive shift of the threshold voltage. The other types of degradation, i.e., changes in leakage current, on-current, and hysteresis in the on-current, are not observed. Figure 3.31b shows the results of device degradation under a stress of high drain bias, where the hysteresis in the

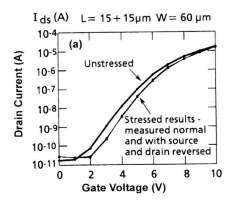

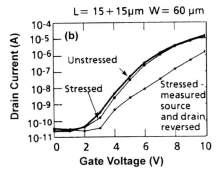

Fig. 3.31a,b. Transfer characteristics for an unstressed TFT and a TFT stressed for 4 h with (a) $V_g = 15$ V and $V_{sd} = 10$ V and (b) $V_g = 10$ V and $V_{sd} = 15$ V [3.85]

on-current is observed. Overall, the degradation of this device is much less than that of the device with single gate and planar structure.

In general, poly-Si TFTs have better stability under stress than a-Si TFTs. With proper defect passivation, such as hydrogenation, the change in the device properties for poly-Si TFTs are sufficiently small that they can be used for various integrated circuit applications.

3.4 Integration of a-Si and Poly-Si TFTs

A-Si:H thin film transistors possess low off-state currents, making them ideal pixel switches in large area arrays. However, the carrier mobility of a-Si:H is too low for high-speed peripheral driver circuits so that current technology relies on connecting the a-Si:H arrays to external high-speed, crystalline-silicon circuits. In contrast, the fabrication of large-area arrays with poly-Si pixel switches and peripheral drivers requires poly-Si TFTs to have uniform performance and low off-state currents. The alternative discussed here is a hybrid system combining a-Si and poly-Si TFTs as illustrated in Fig. 3.1.

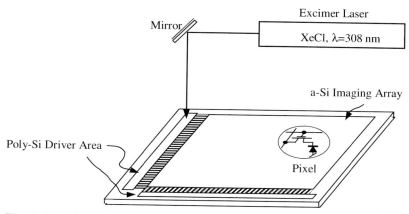

Fig. 3.32. Schematic of a laser crystallization system for converting a-Si:H material to poly-Si films for the drive circuits in the peripheral areas of an imaging array

The hybrid approach utilizes the low leakage current a-Si:H TFTs to fulfill the requirements for the pixel switches and high-mobility poly-Si for the peripheral circuits. Poly-Si devices have the high mobility needed for peripheral drive circuitry, and the requirements on the TFT off-state currents are not as stringent as for the pixel switches. In addition, since the area of the peripheral circuits is generally smaller than the laser beam width (Fig. 3.32), uniformity of the poly-Si TFTs is easier to achieve with a one-dimensional laser scan for the crystallization process than for a two-dimensional scan to crystallize the entire plate. This approach reduces the number of external connections and thereby lowers the packaging costs of large-area electronics. The cost advantage is more pronounced for high-resolution arrays when the pitch of the pixels becomes less than the current bonding pitch for the peripheral circuits, about 100 μm.

3.4.1 Development of Hybrid a-Si and Poly-Si Devices

Integration of a-Si and poly-Si TFTs relies on simple modifications of the existing a-Si device fabrication process. The attempt has been focused on selective laser crystallization of a-Si:H in order to have the same process steps for most of the fabrication of a-Si and poly-Si TFTs. An early study by Sera et al. in 1989 [3.3] demonstrated the principle of a staggered top-gate TFT structure. Figure 3.33 describes the fabrication process. In general, a top-gate poly-Si TFT has a higher mobility compared with that of a bottom-gate TFT, as described in Sect. 3.3.1. However, the performance of a top-gate a-Si TFT is inferior to that of the bottom-gate a-Si:H TFTs [3.86]. The mobility of the top-gate a-Si:H TFT [3.1] is about $0.2\,\text{cm}^2/\text{Vs}$, which is much less than that of a conventional bottom-gate a-Si:H TFT, $\approx 1\,\text{cm}^2/\text{Vs}$.

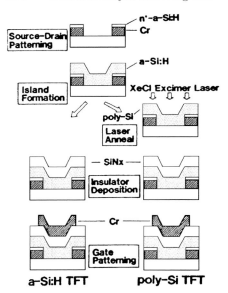

Fig. 3.33. Fabrication process for top-gate a-Si:H and poly-Si TFTs using the selective excimer laser annealing method [3.3]

This work on the top-gate TFTs showed that the inverse staggered (bottom-gate) TFT structure is more suitable for the integration of a-Si and poly-Si devices. An early demonstration of the monolithic fabrication of bottom-gate hybrid a-Si:H and poly-Si TFTs was reported by Shimizu et al. in 1990 [3.6]. The Si film is grown by the CVD process, and the poly-Si material is created by selective pulsed laser crystallization. The fabrication procedure is shown in Fig. 3.34. The discrete a-Si:H and poly-Si TFTs have field effect mobilities of 1.3 and 60 cm^2/Vs, respectively. A demonstration liquid crystal display with a-Si:H pixel switches and poly-Si peripheral circuits was reported by Tanaka in 1993 [3.87], with circuit diagram as shown Fig. 3.35. The poly-Si TFTs connect each output terminal of the driver IC to two drain lines in the pixel array to reduce the number of driver ICs by one half. This work shows the simplicity of the fabrication with an all n-type TFT design.

The major technical difficulties of monolithic fabrication of both a-Si and poly-Si TFTs on the same low temperature substrate are achieving selective laser crystallization, and control of the TFT threshold voltage. Selective laser crystallization of PECVD a-Si:H has to deal with the problem of hydrogen evolution from the hydrogenated a-Si:H film. The threshold voltage issue is related to an appropriate gate insulator for both a-Si and poly-Si TFTs. Both issues are addressed in the following section.

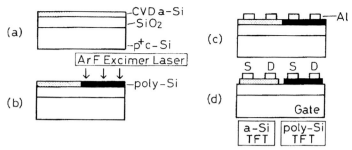

Fig. 3.34a–d. Fabrication flowchart for the on-chip bottom-gate a-Si and poly-Si TFTs with selective laser crystallization process. (**a**) deposition of CVD a-Si film, (**b**) local crystallization by an excimer laser, (**c**) TFT source and drain patterning, and (**d**) TFT island formation [3.6]

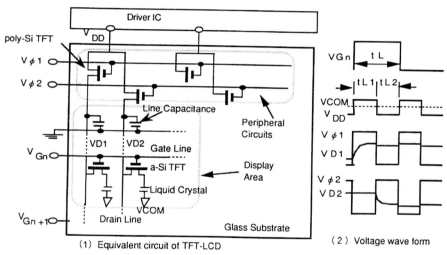

Fig. 3.35. Schematic of the poly-Si peripheral circuitry used to reduce the number of driver ICs in a display [3.87]

3.4.2 Hybrid Materials Processing

To maintain high performance a-Si TFTs, a low temperature PECVD process is used to deposit a-Si:H films for the fabrication of the hybrid a-Si and poly-Si devices. The low-temperature PECVD a-Si:H films contain a large amount of atomic hydrogen, typically 10 at. % which passivates Si dangling bonds and determines the electronic properties of the a-Si:H [3.88].

Whereas H is essential for good electrical performance of a-Si devices, it is problematic for a poly-Si precursor since it causes film ablation and cracking during laser crystallization due to the rapid hydrogen out-diffusion. To prevent this explosive H evolution, a 450°C furnace anneal for several

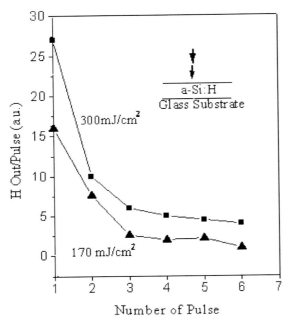

Fig. 3.36. Hydrogen out-diffusion per laser pulse as a function of the number of the laser pulses. The a-Si:H sample was first irradiated by a excimer laser at $170\,\text{mJ/cm}^2$ with a number of pulses. The same sample was then irradiated by the laser at a higher energy density ($300\,\text{mJ/cm}^2$) [3.90]

hours is usually required to remove most of the H from the a-Si:H films prior to laser crystallization. This anneal makes it impossible to fabricate high quality a-Si devices because of the significant loss in H in the amorphous devices at 450°C.

There have been several approaches to make high quality materials for both a-Si and poly-Si devices on the same substrate. Shimizu et al. [3.6] describe an approach in which a-Si deposition is carried out at 450°C using disilane, resulting in a smaller amount of H (3 at. %) in the material. Because of the small hydrogen content in the starting material, the laser ablation is avoided during selective crystallization. This small amount of hydrogen, however, is detrimental to the a-Si TFT performance. In order to achieve high performance, a crucial hydrogenation procedure had to be performed on both a-Si and poly-Si devices.

An alternative approach is reported by Aoyama et al. [3.89]. In their work, the precursor for the poly-Si film is a 40-nm thick hydrogenated a-Si:H. The thin a-Si:H film allows thorough laser crystallization without hydrogen induced film ablation. A restriction in this process, however, is a limit on the laser energy density to a value below that required to get a large poly-Si grains.

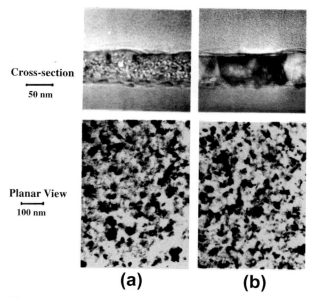

Fig. 3.37a,b. TEM cross sections and planar views of samples after laser dehydrogenation at (**a**) 150 mJ/cm^2 and (**b**) 300 mJ/cm^2 [3.90]

The hydrogen out-diffusion from a-Si:H is also controlled by careful choice of the laser fluence [3.90]. Pulsed laser crystallization of a-Si:H films with moderate thickness (40–100 nm) can be performed by a series of laser exposures with increasing laser energy density. One scheme is to have the laser beam shaped as a staircase in space [3.56]. With this beam profile, the a-Si:H film receives laser irradiations with a gradual increase in energy density as the beam is scanned. Another approach is to use three laser irradiations with increasing laser energy density [3.90]. Figure 3.36 shows the experimental data on the amount of hydrogen that out-diffuses for different laser fluences and number of laser shots.

During a pulsed laser irradiation, hydrogen diffuses rapidly in the molten silicon layer. In the three-step process, the laser energy density of the first irradiation is chosen to be slightly above the Si surface-melting threshold. The hydrogen atoms near the surface are removed during a multiple-shot irradiation at this low energy density. The second laser energy density is chosen to be near the melt-through fluence, which depends on the various factors discussed in Sect. 3.2.2 and includes the film thickness. This irradiation removes most of the hydrogen remaining in the film. The final scan at higher laser fluence produces poly-Si films with large grains.

Figure 3.37 shows cross-sectional and planar-view TEM images of the films processed by the three-step process. After the first laser scan, the film is partially converted to microcrystalline Si with fine grains (Fig. 3.37a). The

(a)   **(b)**   **(c)**   **(d)**

Fig. 3.38a–d. TEM planar views of poly-Si films produced by the three-step laser processing of PECVD a-Si:H. The laser energy densities for the third and final step were (**a**) 425 mJ/cm^2, (**b**) 445 mJ/cm^2, (**c**) 465 mJ/cm^2, and (**d**) 485 mJ/cm^2 [3.47]

second scan at an energy density near the melt-through threshold produces columnar poly-Si grains (Fig. 3.37b). The average grain size, however, is small. The desired film crystallinity is mainly determined by a third and final scan.

Figure 3.38 shows the film crystallinity from samples processed by the three-step laser crystallization process at various laser energy densities for the last step. The behavior of the average grain size versus the laser fluence is similar to that for laser crystallization of LPCVD deposited Si films described in Sects. 3.2.2 and 3.2.3. The grain size increases with increasing laser energy density up to a peak value of a few microns. The grain size then decreases with further increases in the laser energy density. The transistor field effect mobility is correlated with this grain size. At the optimized grain size, however, the film suffers from damage at the interface between the Si film and the gate insulator below. The damage results in non-uniform device performance and a significant leakage current through the gate insulator. Uniform device performance is obtained by choosing a laser energy density near to, but below, the critical value.

3.4.3 Device Fabrication and Performance

Figure 3.39 illustrates the monolithic fabrication process for making a-Si and poly-Si bottom-gate TFTs on the same substrate. The process starts with the gate electrode patterning, followed by PECVD deposition of a dual dielectric of Si_3N_4 and SiO_2 and a-Si:H. Selective laser dehydrogenation/crystallization is then performed. The film is hydrogenated after the crystallization process to passivate the defects in the poly-Si film. The source/drain contacts are made from deposited, doped a-Si:H, followed by metal deposition and patterning.

The dual dielectric gate insulator is used to optimize the threshold voltage simultaneously for both a-Si and poly-Si TFTs. The TFT threshold voltage depends on the fixed charge in the dielectric film, the defect density, the dop-

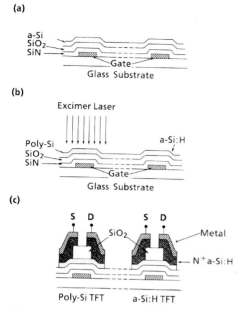

Fig. 3.39a–c. Fabrication process for hybrid a-Si and poly-Si TFTs; (**a**) PECVD deposition of the gate insulator and an a-Si:H film; (**b**) selective laser dehydrogenation and crystallization; (**c**) the final device structure with source and drain contacts [3.48]

ing level in the Si film, and the work function of the gate electrode. a-Si:H TFTs show superior performance with a silicon nitride (SiN) gate insulator over that for a SiO_2 gate insulator. According to the defect pool model [3.91], the superior performance of a-Si TFTs with a nitride gate insulator is attributed to the positive fixed charge in the nitride dielectric. The positive charge pulls the Fermi level up in the band gap, causing a redistribution of the defect density of states. This lowers the defect densities located above midgap in the a-Si:H. Therefore, the threshold voltage is small and positive for SiN. However, the threshold voltage of poly-Si TFTs is just the reverse; namely, it is large and negative with a SiN gate insulator and is small and positive for a SiO_2 gate insulator. Figure 3.40 describes the general behavior of a-Si and poly-Si TFTs with nitride or oxide as the gate insulator.

Large-area display and imaging applications require small and positive threshold voltages for a-Si pixel switches and poly-Si TFTs in the peripheral circuits. A simple approach to obtain small and positive threshold voltages for both a-Si and poly-Si TFTs is to combine SiN and SiO_2 for the gate insulator. Figure 3.41 shows simulation results of the transfer characteristics of poly-Si TFTs with various film thickness of nitride and oxide, keeping the total dielectric film thickness at 100 nm. In this simulation, a uniform

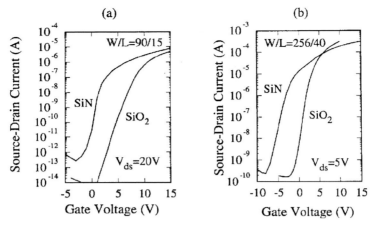

Fig. 3.40a,b. Transfer characteristics of (**a**) a-Si TFTs and (**b**) poly-Si TFTs with nitride and oxide as the gate insulator, at the specified device width-to-length ratio (W/L) and source-drain voltage, V_{ds} [3.47]

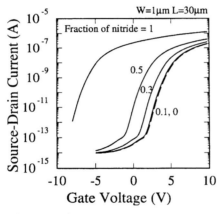

Fig. 3.41. Simulated transfer characteristics of poly-Si TFTs with a dual dielectric of nitride and oxide. The total dielectric thickness is 100 nm [3.47]

density of fixed charge is introduced into each layer of the two-layer gate dielectric. It is assumed that the fixed charge density in nitride is $+4 \times 10^{17}$ charges/cm^3 while it is -10^{17} charges/cm^3 in the oxide. The two-dimensional device simulator is based on one specifically developed for modeling a-Si TFTs wherein the defects and grain boundaries are treated as a spatially uniform density of localized states in the band gap. The effective density of states spectrum for poly-Si has been obtained by fitting the model to both the low drain bias transfer data as well as to the activation energy of the channel conductance as a function of gate bias. This set of calculations demonstrates

3 Laser Crystallization for Polycrystalline Silicon Device Applications 141

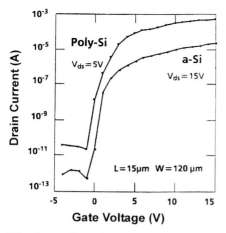

Fig. 3.42. Transfer characteristics of an a-Si and poly-Si TFTs fabricated on the same substrate by the selective laser de-hydrogenation and crystallization process and with a dual dielectric [3.47]

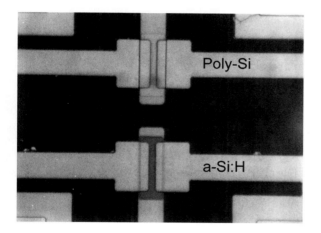

Fig. 3.43. Optical microscope photograph of an a-Si TFT and a poly-Si TFT built side by side by a monolithic fabrication process. The distance between the two devices is about 10 µm [3.92]

that the threshold voltage can be controlled by adjusting the film thickness of the oxide and nitride.

Figure 3.42 shows the transfer characteristics from a-Si and poly-Si TFTs fabricated on the same glass substrate. The gate insulator consists of a dual-layer oxide and nitride dielectric. The performance of the poly-Si TFTs are uniform, although the mobility is lower than the maximum of about

$100\,\text{cm}^2/\text{Vs}$ that one is able to obtain with a monolithic fabrication process. The a-Si TFT exhibits a low leakage current of less than $1\,\text{fA}/\mu\text{m}$ under a drain bias of 10 V, which is sufficiently low for the pixel switch.

With proper light shielding during the laser crystallization process, an a-Si TFT and a poly-Si TFT can be made side by side. Digital or analog circuits may utilize the low mobility and low leakage currents in a-Si for hybrid circuits. The minimum device separation is determined by the lateral thermal diffusion. Figure 3.43 shows a photograph of an a-Si and a poly-Si TFT built side-by-side with a separation of about $10\,\mu\text{m}$ [3.47].

3.5 Conclusion

Materials studies show that laser-crystallized polysilicon contains larger grains with fewer defects than polysilicon prepared by other techniques. This is of particular significance for devices since large grains and few defects imply improved performance, such as higher electron mobility and lower leakage current for thin-film transistors. As a result, laser processing yields good devices and also allows for the creation of CMOS polysilicon on glass substrates, the integration of polysilicon silicon and amorphous devices on the same glass substrate, and the fabrication of self-aligned amorphous silicon thin-film transistors via laser doping. Thus, this processing technique provides a wide range of opportunities for large-area polysilicon electronics. One of the attractive areas is in the application to active matrix liquid crystal displays. The high carrier mobility of poly-Si TFTs allows the dimensions of the pixel switches to be scaled down. This leads to an improvement in the pixel aperture ratio and, therefore, to a brighter and lower-power display. The smaller size and improved performance of poly-Si TFTs also make them suitable for projection displays and high definition displays. The construction of CMOS poly-Si drive circuitry on the glass substrate reduces the number of off-substrate connections, leading to greater reliability, lower cost, and compact construction. With more than 20 years of development invested in low temperature poly-Si technology, new products, such as camera view-finders and small poly-Si displays, are now available in the market place.

Another application area of poly-Si devices on glass is in the combination of amorphous silicon photo sensors with poly-Si pixel switches and drive circuitry for large-area imaging. Similar to display applications, poly-Si TFTs improve the pixel fill factor (aperture ratio) and the packaging for flat panel imagers. In addition, the availability of both n-channel and p-channel TFTs enables digital and analogue circuit design. For example, with the high drive power of CMOS poly-Si TFTs, pixel amplifiers can be built to improve the imaging signal-to-noise ratio.

Other applications of low temperature poly-Si TFTs include printers, smart sensors, memory devices, and neural networks. Considerable development in manufacturing yield, device uniformity and stability, and device leakage current is still required to fully realize all these applications. Nonetheless, several major display manufacturers have prototype poly-Si display fabrication facilities and have plans to scale these up to mass production capacities in the near future.

References

3.1 T. Sameshima and S. Usui, Mat. Res. Soc. Symp. Proc. **71**, 435 (1986).
3.2 T. Noguchi, H. Hayashi, and T. Ohshima, Mat. Res. Soc. Symp. Proc. **106**, 293 (1988).
3.3 K. Sera, F. Okumura, H. Uchida, S. Itoh, S. Kaneko, and K. Hotta, IEEE Trans. Electron Devices **36**, 2868 (1989).
3.4 I.-W. Wu, Mat. Res. Soc. Proc. **182**, 107 (1990); I-W. Wu, A. G. Lewis, T-Y. Huang, and A. Chiang, Proc. of Society for Information Display **31**, 311 (1990).
3.5 K. Winer, G. B. Anderson, S. E. Ready, R. Z. Bachrach, R. I. Johnson, F.A. Ponce, and J. B. Boyce, Appl. Phys. Lett. **57**, 2222 (1990).
3.6 K. Shimizu, O. Sugiura, and M. Matsumura, Jpn. J. Appl. Phys., **29**, L1775 (1990).
3.7 H. Kuriyama, et al, Jpn. J. Appl. Phys. **31**, 4550 (1992).
3.8 S. Chen, J. B. Boyce, I-W. Wu, A. Chiang, R. I. Johnson, G. B. Anderson, and S. E. Ready, SID Proc. Of Active Matrix Liquid Crystal Displays Symp., p. 26 (1993).
3.9 I. Asai, N. Kato, M. Fuse, and T. Hamano, Jpn. J. Appl. Phys. **32**, 474 (1993).
3.10 E. Forgarassy, H. Pattyn, M. Elliq, A. Slaoui, B. Prevot, R. Stuck, S. de Unamuno, and E. L. Matthe, Appl. Surface Science **69**, 231 (1993).
3.11 S. D. Brotherton, D. J. McCulloch, J. B. Clegg, and J. P. Growers, IEEE Trans. Electron Devices **40**, 407 (1993).
3.12 H. Tanabe, K. Sera, K. I. Nakamura, K. Hirata, K. Yuda, and F. Okumura, Mat. Res. Soc. Symp. Proc., **677**, 305 (1994).
3.13 S. D. Brotherton, Semiconductor Sci. Tech., **10**, 721 (1995).
3.14 T. J. King, AMLCDs, 80 (1995).
3.15 J. B. Boyce, P. Mei, D. K. Fork, G. B. Anderson, and R. I. Johnson, Mat. Res. Soc. Proc. **403**, 305 (1996).
3.16 J. S. Im and R. S. Sposili, MRS Bulletin **21**, 39 (1996).
3.17 I-W. Wu, Mat. Res. Soc. Proc. **471**, 125 (1997).
3.18 J. B. Boyce, P. Mei, R. T. Fulks, and J. Ho, Phys. Stat. Sol. **166**, 729 (1998).
3.19 P. Mei, G. B. Anderson, J. B. Boyce, D. K. Fork, R. Lujan, Electrochem. Soc. Proc, **96-23**, 51 (1996).
3.20 M. Matsumara, , Phys. Stat. Sol. **166**, 715 (1998).
3.21 G. K. Giust, T. W. Sigmon, J. B. Boyce, and J. Ho, IEEE Electron Devices Letters, **20**, 77 (1999).
3.22 Corning Product Specification, Display Grade Product, Glass 1737F; Corning Inc., Corning, NY.
3.23 T. Kamins, Polysilicon for Integrated Circuit Applications, Kluwer Academic Publications, New York (1988).
3.24 Polycrystalline Semiconductors: Physical Properties and Applications, Ed. G. Harbeke, Springer-Verlag, Berlin (1985).

3.25 See, for example, references 3.7, 3.9, 3.15, 3.18, and references contained therein.
3.26 See, for example, N. D. Young, D. J. McCulloch, and R. M. Bunn, Digest of Technical Papers, AM-LCD **97**, 47 (1997).
3.27 A. T. Voutsas and M. K. Hatalis, J Electrochem. Soc. **139**, 2659 (1992).
3.28 A. Mimura, N Konishi, K. Ono, J.-I. Ohwada, Y. Hosokawa, Y. A. Ono, T. Suzuki, K. Miyata, and H. Kawakami, IEEE Trans. Electron Devices **36**, 351 (1989).
3.29 N. Yamauchi and R. Reif, J. Appl. Phys. **75**, 3235 (1994); I-W. Wu, A. Chiang, M. Fuse, L. Ovecoglu. and T. Y. Huang, J. Appl. Phys. **65**, 4036 (1989).
3.30 R. Z. Bachrach, K. Winer, J. B. Boyce, S. E. Ready, R. I. Johnson, and G. B. Anderson, J. Electron. Mat. **19**, 241 (1990).
3.31 R. Z. Bachrach, J. B. Boyce, S. E. Ready, and G. B. Anderson, Polycrystalline Semiconductors II, Eds. J. H. Werner and H. P. Strunk, Springer-Verlag, Berlin (1991), pp. 330-341.
3.32 J. Y. W. Seto, J. Appl. Phys. **46**, 5247 (1975).
3.33 T. I. Kamins and P. J. Marcoux, IEEE Device Letters **1**, 159 (1980): M. J. Thompson, J. Non-Cryst. Solids **137&138**, 1209 (1991).
3.34 N. H. Nickel, G. B. Anderson, and R. I. Johnson, Phys. Rev. B **56**, 12065 (1997).
3.35 N. H. Nickel, N. M. Johnson, and W. B. Jackson, Appl. Phys. Lett. **62**, 3285 (1993).
3.36 J. M. Poate and J. W. Mayer, Laser Annealing of Semiconductors, Academic Press, New York (1982).
3.37 Pulsed Laser Processing of Semiconductors, Vol 23 in Semiconductors and Semimetals, Eds. R. F. Wood, C. W. White, and R. T. Young, Academic Press, New York (1984).
3.38 See, for example, references 3.10, 3.13, 3.15, 3.17 and references contained therein.
3.39 S. E. Ready, J. B. Boyce, R. Z. Bachrach, R. I. Johnson, K. Winer, G. B. Anderson, and C. C. Tsai, Mat. Res. Soc. Proc. **149**, 345 (1989).
3.40 R. I. Johnson, G. B. Anderson, S. E. Ready, D. K. Fork, and J. B. Boyce, Mat. Res. Soc. Proc. **258**, 123 (1992).
3.41 R. I. Johnson, G. B. Anderson, J. B. Boyce, D. K. Fork, P. Mei, S. E. Ready, and S. Chen, Mat. Res. Soc. Proc. **297**, 533 (1993).
3.42 LIMP, for Laser Induced Melting Predictions, is the Harvard simulation program provided by Jeffrey West. The original was written by M. O. Thompson and enhanced by J. West, P. M. Smith and D. Hoglund.
3.43 J. S. Im, H. J. Kim, and M. O. Thompson, Appl. Phys. Lett. **63**, 1969 (1993).
3.44 J. S. Im and H. J. Kim, Appl. Phys. Lett. **64**, 2303 (1994).
3.45 J. B. Boyce, G. B. Anderson, D. K. Fork, R. I. Johnson, P. Mei, S. E. Ready, Mat. Res. Soc. Proc. **321**, 671 (1994).
3.46 P. Mei, J. B. Boyce, M. Hack, R. A. Lujan, R. I. Johnson, G. B. Anderson, S. E. Ready, D. K. Fork, and D. L. Smith, Mat. Res. Soc. Proc. **297**, 151 (1993).
3.47 P. Mei, J. B. Boyce, M. Hack, R. A. Lujan, R. I. Johnson, G. B. Anderson, D. K. Fork, and S. E. Ready, J. Appl. Phys. **76**, 3194 (1994).
3.48 P. Mei, J. B. Boyce, ,D. K. Fork, M. Hack, R. A. Lujan and S. E. Ready, Proc. 4th Int. Conf. Solid State and Integrated-Circuit Technology, **76** (5), 3194 (1994).

3.49 D. K. Fork, G. B. Anderson, J. B. Boyce, R. I. Johnson, and P. Mei, Appl. Phys. Lett., **68**, 2138 (1996).
3.50 M. O. Thompson, G. J. Galvin, J. W. Mayer, P. S. Peercy, J. M. Poate, D. C. Jacobson, A. G. Cullis, and N. G. Chew, Phys. Rev. Lett. **52**, 2360 (1984).
3.51 J. Narayan and C. W. White, Appl. Phys. Lett. **44**, 35 (1984).
3.52 D. H. Lowndes, G. E. Jellison, S. J. Pennycook, S. P. Withrow, and D. M. Mashburn, Appl. Phys. Lett., **48**, 1389 (1986).
3.53 S. E. Ready, J. H. Roh, J. B. Boyce, and G. B. Anderson, Mat. Res. Soc. Proc. **258**, 111 (1992).
3.54 S. R. Stiffler, M. O. Thompson, and P. S. Peercy, Phys. Rev. B **43**, 9851 (1991)
3.55 A. G. Cullis, H. C. Weber, N. G. Chew, J. M. Poate, and P. Baeri, Phys. Rev. Lett. **49**, 219 (1982).
3.56 T. Sameshima and S. Usui, J. Appl. Phys. **74**, 6592 (1993).
3.57 G. B. Anderson, J. B. Boyce, D. K. Fork, R. I. Johnson, P. Mei, and S. E. Ready, Mat. Res. Soc. Proc., **343**, 709 (1994).
3.58 D. J. McCulloch and S. D. Brotherton, Appl. Phys. Lett. **66**, 2060 (1995).
3.59 L. L. Kazmerski, Electrical Properties of Polycrystalline Semiconductor Thin Films, Chap. 3 in Polycrystalline and Amorphous Thin Films and Devices, L. L. Kazmerski, Ed., Academic Press, New York (1980).
3.60 P. G. Carey and T. W. Sigmon, Appl. Surface Sci., **43** 325 (1989).
3.61 K. H. Weiner, P. G. Carey, A. M. McCarthy, and T. W. Sigmon, IEEE Elec. Dev. Lett., **13**, 369 (1992).
3.62 K. H. Weiner, P. G. Carey, A. M. McCarthy, and T. W. Sigmon, Microelectronic Eng, **20** 107 (1993).
3.63 T. Sakoda, C. Kim, and M. Matsumura, Mat. Res. Soc. Proc., **345**, 59 (1994).
3.64 J. B. Boyce, G. B. Anderson, P. G. Carey, D. K. Fork, R. I. Johnson, P. Mei, S. E. Ready and P. M. Smith, Mat. Res. Soc. Proc. **358**, 909 (1995).
3.65 P. Mei, J. B. Boyce, D. K. Fork, M. Hack, R. Lujan, and S. E. Ready, Proc. of the 4th Intl. Conf. On Solid-State and Integrated Circuit Tech., p. 721 (1995)
3.66 K. H. Weiner, Fabrication of Thin Base Bipolar Transistors Using Pulsed UV Laser Processing, PhD Thesis, Stanford University, June (1989).
3.67 Figure 3.14 is from Fig. 4.2 of reference 3.66 where the laser energy density has been estimated from the melt depth versus laser fluence data of this reference's Fig. 4.8.
3.68 H. Kumomi and T. Yonehara, Mat. Res. Soc. Symp. Proc., **202**, 645 (1990).
3.69 H. J. Kim and J. S. Im, Mat. Res. Soc. Symp. Proc., **397**, 401 (1995).
3.70 D. H. Choi, E. Sadayuki, O. Sugiura, and M. Matsumura, Jpn. J. Appl. Phys., **33**, Pt. 1, 70 (1993).
3.71 A. B. Y. Chan, C. T. Nguyen, P. K. Ko, M. Wong, A. Kumar, J. Sin, and S. S. Wong, IEEE Electron Dev, Lett., **17**, 518 (1996).
3.72 I-Wei Wu, A. G. Lewis, T. Y. Huang, W. B. Jackson, and A, Chiang, IEEE, IEDM 867 (1990).
3.73 S. D. Brotherton, J. R. Ayres, and M. J. Trainor, J. Appl. Phys. **79**, 895 (1996).
3.74 M. S. Shur, M. D. Jacunski, H. C. Slade, and M. Hack, J. SID, **3/4**, 223 (1995).
3.75 M. Hack, I-Wei Wu, T. J. King and A. G. Lewis, IEEE IEDM, 385 (1993).
3.76 M. Rodder, IEEE Electron Dev, Lett., **6**, 570 (1985).
3.77 G. P. Pollack, IEEE Electron Dev, Lett., **5**, 468 (1984).

3.78 H. N. Chern, C. L. Lee, and T. F. Lei, IEEE Tran. Electron Dev., **41**, 698 (1994).

3.79 H. C. Cheng, F. S. Wang, C. Y. Huang, IEEE Tran. Electron Dev., **44**, 64 (1997).

3.80 S. Seki, O. Kogure, and B. Tsujiyama, IEEE Electron Dev, Lett., **8**, 425 (1987).

3.81 I-Wei Wu, W. B. Jackson, T. Huang, A. G. Lewis, and A. Chiang, IEEE Electron Dev, Lett., **11**, 167 (1990).

3.82 M. Hack, A. G. Lewis, and I-Wei. Wu, IEEE Tran. Electron Dev., **40**, 890 (1993).

3.83 N. D. Young, IEEE Tran. Electron Dev., **43**, 450 (1996).

3.84 V. Suntharalingam and S. J. Fonash, Appl. Phys. Lett. **68**, 1400 (1996).

3.85 M. Hack, P. Mei, R. Lujan and A. G. Lewis, J. Non-Crystalline Solids, **164–166**, 727 (1993).

3.86 M. J. Powell, Insulating Films on Semiconductors, J. F. Verweij and D. R. Wolters (editors), Elsevier Science Publishers B. V. (North-Holland), p. 245 (1983)

3.87 T. Tanaka, H. Asuma, K. Ogawa, Y. Shinagawa, K . Ono, N. Konishi, International Electron Devices Meeting 1993. Technical Digest, 389 (1993).

3.88 W. B. Jackson, Phys. Rev. B., **41**, 1059 and 10257 (1990).

3.89 T. Aoyama, K. Ogawa, Y. Mochizuki, and N. Konishi, Appl. Phys. Lett., **66**, 3007 (1995).

3.90 P. Mei, J. B. Boyce, M. Hack, R. Lujan, R. I. Johnson, G.B. Anderson, D.K. Fork, S.E. Ready, Appl. Phys. Lett., **64**, 1132 (1994).

3.91 M.J. Powell, S.C. Deane, I.D. French, J.R. Hughes, W.I. Milne, Philosophical Magazine B (Physics of Condensed Matter, Electronic, Optical and Magnetic Properties), **63**, 325 (1991).

3.92 P. Mei, J. B. Boyce, D.K. Fork, G.B. Anderson, J. Ho, M. Hack, R. Lujan, Mat. Res. Soc. Symp. Proc. **507**, 3 (1998).

4 Large Area Image Sensor Arrays

Robert Street

Xerox Palo Alto Research Center, 3333 Coyote Hill Road, Palo Alto, CA 94304, USA; E-mail: street@parc.xerox.com

Abstract. The development of large area 2-dimensional amorphous silicon image sensor arrays has taken nearly ten years. At the time of writing, these devices are being introduced into the market for digital X-ray imaging applications such as medical radiography and non-destructive testing. This chapter describes the arrays, the range of applications and why the amorphous silicon technology is well matched to the particular needs of X-ray imaging.

4.1 Introduction

Amorphous silicon 2-dimensional (2-d) image sensor arrays owe their development to earlier research on linear image sensors and to the processing technology developed for active matrix addressed liquid crystal displays. Active matrix addressing of LCDs using amorphous silicon thin film transistors (TFT) was first proposed by LeComber et al. in 1979 [4.1], and was demonstrated by the same group in 1981 [4.2], although the idea of matrix addressing is considerably older. Shortly afterwards, a number of groups applied the matrix addressing technique to linear image sensors, combining the TFTs with amorphous silicon sensors to make arrays for scanners and FAX machines [4.3–5]. The linear sensors use the same addressing structures as is used in the 2-d arrays, but configured as a single line of sensors. The first sensors were Schottky barrier devices, but the better performance of p-i-n photodiodes soon became apparent.

The processing technology to create 1-d and 2-d matrix addressed arrays for scanning and display was also created in the early 1980's and has since been greatly refined due to the rapid development of the AMLCD market for lap-top computer displays. The sensor arrays use the same processing technology, and made the transition from 1-d to 2-d arrays of increasing size starting at the end of the 1980's [4.6].

This chapter begins with a description of the important devices – the photodiodes and TFTs – that make up the sensor arrays, and goes on to describe the arrays, the electronic systems that provide the control and readout, and the various applications in medicine, non-destructive testing, scientific imaging and document capture.

4.2 Devices

Thin film transistors, p-i-n diodes and photoconductors are the active devices in most image sensor arrays. The structure and properties of the key amorphous silicon devices are described before going on to discuss the arrays. Further information about photodiodes is contained in Chap. 7 on color sensors and Chap. 8 on position sensors, and closely related p-i-n devices are used in the solar cells described in Chap. 6. This chapter focuses on those properties that are of importance to the image sensors.

4.2.1 P-i-n Photodiodes

The ideal photosensor has a high sensitivity to illumination and negligible dark current. Either a photoconductor or a photodiode structure can be used, the distinction being that the photoconductor has ohmic electrical contacts while the photodiode has rectifying contacts, either a doped layer or a metal Schottky barrier. The merit of the photoconductor is that gain may be achieved, resulting in high photocurrent, although at the expense of a long response time. On the other hand, photodiodes exhibit exceptionally low leakage currents. Avalanche gain is a desirable property for high sensitivity photodiodes but has not been reported in aSi:H, although it is observed in amorphous selenium [4.7].

The a-Si:H p-i-n photodiode is used as the light detector in image sensor arrays primarily to take advantage of its high quantum efficiency, low reverse bias leakage current and thin film structure. The diode is also used instead of the TFT as the switching element in some arrays, because of its good rectifying properties. The photodiode design is illustrated in Fig. 4.1 and consists of very thin p-doped and n-doped layers with a thicker undoped (i) layer in between. The doped layers provide rectifying contacts but do not contribute to the light sensitivity because doping causes a high density of charged dangling bond defects in a-Si:H – positive in n-type material and negative in p-type material [4.8]. The minority carrier lifetime in doped a-Si:H is therefore so small that most holes generated in the n-layer and electrons in the p-layer recombine before they can cross the reverse bias junction. On the other hand, the high defect density results in a very small depletion width, so that a doped layer thickness of 100 Å or less is sufficient to form the junction.

Incident light which is absorbed in the doped layer reduces the sensitivity to short wavelength illumination. Hence, particular care is needed to minimize such absorption if high sensitivity to blue light is required. Microcrystalline silicon or a-Si:C:H both have a lower absorption than a-Si:H and are often used in solar cells to enhance blue sensitivity [4.9]. This approach also works for sensor arrays, but has not been widely implemented because the extra blue sensitivity has not yet proved to be sufficiently important.

In many sensor arrays, the p-i-n photodiodes are etched to isolate the sensor from the neighboring pixels, as in the structure shown in Fig. 4.1. The

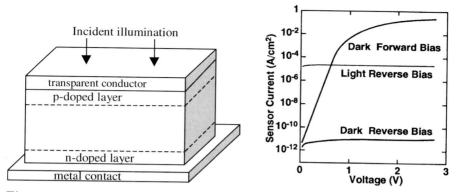

Fig. 4.1. Illustration of the structure of an a-Si:H p-i-n photodiode (*left*), and the forward and reverse bias electrical characteristics (*right*)

exposure of the edge of the diode can affect the leakage current, as is described shortly, particularly when the sensor area is small. The diode requires metal contacts on both top and bottom because the thin doped layers are not sufficiently conductive by themselves. For example, a 100 Å thick p-type a-Si:H layer has a sheet resistance of $10^{11}\,\Omega/\square$, and even the more conductive n-type layer is only $10^8\,\Omega/\square$. A transparent conductor is required for the illuminated surface and indium tin oxide is commonly used.

The basic electrical characteristics of a p-i-n photodiode are shown in Fig. 4.1. The exponentially increasing forward bias current is typical of a diode, as is the low reverse bias current. The large increase in reverse current under illumination is the basis for its use as a light sensor.

The Sensor Photocurrent. Absorption of light in the sensor generates electron–hole pairs in the i-layer. The charges separate under the action of the electric field and cause a current to flow in the external circuit. The electric field includes the internal built-in potential and the applied reverse bias. The photocurrent, $I_{\rm PH}$, resulting from a light flux, $G_{\rm L}$, is given by

$$I_{\rm PH} = G_{\rm L}\eta_{\rm QE}(V,\lambda)\,. \tag{4.1}$$

The quantum efficiency $\eta_{\rm QE}(V,\lambda)$ comprises a wavelength-dependent absorption that generates electron–hole pairs, and a bias-dependent charge collection efficiency. In the absence of photoconductive or avalanche gain the quantum efficiency is always less than unity.

A typical quantum efficiency spectrum is shown in Fig. 4.2. The decrease at long wavelength is due to the low absorption coefficient of a-Si:H below the band gap energy, and the short wavelength decrease mostly arises from absorption in the upper doped layer and perhaps the transparent conductor. The peak quantum efficiency of 80–95% in optimized diodes occurs in the range 500–600 nm. The interference fringes seen in Fig. 4.2 are due to

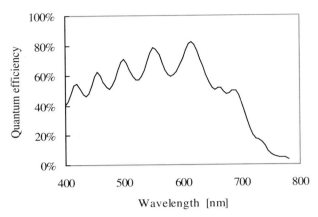

Fig. 4.2. Typical quantum efficiency spectrum of a p-i-n photo-diode with an ITO contact

the thin transparent overlayers to the photodiode, and the choice of materials and thickness can be designed as an antireflection coating for a selected wavelength range.

The charge collection efficiency depends on the applied bias and on the carrier recombination lifetimes. Dangling bond defects act as traps and limit the carrier lifetime, τ, such that the mean free path, L, of a carrier with drift mobility, μ, in a uniform applied electric field, F, is,

$$L = \mu\tau F \ . \tag{4.2}$$

Trapping will occur with high probability when L is smaller than the thickness of the sample. Hence, the collection efficiency is increased by large values of $\mu\tau$ and a high applied bias. In a device of thickness, t_F, the fraction, P, of charge that is collected is given by the Hecht formula,

$$P = \frac{\mu\tau F}{t_F}\left[1 - \exp(-t_F/\mu\tau F)\right] \approx 1 - \frac{t_F}{2\mu\tau F} \ , \tag{4.3}$$

where the approximation on the right applies when $\mu\tau$ is large [4.10].

The values of $\mu\tau$ are usually different for electrons and holes and depend on the density of traps, N_D, and the capture cross-section, σ, as [4.8],

$$\mu\tau = \frac{\text{const}}{\sigma N_D} \ . \tag{4.4}$$

Neutral dangling bonds in undoped a-Si:H can trap both electrons and holes, but the cross section for holes is greater, so that $\mu\tau$ is correspondingly 5–10 times smaller than for electrons. The high density of charged defects in doped a-Si:H strongly trap minority carriers. In low defect density a-Si:H, typical values are $(\mu\tau)_{\text{electron}} = 3 \times 10^{-7} \text{ cm}^2/\text{V}$ and $(\mu\tau)_{\text{hole}} = 5 \times 10^{-8} \text{ cm}^2/\text{V}$ [4.8].

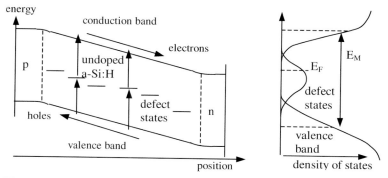

Fig. 4.3. *Left:* band model for bulk thermal generation through defect states in the band gap. *Right:* density of states model for amorphous silicon

Reverse Bias Dark Current. The photodiode reverse bias dark current may have components from the bulk, the doped contacts or the exposed diode perimeter. In optimized devices the bulk current dominates and is attributed to the thermal generation of electrons and holes via mid-gap defect states, as illustrated in Fig. 4.3 [4.11, 12]. Electrons are thermally excited from the valence band to empty or singly occupied defect states, and are later excited from these states to the conduction band, resulting in electron-hole pairs which are collected by the applied field. The equilibrium excitation rates of electrons and holes must be equal, which ensures that the dominant contribution is from states at the middle of the band gap. The thermal generation current, J_{th}, is given by [4.11],

$$J_{\text{th}} = eN(E)kT\omega_0 \exp\left(-E_M/2kT\right) vol. \,, \tag{4.5}$$

where $N(E)$ is the density of defect states in the middle of the gap, E_M is the mobility gap energy, $vol.$ is the volume of the sample, and unity collection efficiency is assumed for the generated charge.

Equation (4.5) predicts that the current is thermally activated with half the mobility gap energy and experiments indeed find an activation energy of about 0.85–0.9 eV, in support of this model (see Fig. 4.4). This activation energy causes the current to increase by an order of magniture with a temperature increase of 20°C near room temperature. Thus, the leakage current of an image sensor array may be reduced by modest cooling, but the array will not operate well at elevated temperature.

The absolute magnitude of the current is also consistent with the thermal generation model. Figure 4.4 shows the reverse current at room temperature after the application of a field. The current decreases to a steady state value of $1-2 \times 10^{-11}$ A/cm^2, which is consistent with a midgap defect density of 10^{16} cm^{-3}eV^{-1} and a mobility gap of 1.8 eV in (4.5), both of which are within the accepted range for amorphous silicon [4.8]. The slow increase of the steady state current with applied field is most probably explained by a field lowering

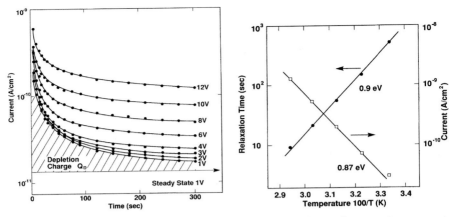

Fig. 4.4. *Left:* Measurements of the room temperature thermal generation current, showing the transient decay after the bias is applied. *Right:* Temperature dependence of the current and relaxation time showing an activation energy of about half the band gap

of the excitation energy [4.12]. The room temperature leakage current in a typical device is indeed remarkably small – the data in Fig. 4.4 indicates a current below 2×10^{-15} A in a photodiode of size $100 \times 100\,\mu\text{m}$.

Figure 4.4 also shows that the current is larger when the field is first applied and decreases quite slowly to the steady state value. The excess current is explained by the depletion of charge from mid-gap defect states [4.11]. Undoped a-Si:H is slightly n-type, with its Fermi energy above the middle of the band gap (Fig. 4.3). After the bias is applied, the occupied states above mid-gap are depleted of charge, and depletion continues until the steady state thermal generation is established. The amount of depletion charge is approximately,

$$Q = eN(E)\Delta E \, vol. \cong 1.6 \times 10^{-8}\,\text{C/cm}^2 , \qquad (4.6)$$

where ΔE is the position of the equilibrium Fermi energy above mid-gap, estimated to be $0.1\,\text{eV}$, and the diode thickness is $1\,\mu\text{m}$. The estimated depletion charge is consistent with the measured decay and the slow decay time is due to the low excitation probability from mid-gap states, and the thermal activation energy of the relaxation time is also consistent with this explanation.

The exposed periphery of the photodiode provides a different source of leakage current which is particularly significant in small devices. In the presence of both bulk and peripheral components, the total current in a square diode of edge length d_S, is given by,

$$I_{\text{total}} = I_B d_S^2 + 4 I_P d_S , \qquad (4.7)$$

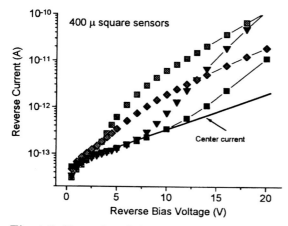

Fig. 4.5. Examples of the bias dependence of p-i-n photodiodes. The variation is largely due to peripheral currents, while the bulk leakage current is similar in each device

where I_B (cm^{-2}) and I_P (cm^{-1}) are the bulk and peripheral conductance. The peripheral current increasingly dominates when d_S is sufficiently small, and the different dependence on the pixel size allows an easy experimental method to determine the relative magnitude of the two components.

Figure 4.5 illustrates the bias dependence of the reverse current in sensors of size 400 × 400 μm [4.13]. The group of samples exhibits much variability in the current when the bias is large. Measurements with different pixel sizes confirm that the excess current is due to the peripheral contribution and that the bulk current follows the line indicated. The mechanism of peripheral current is not definitely identified, but is probably related to band bending at the a-Si:H surface. The different edge leakage currents are no doubt explained by the process variations in the etching and passivation during fabrication of the device. Unpassivated sensors are very sensitive to the ambient atmosphere and generally have larger edge leakage currents.

The Forward Current. The characteristics of the photodiode are completed by the forward bias current which is illustrated in Fig. 4.6 [4.14]. Although photodiode sensors operate in reverse bias, the forward current characteristics are a useful tool for understanding the device, and a high forward current is needed for diode addressed arrays. The p-i-n device exhibits diode behavior with a current increasing exponentially with bias, V;

$$I_F = I_0 \exp(eV/nkT) \,. \tag{4.8}$$

The ideality factor, n, is generally 1.3–1.5, depending on the thickness, deposition conditions and quality of the contacts [4.15]. Schottky diodes using Pt or Pd contacts instead of the p-type contact follow a similar form except that the current is larger and the ideality factor is close to unity.

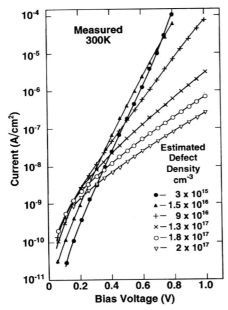

Fig. 4.6. Examples of the forward bias characteristics of an a-Si:H p-i-n diode, showing the effects of a high defect density in the undoped layer

The departure from the exponential slope at larger bias is due to a series resistance, which can originate in the bulk a-Si:H material if the device is thick or defective, or in the contacts, which are often thin films of significant resistance. Figure 4.6 illustrates the effect of high defect density in the undoped layer.

Semiconductor theory attributes an ideality factor $n = 1$ to thermionic emission and $n = 2$ to a generation-recombination mechanism [4.16]. The relative contribution of the two mechanisms depends on the height of the barrier and the density of recombination centers. The thermionic emission rate decreases with a larger barrier height, while generation-recombination increases with defect density. A-Si:H diodes with high work function metal contacts such as Pt or Pd, exhibit thermionic emission [4.17]. However, the p-type contact has a large enough barrier to suppress thermionic emission in favor of the generation-recombination mechanism. Modeling studies show that even though the ideality factor is < 2, there is essentially no contribution from thermionic emission and the intermediate value of n is entirely due to the distribution of localized states in a-Si:H, particularly the band tails [4.14]. Furthermore, the fact that the reverse current of the p-i-n device is due to thermal generation is obviously consistent with the identification of recombination-generation as the forward current mechanism.

Current Stability. The signal of interest in an image sensor is the difference between the photocurrent and the dark current. An image acquired in the dark is subtracted from the measured image to correct for any leakage current or other fixed signals. The long term stability of the dark current therefore determines whether the correction data can be obtained only once or whether it must be obtained frequently to reflect a time-varying current. Clearly, current stability is less important when the dark current is small, and therefore low leakage devices are desirable.

There are several potential sources of instability in a-Si:H photodiodes. Prolonged illumination of amorphous silicon is well known to cause a reversible metastable defect generation process, known as the Staebler-Wronski effect [4.18]. However, this mechanism plays almost no role in the sensor arrays because the total illumination of sensors under normal conditions is much less than the magnitude needed to create defects. A typical sensor array saturates with an illumination of $< 10^{-7}$ W/cm^2, whereas sunlight illumination at 10^{-1} W/cm^2 is 6 orders of magnitude more intense and even then takes hours or days to cause observable defect creation.

The actual sources of leakage current instability are all related to the mechanism of conduction. Figure 4.4 shows that the thermal generation current decreases slowly after the application of a bias, and consequently several seconds are needed for stable operation. The peripheral leakage current is often found to increase slowly with time, possibly due to a build-up of space charge in the passivation layer [4.13]. Minimization of this mechanism involves careful passivation and is helped by maintaining a low applied bias. The strong temperature dependence of the leakage current (Fig. 4.4) is another mechanism for current variation if the array temperature is allowed to vary.

The bias dependence of the photodiode current is an indirect cause of leakage current changes in sensor arrays, because the device is usually operated in charge integration mode. As light falls on the sensor, charge is collected at the electrodes and the effective bias voltage across the sensor decreases. Hence the leakage current correction measured in the dark may not apply precisely to an illuminated sensor.

For all the above reasons, it is important that the sensor has a very low leakage current so that the instability effects are negligible. P-i-n diodes satisfy this requirement better than Schottky diodes, and very stable dark currents are obtained near room temperature and at fairly low bias.

Thick Photodiode Sensors. Early in the research on sensor arrays, very thick amorphous silicon p-i-n photodiodes were investigated for detection of high energy radiation [4.19–22]. A thick layer increases the energy loss of an incident energetic charged particle – an electron or alpha particle – within the sensor, and results in a correspondingly larger signal. For reasons given in Sect. 4.3.4, recent research favors the use of other semiconductor materials for

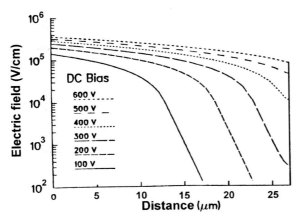

Fig. 4.7. Calculated electric field profiles in a 27 μm p-i-n photodiode showing full depletion at a bias larger than 400V [4.22]

the fabrication of thick sensors to detect ionizing radiation. However, effective amorphous silicon particle sensor diodes have been made with thickness of 20–50 μm. It is essential that the thick device is fully depleted, because the diffusion length of carriers in a-Si:H is very small, and essentially no photocurrent flows in a field-free region.

The depletion lengt, W, is determined by the space charge within the a-Si:H layer,

$$W = \sqrt{2\varepsilon\varepsilon_0 V_B / e N_{DI}} \,, \tag{4.9}$$

where V_B is the applied bias voltage and N_{DI} is the density of ionized defect states. The thick p-i-n device is in a condition of deep depletion, in which the ionization state of defects is governed by the relative excitation rates of electrons and holes in the defect states near the middle of the band gap. Measurements show that the density of ionized states is about 1/3 of the total density of defects, and for good quality a-Si:H, N_{DI} is about $7 \times 10^{14}\,\mathrm{cm}^{-3}$ (4.6) [4.22]. Figure 4.7 shows calculated field profiles for a 27 μm thick p-i-n sensor based on the measured ionized defect density [4.23]. The depletion layer increases with bias and measurements confirms the prediction that full depletion occurs at 300–400 V.

The depletion length increases with applied bias, and is ultimately limited by breakdown or an excessive leakage current, due to the high field at the electrical contact. The field at the contact can be reduced by growing the films with buried doped layers, giving a p-i-p-i-n-i-n structure [4.23]. The internal lightly-doped layers are ionized and cause a reduction of the field at the contact. Such devices are indeed found to have higher breakdown fields, and higher sensitivity to alpha particles.

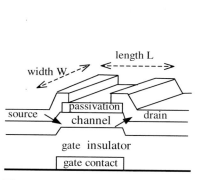 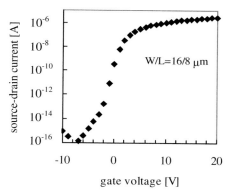

Fig. 4.8. Bottom gate TFT structure with source and drain contacts overlapping the channel (*left*). The transfer characteristics (*right*) show low leakage and a high on/off ratio

4.2.2 Thin Film Transistors

Thin film transistors, made of both amorphous and polycrystalline silicon are discussed in more detail in Chaps. 2 and 3, and only those properties of most interest for the image sensor arrays are discussed here. The most common a-Si:H TFT structure, shown in Fig. 4.8, has a bottom gate with a silicon nitride gate dielectric, an undoped a-Si:H channel, and n-type source and drain contacts. This structure has proved to be less prone to charge trapping effects than top gate designs, in part because the silicon nitride layer can be deposited at about 350°C. The TFT operates in the n-channel accumulation mode because of the particular material properties of a-Si:H. Electrons have higher mobility than holes (hence n-channel operation), and the high defect density of doped a-Si:H precludes a depletion mode of operation.

The silicon nitride gate dielectric, the a-Si:H channel, and a top nitride passivation layer are all deposited by plasma enhance chemical vapor deposition. In this fabrication process, the source-drain contacts are deposited over the protecting passivation layer which is self-aligned to the gate. The overlap of the source and drain contacts over the TFT channel ensures reliable contacts, but provides some coupling capacitance to the gate which has a significant effect on the performance of the sensor array.

The principal attributes needed for image sensors are an on-resistance of a few MΩ, a very low off-resistance of $> 10^{14}\,\Omega$, small size, and low parasitic capacitance. The reasons for these features are developed further in Sect. 4.3.2 describing the array operation. The TFT electrical characteristics obey the usual semiconductor relations for MOS transistors. The image sensor mostly operates in the linear regime for which the on-resistance, R_{ON}, the ratio of the source-drain voltage and current, is [4.8],

$$R_{ON} = V_{SD}/I_{SD} = [C_G \mu_{FE}(V_G - V_T)W/L]^{-1}\,, \tag{4.10}$$

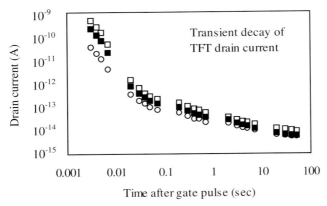

Fig. 4.9. Transient response of the TFT leakage current showing the slow decay to the steady state value. The TFT has $W/L = 75/8.5\,\mu\text{m}$ and the Source-drain voltage is 5 V (*circles*), 7.5 V (*closed squares*) and 12.5 V (*open squares*) [4.26]

where V_G is the gate voltage, V_T is the threshold voltage, W is the TFT width and L is the length. A nitride thickness of 3000 Å has a gate capacitance, C_G of $5 \times 10^{-8}\,\text{F/cm}^2$, and a field effect mobility, μ_FE, of $0.8\,\text{cm}^2/\text{Vs}$, yields,

$$R_\text{ON} = 25[(V_\text{G} - V_\text{T})W/L]^{-1}\,\text{M}\Omega\,. \qquad (4.11)$$

An on-resistance of a few MΩ is achieved with a width-to-length ratio, $W/L \sim$ 1–5 and $V_\text{G} - V_\text{T} \sim 10\,\text{V}$, and the threshold voltage is usually 0–2 V.

The TFT transfer characteristics are shown in Fig. 4.8 [4.24]. In the sub-threshold region with V_G below V_T, the current increases exponentially. The minimum leakage current is attained with V_G between $-5\,\text{V}$ and $-10\,\text{V}$, and the corresponding off-resistance is extremely large, $> 10^{15}\,\Omega$. Contributing to the low current are the barrier formed by the n-type source and drain contacts, the low density of states at the a-Si:H/nitride interface and the low mobility of a-Si:H. The leakage current is much lower than in polysilicon devices and make the a-Si:H device preferred for image sensors. The leakage current is probably a bulk generation current in the channel [4.25], similar to the mechanism of photodiode leakage described in Sect. 4.2.1.

There is a significant slow transient decay to the TFT leakage current when the device is turned off, as indicated in Fig. 4.9 [4.26]. The slow decay is similar to that of the photodiode in Fig. 4.4 and reaches a steady state only after several seconds. The effect is most likely due to the release of trapped charge and affects the image sensor by providing a source of fixed charge that can be subtracted from the signal provided that the signal and offset images are acquired with the same timing conditions.

The capacitance of the TFT is a combination of the channel capacitance, $C_\text{G}WL$, and the parasitic capacitance of the source and drain overlaps, which are also proportional to W. Smaller feature sizes greatly reduce the capacitance, and since this is a major contribution to the electronic noise of the imager, there are immediate benefits in reducing the photolithography feature sizes for the large area arrays.

Undoped a-Si:H is a highly photoconductive material, so that the TFT leakage current is sensitive to illumination even when the channel is very thin. It is therefore usually necessary to provide a light shield for the TFT to obtain the very low leakage currents that are required. Shielding is readily accomplished with a metal line placed over the TFT, and the gate metal already provides a light shield from below.

Other TFT Materials. Arrays are also fabricated with different active semiconductor materials for the TFTs. The two most widely explored are polycrystalline silicon and CdSe. The use of polycrystalline silicon in active matrix LCDs has been investigated extensively, and aspects of the technology are described in Chap. 2. The interest in poly-silicon is to provide higher performance devices to allow the drive circuitry to be integrated into the array, and LCDs with completely integrated drivers have recently been introduced onto the market. The main issue with poly-silicon is to fabricate TFTs on glass substrates. High temperature processing that closely follows the silicon IC technology uses thermal oxidation to fabricate the TFT gate dielectric and annealing to activate dopants. This process results in excellent TFTs, but cannot be performed on glass substrates. Efforts to use intermediate temperatures near the glass softening point have given way to laser crystallization that allows the silicon films to be melted without heating the glass substrate. The laser crystallized material has a mobility of typically 20–100 cm^2/Vs, which is high enough to fabricate integrated drivers [4.27]. However, the off-current is significantly higher than for a-Si:H, and as is discussed in Sect. 4.3.2, a low off-current is essential for the pixel switch of image sensor arrays. This requirement has led to the development of a hybrid technology that allows selective crystallization of some TFTs so that both devices can be made on the same array, and this is the subject of Chap. 3.

Polycrystalline CdSe has also been studied as a TFT material [4.28]. The devices are usually fabricated with a bottom gate and an SiO_2 gate dielectric. Low temperature deposition is consistent with the use of glass substrates. The carrier mobility is usually in the range 50–150 cm^2/Vs, and sometimes even higher. The leakage current can be made to be almost as low as a-Si:H, with the low values helped by the use of a double gate structure. Despite the good properties, CdSe TFTs are not widely used, evidently because the processing is difficult. Amorphous and poly-crystalline silicon benefit greatly from the fact that the processing chemistry is very similar to that used for the single crystal silicon IC industry and so can benefit from the extensive developments that have taken place.

4.3 Sensor Array Designs

The image sensors are two-dimensional arrays of pixels. Each pixel senses the radiation that impinges on it and stores the corresponding charge until it is transferred to the external electronics. The general architecture of the pixel, shown in Fig. 4.10, illustrates the various elements. The pixel contains a sensor to respond to incident radiation, a charge storage capacitor to hold the charge until it is read out, and a pixel switch to control the readout. Future imagers may also contain some signal processing capability in the pixel. The pixels are connected together with common address lines that cross the array – a bias contact for the sensor, address lines to trigger the pixel switch and data output lines to transfer the charge to the external electronics.

The array design attempts to optimize the most important performance attributes of the imager which are the size, spatial resolution, sensitivity and dynamic range. The fact that the performance of these parameters tends to conflict with each other represents a challenge for design optimization. For arrays that are typically 20–40 cm in size, the number of pixels ranges from 100 000 to 10 000 000, depending on the pixel pitch. The spatial resolution is ultimately limited by the pixel size, and the interesting range is from 50 μm up to 300–500 μm, depending on the application. Sensitivity, in terms of the signal from a pixel, depends on the quantum efficiency of the sensor, and is also proportional to the sensor area in the pixel. Thus, high resolution arrays are intrinsically less sensitive than low resolution arrays, in the sense that for a fixed illumination flux, the charge developed per pixel is smaller.

The dynamic range is the ratio of the maximum signal to the noise, and highlights the need for low noise performance. However, the overall performance of the image sensor depends on a great many factors relating to the specific imaging environment.

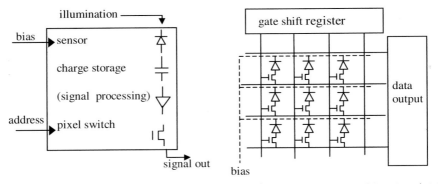

Fig. 4.10. The general design of the pixel (*left*), and the matrix addressing (*right*) of an image sensor array

4.3.1 Matrix Addressed Readout

The read-out of the image information is performed by matrix addressing, similar to that used for liquid crystal displays (Chap. 2). Matrix addressing reduces the number of individual contacts that must be made to the array, and consists of intersecting address lines and data sensing lines (Fig. 4.10). The address lines activate the pixel switches along one column of the array and those pixels transfer their signal to the data sensing lines, which are connected to the external electronics. After the measurement of the signal is complete, the next address line is activated. Thus the array is read out one line at a time until the image transfer is complete. Normally the gate lines are addressed in sequential order across the array, but in principle the columns of pixels could be sampled in any arbitrary order. Several different embodiments of this general design have been proposed and tested. In most cases the pixel switch is a thin film transistor (TFT), but diode-addressed imagers have been successfully fabricated, as they have for liquid crystal displays. Although amorphous silicon TFTs have been widely used, both poly-silicon and CdSe devices are alternatives. The a-Si:H p-i-n photodiodes form excellent light sensors (and X-ray imagers when coupled to a scintillator), while selenium and other photoconductors are also in use for X-ray sensitive imagers. Some more complex a-Si:H light sensor designs that could be used in image sensor arrays are described in Chap. 7.

4.3.2 TFT Addressed, p-i-n Photodiode Arrays

Much of the a-Si:H image sensor data is obtained from array designs which have an a-Si:H p-i-n photodiode together with a TFT pixel switch, and the following description of this device is used to develop many general aspects of image sensors which apply to other designs. A photograph of one such pixel design with a size of 127 μm is shown in Fig. 4.11, along with the pixel circuit [4.28]. The pixel is bounded on each side by the address lines. In the lower left corner of the pixel is the TFT whose gate is connected to the vertical address line to the left. The TFT source and drain contacts are connected to the photodiode and to the data line respectively. The bias contact runs horizontally across the pixel and in this particular design is placed directly over the TFT to provide a light shield. The sensor occupies the available space in the pixel between these other elements.

A schematic cross-sectional view of the pixel is shown in Fig. 4.12 and illustrates one particular structure of the array. The TFT is a bottom gate device as described in Sect. 4.2.2. The sensor is deposited directly on top of the source contact metal. For the design under consideration, the n-type layer is deposited first and the top junction is p-type, which is covered with a transparent conducting film, usually indium tin oxide. The sensor is patterned to provide isolation from the adjacent pixels, and a passivation layer is applied to protect the side walls of the sensor and to provide an inter-layer insulator

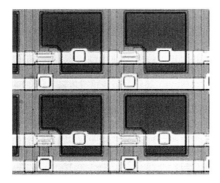 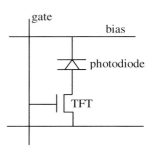

Fig. 4.11. Photograph of sensor array pixels showing the vertical gate line, the horizontal data and bias lines and the TFT in the *bottom left corner* [4.24]. The pixel size is 127 µm and the sensor fill factor is 57%. (*Right*) The pixel circuit

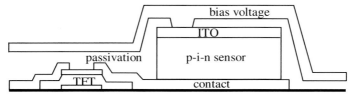

Fig. 4.12. Cross-sectional view of a pixel showing the a-Si:H TFT and p-i-n photodiode sensor

for the next metal layer. The top metal layer is used for the data address line and the bias voltage contact to the sensor, as well as the light.

Overview of Array Operation. The array performs three operations – signal acquisition, signal readout, and reset. Generally, the array is repetitively cycled though these operations, and the time for each cycle is referred to as the *frame time*. The signal acquisition begins immediately after the pixel is reset. The TFT is turned off and the sensor is charged to the full bias voltage. Incident light absorbed in the photodiode, creates electron–hole pairs which drift to the contacts and discharge the sensor. The thin sensor has a significant capacitance which stores the charge – the pixel circuit in Fig. 4.11 shows the sensor as both a diode and a capacitor.

When n photons are incident on the photodiode, it discharges by an amount $Q_S = \eta_{QE} n e$, and the voltage across the sensor is reduced to V_{SIG}, where,

$$V_{SIG} = V_B - Q_S/C_S, \qquad (4.12)$$

provided that the charge is insufficient to discharge the sensor capacitance completely. Although the sensor is discharged by the light, it is convenient to think of Q_S simply as the signal charge, and it is frequently referred to in this way. Thus, the signal charge is initially zero, and increases in proportion to the illumination flux until either pixel saturation is reached or readout occurs.

Readout is initiated when the TFT is switched on, which causes the lower contact of the sensor to be restored to the data line voltage. The charge Q_S passes from the data line to the photodiode and this amount of charge required to reset the sensor is recorded by the external electronics as the signal. Thus, the measurement of the signal charge and the reset of the photodiode are performed simultaneously.

The basic operational requirements of the pixel are therefore to covert the light flux to charge as efficiently as possible, to store the charge without loss during signal acquisition and to transfer it completely and quickly during readout. The next sections discuss the details of these operations and the design issues that allow the array to acquire images at high speed and with minimum distortion or loss of signal.

Sensitivity and Fill Factor. The design of Fig. 4.12 is one in which the sensor occupies the space remaining between the address lines and TFT. The fraction of the pixel area which is occupied by the exposed sensor surface is known as the *fill factor*, F_F, and it affects the overall sensitivity of the array. The photodiode charge that results from an incident light flux of G_L photons/cm^2 on a square pixel of edge size, d, is,

$$Q_S = eF_F\eta_{QE}G_Ld^2 \,. \tag{4.13}$$

High sensitivity is obtained with low electronic noise together with high fill factor, collection efficiency and pixel size. The minimum detectable charge is defined to be equal to the electronic noise, q_N, so that the sensitivity can be given in terms of a noise equivalent power, expressed as the light flux, G_{NE}, given by,

$$G_{NE} = q_N/(eF_F\eta_{QE}d^2) \,. \tag{4.14}$$

For the pixel design under consideration, the fill factor is determined by the size of the pixel and the processing design rules. A simple, illustrative model for the fill factor is,

$$F_F = (d-a)(d-b) - c \,, \tag{4.15}$$

where d is the pixel size, a and b are respectively the space occupied by the address lines in the gate and data directions, and c is the additional area taken up by the TFT. Figure 4.13 shows the fill factor according to (4.15) as a function of pixel size for design rules that approximately correspond to present arrays, as well as with each of the values, a, b and c, reduced by a factor 2. The fill factor approaches unity at large pixel sizes, and drops rapidly to zero with sufficiently small pixels. Present photolithigraphic manufacturing processes allows a fill factor of about 50% with 100 μm pixels and such designs

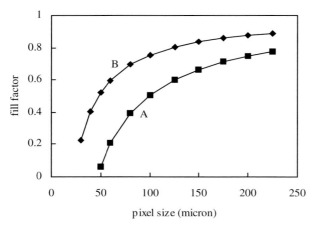

Fig. 4.13. Calculated fill factor vs pixel size for the model of (4.15). Curve A uses values, $a = 20\,\mu\text{m}$, $b = 30\,\mu\text{m}$, $c = 400\,\mu\text{m}^2$, roughly corresponding to present design rules, and curve B has roughly half these values

are therefore quite adequate for the larger pixel arrays [4.30]. However, this design has insufficient fill factor for higher resolution devices, and the solution for such arrays is either improved design rules – which will undoubtedly come with time – or the high fill factor designs that are discussed in Sect. 4.3.3.

Table 4.1 shows the sensitivity, G_{NE}, for various pixel sizes, calculated from the fill factor model. The results assume a minimum detectable charge of 1000 electrons (0.16 fC), which is in line with current imaging systems, and a typical sensor quantum efficiency, η_{QE}, of 0.8. The rapid decrease in G_{NE} with pixel size is due to a combination of the increased fill factor and pixel size. The value of $2.5 \times 10^{-12}\,\text{W/cm}^2$ at a pixel size of 150 μm illustrates the extraordinary ability of these devices to detect a very weak light flux.

Figure 4.14 shows how the maximum sensor charge, Q_{SMAX} ($= C_{\text{S}} V_{\text{B}}$), increases with pixel size, again based on the same fill factor model and a bias voltage of 5V. For the 1000 electron minimum detectable charge, the dynamic range is in the range 10 000–100 000 for pixel sizes above 100 μm. These devices therefore have excellent dynamic range, and in most cases it

Table 4.1. Calculated values of the noise equivalent power for different pixel sizes

Pixel size (μm)	Noise equivalent power, G_{NE} (W/cm^2)
50	3×10^{-10}
100	8×10^{-12}
150	2.5×10^{-12}
200	1.3×10^{-12}

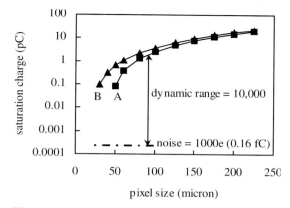

Fig. 4.14. The maximum sensor charge vs pixel size based on the fill factor model of (4.15). A and B refer to the same parameter values as in Fig. 4.13. The dynamic range of the sensor is indicated

is the readout electronics rather than the array itself that limits the overall dynamic range of the complete imaging system.

Readout Timing. The TFT and the photodiode in the pixel determine the time required to read out the image to the external electronics. During the time that the TFT is switched on, the signal charge stored on the sensor capacitance, C_S, is transferred through the TFT to the data lines. The fraction of charge transferred in time t, is $\exp[-t/R_{ON}C_S]$ when the TFT has an on-resistance of R_{ON}. Several time constants are required to transfer the charge completely (i.e., 99.3% of the charge is transferred after 5 time-constants and 99.995% after 10). The minimum time, τ_{RO}, required to read out the complete image of N gate lines, is therefore,

$$\tau_{RO} = 10\,N\,R_{ON}C_S\,, \qquad (4.16)$$

where 10 time constants are assumed for definiteness. For example, an array with 1000 lines, 1 pF pixel capacitance and 3 MΩ TFT on-resistance, can be read out in 30 ms. The value of τ_{RO} sets a lower limit on the imager frame time.

It is of interest to understand how the minimum readout time varies with the size and resolution of the array, for which a simple scaling analysis can be given [4.29]. The sensor capacitance is proportional to the pixel area and the fill factor ($C_S \approx C_{SO}F_F d^2$), while the TFT is limited in size by the pixel dimension ($R_{ON} \approx R_{TO}d$), where C_{SO} is the sensor capacitance per unit area, R_{TO} is the TFT resistance for unit width. Equation (4.16) therefore becomes,

$$\tau_{RO} = 10\,N\,R_{TO}C_{SO}F_F\,d = 10\,R_{TO}C_{SO}F_F\,L_D = 10^{-3}\,F_F\,L_D\,, \qquad (4.17)$$

where $L_D (= Nd)$ is the size of the array in centimeters. For this simple analysis, the minimum readout time is independent of the pixel size, and increases with the linear dimension of the array. The reason is that the shorter time constant of small pixels is offset the larger number of gate lines in such an array. High resolution arrays can, in principal, be read out faster than lower resolution arrays for the same area because the sensor fill factor is smaller. The value of 10^{-3} given on the right-hand side of (4.17) uses typical values for amorphous silicon sensors and TFTs (a sensor thickness of 1 μm, TFT channel length of 10 μm, mobility $0.8\,\mathrm{cm}^2/\mathrm{Vs}$, gate voltage 10 V etc). Video rate output can be sustained for array sizes up to 20–30 cm, and even larger by optimizing the design, and this speed is required for X-ray fluoroscopy applications described in Sect. 4.5.1. In practice, and particularly for a large high resolution array, the speed is more constrained by the performance of the external electronics than by the array itself.

High readout speeds also require that the time constants for the gate and data lines are adequate. The address line capacitance for a typical large area array (> 1000 lines) is approximately 100 pF, so that a line resistance less than about $3 \times 10^4\,\Omega$ is needed for a time constant $< 3\,\mu\mathrm{s}$. Although this seems a large value for a metal, the line is so long and thin that it corresponds to a resistance of about $0.5\,\Omega/\square$, which is readily obtained by thin films of aluminum, but not by some other metals such as chromium. The choice of metals for the address lines is an important factor determining the performance of high speed arrays.

For the detection of very weak illumination, it is often desirable to integrate the signal for a long time. The limitation on the maximum integration time is the leakage current of the sensor (I_{LS}) and the TFT (I_{LT}). The sensor leakage causes a build-up of sensor charge on the pixel, which eventually exceeds the maximum sensor charge, while the TFT leakage causes the loss of sensor charge and therefore a non-linear image response. The leakage currents depend on the bias on the devices, which in turn depends on the sensor charge [4.30]. The sensor charge accumulating in the dark is given by,

$$\frac{dQ_S}{dt} = I_{LS}(Q_S) - I_{LT}(Q_S) , \tag{4.18}$$

with,

$$Q_S = C_S(V_B - V_0) , \tag{4.19}$$

where V_0 is the small bias on the TFT source after reset, due to the feed-through capacitance discussed below. Experimental observations indicate that that the sensor leakage current is nearly independent of bias while the TFT is ohmic with resistance R_{OFF}. These approximations give,

$$\frac{dQ_S}{dt} = I_{LS}(Q_S) - V_0/R_{OFF} - Q_S/R_{OFF}C_S , \tag{4.20}$$

which has the solution,

$$Q_S = R_{OFF}C_S(I_{LS} - V_0/R_{OFF})[1 - \exp(-t/R_{OFF}C_S)] . \tag{4.21}$$

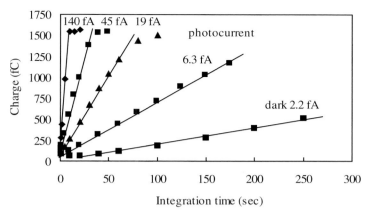

Fig. 4.15. Sensor charge in the dark and with weak illumination, at long integration times [4.30]

When $t \ll R_{\text{OFF}} C_{\text{S}}$,

$$Q_{\text{S}} = (I_{\text{S}} - V_0/R_{\text{OFF}})t \,. \tag{4.22}$$

The sensor charge initially increases at a rate corresponding to the sensor leakage currents (because V_0 is usually small) and eventually approaches a steady state value with a time constant of $R_{\text{T}} C_{\text{S}}$ provided that the sensor does not saturate first. Figure 4.15 shows long integration time data obtained with weak illumination and in the dark using a room temperature imager with 127 µm pixel size, 35% fill factor and sensor capacitance 0.5 pF. The dark signal increases linearly with integration time, apart from a small deviation at short times, and corresponds to a dark current of 2.4 fA, while weak illumination causes a more rapid increase in signal charge as is expected. The dark current is consistent with the sensor leakage measurements in Fig. 4.5 scaled to the area of the sensor. There is no indication of the signal reaching a steady state value even after 250 s, which implies that $R_{\text{OFF}} C_{\text{S}} > 750$ s. This value places a lower bound on the TFT resistance of 1.5×10^{15} Ω for a device with $W/L = 1.5$. Again, the results are consistent with the TFT measurements in Fig. 4.8. Since both the sensor and TFT leakage currents depend strongly on temperature, cooling by a few degrees should allow integration times well in excess of 1000 s.

The deviation from a linear increase in dark signal seen at short times in Fig. 4.15 is explained by the transient TFT response [4.30]. The data in Fig. 4.9 show that the TFT current decays quite slowly to its very low steady state value, and the excess leakage current in this transient causes a loss of signal charge. Thus, at short times there is a small decrease in dark signal charge with increasing time before the linear growth of the signal sets in. This transient charge may have an effect on the imager performance when two images are compared with different integration times.

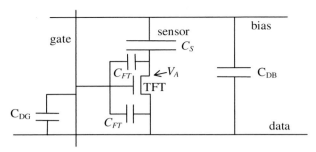

Fig. 4.16. Diagram showing the various sources of capacitance within the array

Capacitance Effects. A full understanding of the charge storage and transient readout properties of the array requires a more detailed analysis of capacitance within the array [4.31]. As described in more detail in Sect. 4.4.2, the electronic noise and much of the array performance is determined by various capacitance effects. Figure 4.16 shows an equivalent circuit that indicates the important parasitic and coupling capacitance. It has just been noted that the high address line capacitance sets limitations on the metal resistance, and in addition, a high data line capacitance is the dominant source of electronics noise in most systems. The gate and data line capacitance arises from several sources within the pixel;

- the cross-over of the gate and data lines which form parallel plate capacitors;
- the TFT channel capacitance and the parasitic gate to source and drain capacitance;
- the capacitance between the TFT source or drain and the light shield (if used);
- the fringing capacitance between the data line and the sensor.

For the type of array under discussion, these effects usually amount to about 20–50 fF per pixel, but the values can be selected by the specific design of the array.

These sources of capacitance have several effects on the readout. When the TFT gate is switched, there is a capacitative feed-through of charge, Q_{FT}, to the source and drain contacts, which may be a significant fraction of the maximum signal

$$Q_{FT} = C_{FT}\Delta V_G \approx 10\,\text{fF}\ 20\text{V} = 0.2\,\text{pC} \,, \tag{4.23}$$

where typical values are assumed for illustration. The charge on the source alters the bias on the sensor by ΔV_B, which must be taken into account in calculating the saturation charge of the pixel and other parameters, where,

$$\Delta V_B = Q_{FT}/C_S = C_{FT}\Delta V_G/C_S \approx 0.2\,\text{pC}/1\,\text{pF} = 0.2\,\text{V} \,. \tag{4.24}$$

The charge on the TFT drain is transferred to the readout electronics and provides a background charge that must be removed to obtain the correct image.

These effects are described in Fig. 4.17 which illustrates the voltages and charge during the readout of one pixel. The response is shown with and without exposure to illumination. The data line is attached to a charge sensitive amplifier, which holds the voltage constant. Consider first the voltage, V_A, on the TFT source (see Fig. 4.16). Near the end of the gate pulse all the charge has been transferred and the voltage V_A is equal to the data line voltage. As the gate is turned off, the parasitic capacitance causes V_A to decrease by ΔV_B. This voltage drop can be significant, and must be taken into account to obtain the actual sensor bias voltage. In the absence of illumination, V_A changes during the frame time by an amount dependent on the time-integrated sensor leakage current. When the gate is next turned on the feed-through charge raises V_A, and the current flow through the TFT resets V_A to the data line voltage. Exposure to illumination during the integration time causes V_A to decrease as extra charge is generated by the sensor. The read out again resets V_A and transfers sensor charge to the data line. If the exposure is sufficiently large, V_A reaches the sensor bias voltage and prevents further charge from being collected. The sensor is saturated and the readout charge is independent of further increase in illumination.

The charge transferred to the data line is shown in the lower curve in Fig. 4.17. The capacitive coupling between gate and data line causes an abrupt transfer of the feed-through charge when the TFT is switched. The real sensor charge is transferred to the data line more slowly, with the RC time constant $R_{ON}C_S$. When the gate is turned off there is another compensating transfer of feed-through charge. How the charge is detected by the external electronics is discussed in Sect. 4.4.1. Generally the data line charge is sampled twice – once before the TFT is switched on to give a background value, and once afterwards. After the sampling has taken place the data line is reset by the external electronics in readiness for the next measurement.

Image Lag and Charge Trapping in the Photodiode. So far it is assumed that all the sensor charge can be read out quickly and completely. Incomplete transfer that leaves signal charge at the pixel is doubly undesirable because it both reduces the detected signal, and releases the charge at a later time when it can create ghosts in subsequent images. This effect called *image lag* is fairly common in detectors, and is most noticeable under high frame rate conditions.

The simplest origin of image lag occurs when the TFT is not turned on for long enough to transfer the signal, but this situation is readily correctable by changing the timing. Image lag may also occur if the photodiode leakage current tends to drift (see Sect. 4.2.1 for a discussion of diode stability). However, in good quality photodiodes, the main source of image lag arises from

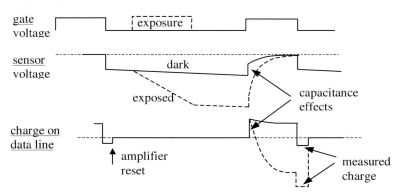

Fig. 4.17. Voltage on the sensor and charge on the data line during readout, with and without illumination, showing the effects of the parasitic capacitance, as described in the text. The illustration is for the case that the feed-through charge has the opposite sign from the signal charge

electronic traps in the undoped layer. Amorphous silicon contains deep defect states which may trap charge for much longer than the readout time. Charge trapping is governed by the mobility-lifetime product, $\mu\tau$, as described in Sect. 4.2.1. Equation (4.3) shows that the fraction of charge trapped is approximately,

$$P_{\mathrm{T}} = 1 - P = t_{\mathrm{F}}^2/2\mu\tau V \qquad (4.25)$$

assuming a constant voltage and uniform field. Thus, trapping is minimized in thin samples with a high bias. In a good quality a-Si:H photodiode of 1 μm thickness with 5 V bias, 1–3% of the electrons are trapped and 5–10% holes, because of their smaller $\mu\tau$ [4.10]. It is therefore best to design the sensor such that the electrons travel the larger distance, and this is accomplished by placing the p-type contact towards the illumination. Provided that the illumination is significantly above the band gap, then the absorption is close to the illuminated p-type contact. Under these conditions, holes do not have a significant distance to travel.

The situation in a sensor array is slightly more complicated because the bias decreases as the light flux on the sensor increases. The charge trapping therefore increases as the sensor approaches saturation. The collected signal charge when there is an illumination intensity, G_{L}, can be expressed as,

$$Q_{\mathrm{S}}(V,t) = \int_0^t P_{\mathrm{T}}(V) G_{\mathrm{L}} \mathrm{d}t \ . \qquad (4.26)$$

Figure 4.18 shows a simulated plot of signal charge versus exposure for different degrees of initial trapping (i.e. the values of P_{T} at the start of illumination). The trapping causes a departure from linearity at increasingly low exposure, and the fraction of carriers which are trapped increases rapidly as saturation is approached due to the decrease in bias. Figure 4.18 also shows

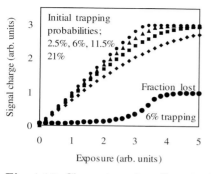 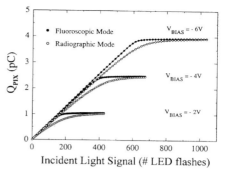

Fig. 4.18. Charge trapping effects in the photodiode simulated (*left*) and measured on an image sensor array [4.31]

that the actual data for an array agrees with the simulations. The charge trapping only reduces the charge collection in radiographic mode operation (see Sect. 4.4.1), so that the different results for the two modes clearly distinguishes the trapping effect. The results indicate that P_T is about 95% for the particular array measured [4.31].

The assumption of a uniform field is an approximation. The internal field of an unilluminated sensor has a depletion layer with a field profile that decreases away from the p-type contact and is determined by the defect density. The field profile becomes further distorted by the charge trapping. This situation has not been fully analyzed but it is clear that any region of the sensor in which the field is low greatly enhances the trapping effects.

4.3.3 High Fill Factor Array Designs

The imager sensitivity is proportional to the sensor fill factor which, as Fig. 4.13 shows, decreases rapidly as the resolution increases for the array design of Figs. 4.11, 12. The fill factor may be increased by using a 3-dimensional array architecture, in which the sensor is placed on top of the addressing matrix. Some designs of this type have been tried and are effective [4.32, 33]. Figure 4.19 illustrates the general approach by showing the cross-section view of one such design. The sensor is a continuous film covering the whole pixel. It is connected to the TFT through a contact pad and isolated from the address lines and the TFT by a passivation layer.

A continuous sensor layer can give unity fill factor, but the absence of an etched region separating pixels raises the possibility that charge created at one pixel can migrate to neighboring pixels. A continuous p-i-n sensor, in particular, allows such cross-talk due to the conductivity of the doped layer. Adjacent pixels with different signal charge have a potential difference between their contact pads and so current can flow taking charge from one pixel to the next. The extent of cross-talk depends on the conductivity of the doped layer and the frame time. In one study, the cross-talk was found

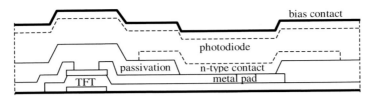

Fig. 4.19. Example of the design of a high fill factor sensor array using a continuous a-Si:H photodiode layer with a patterned n-type doped contact

to be minimal at frame times less than 0.1 s, but extensive if an integration time of 1 s or longer was used. A solution to this source of cross-talk is to eliminate the continuous doped layer by an extra lithography step, and this is the design shown in Fig. 4.19. The fill factor is slightly reduced, but the spatial resolution is restored [4.32].

Placing the sensor over the address lines also causes increased capacitance between these lines and the sensor. High capacitance is generally undesirable because it tends to increase the electronic noise (see Sect. 4.4.2), and the minimization of the capacitance is a matter of design optimization, which can include the use of a thick passivation layer. The capacitive coupling also can introduce unwanted parasitic charge into the system when gate voltage switching takes place.

An alternative high fill factor design to address these concerns is shown in Fig. 4.20 [4.33]. The two innovations are to invert the sensor so that the bias voltage plane lies between the address lines and the sensor to shield it from parasitic capacitive coupling, and to pattern the sensor with small gaps to eliminate crosstalk. This design is less susceptible to image degradation, while maintaining a high but not unity fill factor. The particular array that was fabricated had 192 × 192 pixels of size 200 μm, with a 5 μm gap between the sensors. Together with the contact via to pass the signal from the top contact to the TFT below, the resulting fill factor is 93%. In order to minimize the data line capacitance, the sensors are separated by a 5 μm thick polymer layer with a low dielectric constant, and the resulting capacitance increased by only about 20% compared with the design of Fig. 4.12. However, some degradation of the photodiode was observed due to the polymer film. The capacitance between the sensor and the data line is virtually eliminated on this design and so removes any cross-talk between pixels that arises from this source.

4.3.4 TFT Addressed, X-Ray Photoconductor Arrays

Amorphous silicon p-i-n sensors are efficient detectors of visible light, but thin sensors have little sensitivity to X-rays because their absorption is weak. X-ray sensitivity is achieved by placing a phosphor in close contact with the a-Si:H sensor array [4.6, 34, 35]. The phosphor absorbs X-rays and emits light

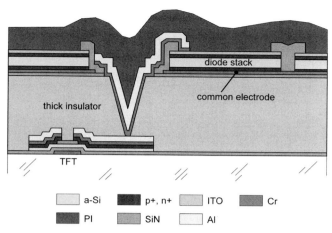

Fig. 4.20. High fill factor array with capacitance transient isolation and sensor isolation [4.33]

that is sensed by the array – the technique is described in Sect. 4.4.3. The use of a phosphor to convert X-rays to visible light is very effective and has been used for the past 100 years with X-ray film. However, an alternative approach to X-ray detection replaces the thin amorphous silicon sensor with a much thicker X-ray sensitive photo-conductor [4.36].

The requirements of the X-ray photoconductor place significant constraints on the choice of materials and the design of the array. The sensor material must absorb a high fraction of incident X-rays in order to make an effective imager. Since X-rays are weakly absorbed, the X-ray photoconductor materials must be thick – from 100 μm to 500 μm – in comparison to the 1 μm a-Si:H sensors. The thick films have several consequences for the array design. It is extremely difficult to pattern such devices into isolated sensors at each pixel because of the high aspect ratio needed for the etching and because of the difficulty of preventing etching of the underlying layers. The array designs use the same approach described above for high fill factor arrays and are fabricated as a continuous film with a contact pad at each pixel. Such thick films have very low pixel capacitance, since the capacitance is inversely proportional to the thickness, and therefore cannot hold much sensor charge. An additional capacitor is provided at each pixel to enhance the charge storage.

Figure 4.21 illustrates the structure of an X-ray photoconductor array. The pixel circuit contains the addressing TFT, the small sensor capacitance and the extra storage capacitor [4.37]. The latter requires a common electrode which must be provided along with the usual gate and data address lines. In the design shown, the storage capacitor is made from the same silicon nitride dielectric layer as for the TFT. The capacitance is comparable to that of the usual amorphous silicon sensor – of order 1 pF for a 100 μm pixel

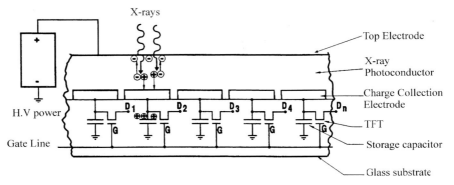

Fig. 4.21. A cross-section view of the structure of a photoconductor array with additional capacitance and matrix addressing [4.37]

– since the nitride layer is usually thinner, but also has a lower dielectric constant than a-Si:H. It can be noted in passing that the additional capacitance could be added to arrays with a-Si:H sensors to increase their charge storage capacitance.

Sensitivity. The increased thickness of the photoconductor requires a high quality material, but has the potential of high sensitivity X-ray detection. Sensitivity depends on the combined efficiency of charge generation and charge collection. Charge generation occurs when an incident X-ray interacts in the material to create a high energy electron which loses energy by ionization of electron–hole pairs. The ionization charge is collected by the applied field, and the collection efficiency is given by (4.3). As an example, a film of 300 μm thickness requires $\mu\tau V > 10^{-3}$ cm^2, which is achieved with a combination of a material with a very low defect density, and by applying a high bias voltage. Furthermore, the X-ray interactions occur throughout the material due to the low X-ray attenuation coefficient, so that efficient collection of both electrons and holes is essential.

Suitable photoconductors must have high atomic mass, large $\mu\tau$ values, and low leakage currents. To form large area arrays, the material must maintain its good properties in thin film form, which usually implies an amorphous or polycrystalline layer. There are a number of candidate materials – examples are selenium, lead iodide and oxide, thallium bromide, mercury iodide, and cadmium telluride. Amorphous selenium is used as an X-ray sensitive layer in other types of detector, and most of the other materials are known to perform well as X-ray detectors in their single crystal form. At the time of writing, Se, PbI$_2$ and PbO have been used in active matrix X-ray imagers and of these, Se is the most extensively studied [4.38–41].

The design of the array can mitigate some of the difficulties associated with high bias voltages and the continuous thick films. Selenium sensors may

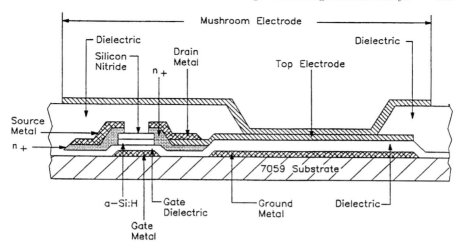

Fig. 4.22. The structure of a pixel for photoconductor detection, showing the "mushroom" electrode designed to maximize the charge collection [4.37]

require in excess of 3000 V, a voltage that would easily destroy the TFT or the charge storage capacitor if it is allowed to drop across the contacts. In principle, excessive exposure can charge up the capacitor and the source contact of the TFT to the bias voltage. One design prevents this by providing a leakage mechanism when the charge stored gets too large [4.39]. The TFT has a dual gate structure in which the second gate is connected to the source of the TFT. When the voltage rises sufficiently, the TFT is turned on and the charge leaks away before breakdown occurs.

A different problem is the build up of space charge in the sensor layer. Selenium has a low dark current and charge can be trapped between the pixels where there is no contact pad. Such charge may be released slowly and provide a source of image lag, or might be trapped near the TFT and cause an increase in its leakage current. A solution is to make the lower contact cover as large an area as possible, and in particular to cover the TFT to isolate it from the charge. The design illustrated in Fig. 4.22 is of this type. An isolation layer covers the TFT and address lines leaving the contact pads to the TFT exposed. Another metal film extends over the isolation layer and forms the bottom contact to the photoconductor. Even with the increased electrode area, there is sufficient trapped charge to cause visible ghosting effects due to image lag. This effect is removed by illuminating the imager between each image capture to reset the array to a uniform state.

4.3.5 Diode Addressed Arrays

Diode addressing of matrix addressed arrays has been applied to both liquid crystal displays and image sensors [4.42–44]. The approach is effective in

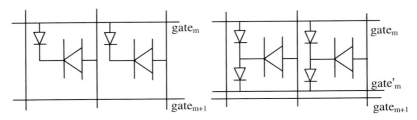

Fig. 4.23. Two alternative designs (single diode – *left*, and double diode – *right*) of diode-addressed image sensor arrays

both cases, but is not as widely used as TFT addressing. Diode addressing is an attractive choice for image sensors using a-Si:H p-i-n sensors, because the same sensor layer can in principle be used for both the sensor and the addressing device, thus simplifying the fabrication.

Figure 4.23 illustrates the pixel circuit used for addressing with a single or double diode arrangement. The single diode is placed back-to-back with the much larger photodiode, and the pair are connected between the address line and the data output line. Similar to the operation of the TFT addressed array, a voltage pulse on the address line serves both to read out and reset the pixel. The voltage pulse applies a forward bias to the addressing diode and transfers the bias voltage to the photosensor. The diode reverts to reverse bias when the pulse is removed, which holds the bias voltage across both diodes. Illumination of the photodiode causes charge collection and reduces the bias. When the voltage is reset by again applying the address pulse, charge of amount $C_S \Delta V$ is transferred from the data readout line to the sensor, where ΔV is the voltage drop on the sensor due to the illumination. This reset charge is exactly the signal developed by the sensor, and is recorded by the external electronics. The development of the signal charge and its read out is therefore generally similar for both TFT and diode addressing, and much of the discussion in Sect. 4.3.2 applies.

In both approaches, the capacitative feedthrough charge must be taken into account. The diode capacitance causes a change in the sensor bias when the address line is switched. The magnitude of the effect depends on the diode capacitance, but is similar to the gate capacitance of the TFT so the effects are of similar magnitude.

The most significant difference between diode and TFT addressing is in the readout time. The resistance of the diode varies exponentially with bias voltage, unlike the resistance of a TFT which is independent of source–drain voltage, at least for low voltages. In the diode exponential region (4.8),

$$R = (V/I_0)\exp(-eV/nkT) \ . \tag{4.27}$$

The diode resistance increases as the sensor voltage approaches V_B, and the reset time constant decreases correspondingly. Consequently, the sensor cannot be reset properly without waiting for a long time. This inability to reset the sensor fully affects the dynamic range only slightly, but can cause significant image lag, because the degree to which the pixel is reset depends on the amount of charge detected in the previous image acquisition cycle.

Another property of the diode also tends to enhance image lag effects. When a diode is switched into forward bias, there is considerable charge trapping in defect states in the band gap. The charge is released slowly upon switching to reverse bias, and modifies the subsequent measurement of the signal. Similar trapping and release occurs in the TFT, but usually involves a smaller charge, because the TFT channel is 20–50 Å thick while the diode is more typically 1 μm thick, with correspondingly more trapping states.

Chabal et al. [4.43], and also Hoheisel et al. [4.44], discuss in detail the design and performance of a large area diode addressed array with 1024×1024 pixels. The arrays have a fill factor of 70% with 196 μm and 43% with 143 μm pixel sizes, using an addressing diode of 20×20 μm^2, but larger fill factors are expected with improved design rules. The two diodes are deposited separately so that they can be individually optimized – the photodiode for high quantum efficiency and low leakage, and the switching diode for high forward current. They overcome the limitations of the diode addressing by using a light flash to fully discharge the sensor charge and the reset of the sensor bias is separated from the readout addressing, using different bias voltages for each. The illumination ensures that all the photodiodes are reset identically by saturating their deep traps. This procedure eliminates the image lag effects but does complicate the read out. The result is an imager that operates up to 12.5 frames/s, and with excellent imaging performance.

Double Diode Addressing. The double diode approach solves some of the problems associated with single diode addressing [4.45]. The reset and readout is performed by two diodes connected to different address lines, as illustrated in Fig. 4.23. The operation is such that both diodes are switched from reverse bias to forward bias during the reset. This immediately solves the problem of a slow reset time, because the diodes maintain a high forward current and therefore have a low resistance with a correspondingly small RC time constant. The effects of the capacitance feedthrough charge are also eliminated in this arrangement because the charge transfer from one diode is offset by an equal and opposite charge from the other. The penalty for these performance improvements is that an extra address line must be included in the array to operate both diodes.

Graeve et al. have developed a double diode imaging system [4.46]. The array has 100 μm pixels in a 512×512 format with a photodiode fill factor of 45%. The readout electronics allows operation at up to 6.5 frames/s, and the imager shows good linearity and sensitivity properties with CsI and GdO_2S_2:Tb phosphors. The image lag of 5–8% arises from the photodiode and is apparently not a result of the double diode addressing.

4.3.6 CMOS Sensors

It is worth giving a brief mention of the emerging field of CMOS sensor arrays, because of their use of a-Si:H photodiodes and their general similarity to the large area devices. CMOS sensor arrays are an alternative to charge coupled devices (CCD), whose widespread use ranges from video cameras to astronomical telescopes [4.47]. The attraction of the CMOS sensor is that it is fabricated with the standard technology of silicon integrated circuits, for which there is an enormous technology development effort. Unlike the CCD, the CMOS sensor usually has a matrix-addressed design based on the general architecture of Fig. 4.10. The pixel size is about 10 μm, to keep the device small and of low cost. Low fill factor limits the sensitivity just as for the large area arrays because, unlike the CCD, significant space in the pixel of a CMOS sensor is needed for address lines and switching transistors. The continuing improvement in the CMOS feature sizes, now down to 0.25–0.35 μm, makes the device possible by allowing more space in the pixel for the sensor.

However, the fill factor can be further increased by the technique described in Sect. 4.3.3 – coating the surface with a thin film sensor. Amorphous silicon p-i-n photodiodes are an obvious choice and have been applied successfully to CMOS sensors [4.48]. The diode provides sensitivity across the visible spectrum, high quantum efficiency, very low leakage current and ease of process integration. The design is essentially that shown in Fig. 4.19 for the high fill factor array – the continuous sensor film is connected to each pixel by a metal contact pad, with a common transparent electrode on top.

One of the attractions of the CMOS sensor array is its ability to integrate electronics on the array. Pixel circuits of considerable complexity can be fabricated for amplification, signal calibration, image processing, etc. In addition, further electronics at the periphery of the sensor array can provide control signals, analog-to-digital conversion and further signal processing. Clearly similar functions could also be integrated into the large area imagers provided that the device feature size is small enough and the TFTs are of sufficient quality. In the future both large and small area imagers may have similar levels of integration.

4.4 Imaging Systems and Their Performance

The complete imaging system comprises the sensor array together with the associated electronics to amplify, digitize, and display the image. The performance of the imager is in large part determined by the electronics, in particular the electronic noise and dynamic range of the amplifiers, which read the signals directly off the data lines. The electronics also determine the bit depth of digitization, and often limits the minimum frame time.

The performance of an image sensor array is a combination of resolution, sensitivity and noise. The analysis and measurement of the detector performance is fairly complicated, particularly for the case of X-ray imaging [4.49],

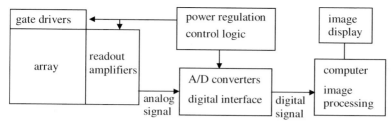

Fig. 4.24. Elements of the imaging system. The gate drivers and readout amplifiers are attached to the array. The other elements can be connected in many different ways

because of the additional effect of the X-ray conversion stage in which each X-ray generates many electronic charges that are detected.

The connection of the array to the external electronics is an important aspect of the technology. A high resolution image sensor requires several thousand such connections, and is a challenging task. The liquid crystal display technology has led the way with tape automated bonding (TAB), which allows a simple adhesive connection between the array and a driver or amplifier chip.

4.4.1 Electronics

The main elements of the imaging system are shown in Fig. 4.24 [4.50]. The gate lines of the array are connected to shift registers that provide sequential addressing pulses – the same devices that are used to address liquid crystal displays. These integrated circuits usually have 100–200 channels, so that several are required for a large area, high resolution imager. The readout amplifiers sense the charge on the data lines and the amplified analog signal is provided to analog-to-digital (A/D) converters [4.51]. The amplifiers are CMOS integrated circuits each with typically 128 channels, and several are used for each array. The signal is then digitized and passed to the host computer for processing and display. There is additionally the control logic that provides timing for all the operations and power regulation which provides filtered dc power to the system.

The task of the electronics is to take the image on the array that is acquired simultaneously at every pixel, and to organize the sequential transfer of digital data to the host computer. The task is accomplished by sequential stages of multiplexing. The first stage is to read out each line in sequence and is provided by the matrix addressing and the gate shift register. The next stage of multiplexing is performed by the readout amplifier which transforms parallel inputs into a single analog output, and further mutiplexing generally takes place either before or after the A/D converter. With each stage of multiplexing, the data rate progressively increases. The design of the elec-

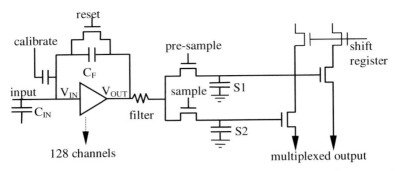

Fig. 4.25. Schematic diagram showing the elements of the charge sensitive amplifiers used to read out data from the array

tronic system matches the multiplexing stages to the speed performance of the various devices in the circuit, to obtain the desired overall readout speed.

Readout Amplifiers. The readout amplifiers are, after the array, the component that most determines the system performance [4.52, 53]. The array generates charge, and therefore a charge sensitive amplifier is the natural readout device, although voltage amplifiers have also been used [4.32]. The usual organization of the amplifier is illustrated in Fig. 4.25 and consists of several stages in sequence. After the high gain input amplifier with its feedback capacitor, there may be additional bandwidth filtering to minimize noise. The sample-and-hold stage, which stores the amplified charge, is followed by an output multiplexer. It is possible to include A/D conversion in the chip, but usually this is performed separately.

The feedback capacitor across the input amplifier determines the maximum input signal. The signal charge, Q_S, is stored on the feedback capacitance, C_F and the input capacitance, C_{IN}, which mostly arises from the data line capacitance of the array,

$$Q_S = C_F(V_{OUT} - V_{IN}) + C_{IN}V_{IN} . \tag{4.28}$$

The output voltage swing, V_{OUT}, is related to the input voltage swing, V_{IN}, by,

$$V_{OUT} = g_A V_{IN} , \tag{4.29}$$

where g_A is the amplifier gain. Hence,

$$Q_S = [C_F(g_A - 1) + C_{IN}] V_{IN} . \tag{4.30}$$

The gain must be large enough that the charge is practically all stored on the feedback capacitance rather than the input capacitance, for which the condition is,

$$C_F(g_A - 1) \gg C_{IN} . \tag{4.31}$$

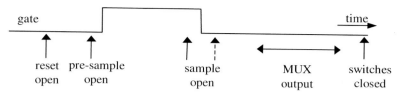

Fig. 4.26. Switching sequence to read out the array signal, using double correlated sampling

A high gain amplifier with $g_A > 1000$ is required, since C_{IN} is typically of order 100 pF, while C_F is about 1 pF.

The amplifier is designed with a single input, which is the gate of a very large FET to achieve low noise, as described in Sect. 4.4.2. The amplifier in Fig. 4.25 has various switches (reset, pre-sample, sample) to allow the measurement of the charge to be synchronized with the gate addressing pulses. In the technique of double correlated sampling, the data line charge is measured twice, a pre-sample reference occurring before the gate line is switched and the signal charge is measured afterwards. Noise is minimized because the measurements occur in the short time interval when signal charge is released from the pixel. The output charge is the difference signal,

$$Q_{OUT} = \text{sample} - \text{presample} = Q_2 + [q_2(t_2) - q_1(t_1)] , \qquad (4.32)$$

where Q and q represent the signal charge and the noise, and t_1 and t_2 are the sample times. Low frequency noise which does not change between the two samples is therefore cancelled out. This double measurement of noise within a short time interval acts as a low frequency filter with a cut-off frequency of $1/(t_2 - t_1)$, which is 30–100 kHz for typical sample intervals of 10–30 µs. In addition to removing low frequency noise, double correlated sampling corrects for any stray leakage charge that gets on to the data line.

Figure 4.26 shows how the switching sequence of the amplifier is synchronized to the gate pulses. When the feedback capacitor reset switch is closed, the amplifier input is clamped and is held at a voltage determined by the design of the amplifier. Opening the reset switch releases the amplifier and shortly afterwards the presample switch is opened to store a signal in S1 which is the data line charge before the gate pulse. The gate pulse is then applied and the signal charge transfers from the pixel to the data line. The sample S2 occurs when the charge transfer is complete, and may be before or after the gate pulse is switched off. The reset switch is closed after the samples are taken, and the complete sequence is repeated for each gate line.

After the sampling is completed by the opening of switch S2, the data are read out to the ADC though the output multiplexer. The amplifier design of Fig. 4.25 contains a shift register that sequentially transfers the two signals, and their difference is the input to the A/D converter. Apart from the low noise requirement, other important attributes of the amplifier are high dynamic range, good linearity and a high speed output shift register. Amplifier chips can be designed to operate at 5–10 MHz on the output multiplexer, and with better than 1% linearity over a dynamic range of 10 000. Further performance improvements can be expected with the ever-continuing development of CMOS integrated circuits.

Timing, Digitization and Display. Most imagers are designed to acquire either single images well separated in time, or sequences of images recorded at video rates. The timing sequence of the data acquisition and the readout is similar in each case and is shown in Fig. 4.27 which illustrates the sequential transfer of data from each line of the array. The time taken for the complete readout is the product of the number of gate lines and the line time. The line time is that required to transfer multiplexed data from the readout amplifier chip, through the ADC and then to the host computer or to a data buffer. A chip with 128 channels takes 26 μs to transfer data at 5 MHz, which represents the minimum line time provided that the readout chip buffers the input from the output stage and the digital system can maintain the speed. Such line times are typical for video rate imaging, while single frame imaging can afford to slow down the transfer rate and have more multiplexing of the readout amplifiers.

Both single frame and video rate imaging requires that the imager is continuously cycled. If the array is not cycled, then the charge accumulates and eventually saturates the pixel sensor. We have seen that the a-Si:H TFTs and photodiodes exhibit some slow transients when the bias is changed, and therefore the first frame after the array is fully saturated may have different dark signal charge compared to subsequent frames. In normal operation,

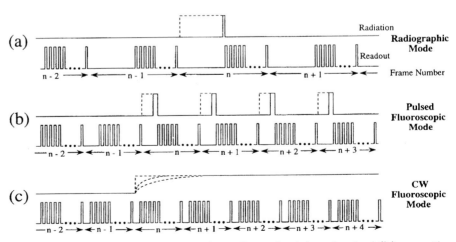

Fig. 4.27a–c. The timing sequences for radiography (**a**), and pulsed (**b**) or continuous (**c**) fluoroscopy imaging, showing how the X-ray radiation is synchronized to the integration time and readout over multiple image frames [4.54]

therefore, the array is cycled through the acquisition and readout cycles, and for single frame image acquisition, the exposure and the readout must be appropriately synchronized.

For the most part the images acquired in single frame or in continuous capture mode are equivalent, but they do have a different response to image lag effects, which can affect the linearity of the response. In single shot mode, some charge is retained by the pixel and does not contribute to the image. However, when readout is continuous, the charge that is retained by the pixel is ultimately readout in a later frame. Thus, although there may be some ghosting, the total signal charge making up the image sequence has no loss due to image lag. A comparison between the signal in the two modes of operation is therefore a convenient measure of the image lag and such data are shown in Fig. 4.18.

In many applications, the image is corrected for gain and offset, which removes variations in the fixed background pattern and the detector sensitivity as well as compensating for illumination non-uniformity. Such a correction is needed when better than 2–5% accuracy in the measurement of intensity is needed, because the offset variations across the imager are often about this value. The correction calculation is made on individual pixels according to the formula,

$$\text{image} = \frac{(\text{measured} - \text{dark})}{(\text{light} - \text{dark})}. \qquad (4.33)$$

The correction requires that the dark and light images are separately measured before the image is acquired. There is a reduction in the digitization bit depth of the corrected image, as this reduces to the difference in the gray levels of dark and light images. Thus the image should be digitized beyond what is actually required for the final image. The correction may also increase the noise because three individual images are involved, but the extra noise can be prevented by using averaged values for correction images.

A further correction is usually made for bad pixels by using a median or average of the neighboring pixel values. The correction is very effective at removing the distracting effects of viewing the bad pixels, and usually a small number in the imager does not detract from its accurate rendering of the signal. The same correction can be made for bad lines, but generally such defects should not be present in an imager in which image fidelity is very important.

Even such simple image processing places a burden on the host computer when large video rate images are processed, and generally such system require specialized digital signal processors. However, current generations of PCs are adequate for single frame imagers for which the acquisition time is 1 s or more.

Interconnect Technology. A high resolution array with 2000×2000 pixels, requires 4000 contacts to the external electronics – 2000 to the gate drivers

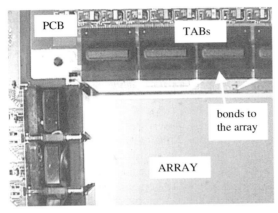

Fig. 4.28. Photograph of TAB-bonded image sensor array

and 2000 to the read-out amplifiers. The challenge of packaging the system increases as technology progress allows the development of higher resolution and larger size arrays. Present interconnect technology uses wirebonding or tape automated bonding (TAB), of which the latter is better suited to volume manufacturing [4.55]. The TAB package contains a silicon integrated circuit bump-bonded to metal traces on a flexible plastic carrier. The contact to the array is made by heat and pressure, using an anisotropic adhesive which contains small metal particles in an insulating matrix. The particles provide a conduction path between the pads bonded together but no lateral conduction between adjacent channels.

The liquid display industry has been the main driver of developments in the TAB technique and an example applied to an imager is shown in Fig. 4.28. The capabilities of the technology are determined by the outer lead bonds connected to the array and the inner lead bonds which are bump-bonded to the IC. The attachment of the TAB part to the circuit board is at lower resolution because there are fewer connections and is done either by soldering or by adhesive bond. The lower limit to the pitch of the outer lead bonds is 60–70 µm, in part determined by the tendency of the plastic carrier to stretch when it is bonded to the array. For example, when using a TAB part with 128 contacts, stretching by as much as 0.5% will cause failure due to mis-alignment of the carrier to the array.

Wirebonding is a well established technique that is presently capable of finer pitch than TAB, but is slower, and more prone to damage. Figure 4.28 shows a photograph of a readout amplifier wirebonded to an image sensor array. The amplifier is on a circuit board placed next to the array, and the pitch of the contacts is 44 µm, which is within the capabilities of wirebonding, although still challenging for an array with several thousand contacts.

It is obviously desirable to reduce the number of contacts from the array to the external electronics, and there are several approaches that can be taken, again mostly pioneered by the liquid crystal display industry. Chip-on-glass technology allows the drive and readout chips to be attached directly on to the array. One approach is to use flip-chip bonding, in which the chip is inverted and attached to metal traces on the array using bump bonds. Alternatively, the readout and drive chips can be eliminated by designing their functions directly into the periphery of the array. It is generally expected that amorphous silicon TFTs do not have the required performance to make fast shift registers and amplifiers, but polysilicon has performance that can approach that of single crystal silicon as is described further in Chap. 3. Liquid crystal displays with integrated drive electronics have been demonstrated [4.56], and although there are no examples of similar image sensors, there is little reason to doubt that the gate drivers could be integrated. Polysilicon readout amplifiers are much more challenging, and most likely would not have the same low noise performance, but this approach may be effective, particularly if pixel-level amplification reduces the noise requirements of the amplifier [4.57].

4.4.2 Electronic Noise

Minimizing noise in the electronic system is essential because the signal-to-noise and dynamic range requirements for many imaging applications are very demanding. However, electronic noise is only one component of the image noise, as discussed further in Sect. 4.4.4. Several sources of electronic noise arise from the array and from the associated electronics. Some of the noise sources are fundamental limits that cannot be reduced, and some are extrinsic sources that are merely difficulty to eliminate. The most important are thermal noise in the pixel TFT and data line, the amplifier input noise, and voltage fluctuations in bias supplies. The RMS values, q_i, of random noise sources add in quadrature,

$$q^2 = \sum_i q_i^2 \,. \tag{4.34}$$

The noise is often dominated by a single component, although this may be a different source in different imaging systems. The following discussion is based on TFT-addressed, a-Si:H sensor arrays, for which there has been the most analysis and measurement. However, the discussion applies to other array types with little modification.

Resistances generate thermal noise, current sources generate shot noise, and generally any device has additional $1/f$ noise [4.58]. The $1/f$ noise is sensitive to the device fabrication, and Fig. 4.29 illustrates the noise spectrum for one a-Si:H TFT [4.59]. There is a low frequency $1/f$ noise that increases with the drain current and a white noise component evident at high frequency. The transition from $1/f$ to white noise occurs at 10^5–10^6 Hz depending on the current. The TFT off-resistance is so high that its noise only occurs at very low frequency and is eliminated by the double correlated sampling. Thick,

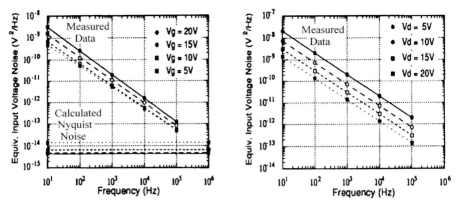

Fig. 4.29. Noise spectrum of an a-Si:H TFT, showing the $1/f$ component and its dependence on gate voltage (*left*) and source–drain voltage (*right*) [4.59]

reverse bias p-i-n photodiodes exhibits $1/f$ noise to about 10^4 Hz at high bias, but the current noise is negligible in thin, low bias sensors, because of the very low leakage currents [4.60, 61]. Photocurrent noise is discussed later and in X-ray imagers reflects the shot noise of the X-ray flux. The sensor may have extrinsic noise due to current instabilities, but these are not considered in the present discussion.

Thermal Noise from the Array. Double correlated sampling acts as a low frequency filter and greatly reduces $1/f$ noise from all the components. To a good approximation, the dominant pixel noise arises from the thermal noise of the single TFT that is switched on [4.62]. The noise voltage fluctuations, with spectral density $\nu(f)$, cause charge fluctuations on the sensor capacitance, C_S. The charge noise is given by,

$$q_{\text{pix}}^2 = \int_0^\infty [C_S \nu(f) g(f)]^2 \, df \, , \tag{4.35}$$

where $g(f)$ is the filter function of the circuit at frequency f. Equation (4.35) represents the sum of noise components over the frequency response of the circuit.

Figure 4.30 shows an approximation to the array circuit, in which the sensor capacitance C_S is in series with the TFT resistance R_{ON} and its noise source, and the components of the much larger data line capacitance are denoted C_D and in parallel with the pixel. The data line capacitance reference voltage is both the gate and bias voltages, but can be considered a single parallel capacitance. The data line resistance, R_D, is shown as a single resistor with its corresponding thermal nose source, although in reality, both C_D and R_D are distributed.

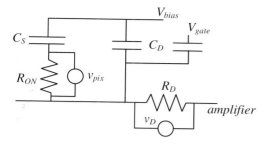

Fig. 4.30. Pixel circuit used for noise analysis, as discussed in the text

The pixel R–C circuit, with the approximation that $C_D \gg C_S$, provides a high frequency cut-off to the noise spectrum with $g(f)$ having the form, $1/[1 + (2\pi f R C_S)^2]$. Resistance noise has a white spectrum with magnitude,

$$\nu^2(f) = 4kTR . \tag{4.36}$$

Neglecting the low frequency filtering provided by the sampled readout, (4.35) yields the well known result,

$$q_{\text{pix}}^2 = kTC_S . \tag{4.37}$$

Although the noise arises solely from the resistance, its value is proportional to the capacitance and independent of the resistance, because the increase in noise from a larger resistance is cancelled out by the reduction in bandwidth. At room temperature the kTC noise from a resistance and a 1pF capacitor is 400 electrons.

There is a low frequency cut-off in $g(f)$ because the TFT is turned on only for a limited integration time of typically 10–50 μs. Usually the integration time is several times larger than the pixel RC time constant, so that the low frequency cutoff removes $1/f$ noise but only reduces the white noise component by a small factor. However, the noise is reduced significantly when a short integration time is used.

From the point of view of the array operation, one must take into account the switching of the gate lines and sampling of the pixel signal to obtain the expected noise. There has been some debate over the correct analysis, but the consensus supports the following argument. When the gate is turned on the resistance noise generates charge fluctuations as described above. While the gate is off, the charge is held on the pixel without further fluctuation until the next time the pixel is addressed (assuming an ideal TFT). Thus each time the pixel is addressed, it starts with a noise charge q_{n1}, from the previous measurement, and finishes with a charge q_{n2}, so that the measured noise charge is,

$$q_{\text{pix}}^2 = q_{n2}^2 + q_{n1}^2 = 2q_n^2 = 2kTC_S . \tag{4.38}$$

This result holds when the two noise measurements are uncorrelated which corresponds to the usual case when the integration time is much longer than $1/R_{ON}C_S$.

In effect, the noise is sampled twice each time the pixel charge is read out, and is therefore increased by a factor of $\sqrt{2}$. Measurements on a single pixel with a 0.7 pF sensor find a value of 620 electrons, compared with the theoretical value of 520 electrons [4.63]. The excess might be due to a small $1/f$ component. The pixel kTC noise values sets the lower limit on the possible performance of the array with a particular sensor capacitance, at least for the usual method of sensing.

The resistance of the data line is another significant source of thermal noise originating on the array. The voltage noise causes charge fluctuations on the data line capacitance. The noise is given by (4.35), but the filter function is usually determined by the high frequency cut-off of the amplifier f_A, rather than the R–C filtering of the data line which typically has a higher frequency cut-off. The noise is therefore approximately,

$$q_D^2 = 4\pi kTR_D C_D^2 f_A . \tag{4.39}$$

For typical array parameters the data line resistance should be no more than a few hundred ohms for this noise to be less than the pixel kTC value. A low amplifier cut-off frequency helps minimize the effect and the double correlated sampling provides a low frequency filter which further reduces noise and eliminates any $1/f$ component.

Thermal noise from the gate line resistance is capacitatively coupled to the data line through the parasitic capacitance of the TFT and the other pixel elements. However, this contribution to the overall noise is sufficiently small to be neglected.

Amplifier Noise. The amplifier input for the integrated CMOS amplifier chips is the gate of a large MOS field effect transistor [4.64]. The primary sources of noise are the thermal and $1/f$ noise of the FET channel. To account properly for the effects of gain, it is conventional to calculate the equivalent noise on the input gate for which the thermal noise power in saturation is [4.64],

$$\nu_{amp}^2 = i_{DN}^2/g_m^2 = 4\beta kTg/g_m^2 = 4\beta kT/g_m , \tag{4.40}$$

where I_{DN} is the FET drain current thermal noise, g_m is the transconductance, g is the channel conductance, and β is a constant whose value depends on the details of the FET but is about 2/3. The factor $1/g_m^2$ refers the current noise in the channel to the gate, and the expression on the right arises because $g = g_m$ for an FET in saturation. $1/f$ noise is mostly eliminated by double sampling, but NMOS FETs have particularly high $1/f$ noise so that PMOS devices are invariably used, in which case the $1/f$ noise can be ignored for the normal array readout speeds.

The input charge noise is $C_{\text{in}}\nu_{\text{amp}}$, and a general expression that is a good approximation for most amplifiers is

$$q_{\text{amp}} = q_0 + \gamma C_{\text{D}}, \quad \gamma = (4\beta kT\Delta f/g_{\text{m}})^{1/2} ; \qquad (4.41)$$

γ is the noise slope and q_0 is the residual noise from the FET gate capacitance and typically has a value of a few hundred electrons.

A low noise amplifier needs a small value of γ, and is therefore designed with as high a transconductance as possible, which occurs in saturation where,

$$g_{\text{m}} = \sqrt{\mu_{\text{FE}} C_{\text{G}} I_{\text{SD}} W/L} . \qquad (4.42)$$

The input FET is therefore designed with large W/L – values of several thousand are commmon – operating at high drain current, and therefore consuming high power.

For low noise operation, the amplifier band width, Δf, should also be as low as possible. There is much theoretical work on the design of optimal filters for single charge detectors, but for the multiple channel integrated circuits the filtering tends to be a simple R–C circuit. The band width is chosen carefully to be the smallest consistent with the desired readout speed of the array [4.65]. Such an amplifier can have an input referred noise of $< 2nV/\sqrt{\text{Hz}}$, and for a bandwidth of 300 kHz, the resulting noise slope is about $\gamma = 10\,e/\text{pF}$ [4.53]. The overall amplifier noise is then in the range 500–1500 for large area arrays.

Extrinsic Noise. The most problematic extrinsic source of electronic noise is the voltage fluctuations in the bias supplies. The TFT addressed, photodiode arrays are provided with the sensor bias voltage, the TFT off-voltage and the TFT on-voltage. Each of these bias supplies is capacitatively coupled to all of the data lines of the array. Hence, a voltage noise, ν_{EXT}, on any of the bias voltages induces a charge noise at the input of the readout amplifier of approximately [4.35, 62],

$$q_{\text{EXT}} = C_{\text{D}} \nu_{\text{EXT}} . \qquad (4.43)$$

The various capacitances that make up the data line capacitance are discussed in Sect. 4.3.2. Since all but one gate line is turned off, the capacitance to the TFT on-voltage is negligible compared to the capacitance to the bias and TFT off-voltage power supplies and these two capacitances are roughly comparable in magnitude. A capacitance of 50 pF with a voltage fluctuation of 3 µV, yields a noise of 950 electrons. Given that the pixel and amplifier noise can be held to this value or less, it is clear that very low noise power supplies are needed. The problem, however, is greatly helped by double correlated sampling which filters out low frequency noise. The system is sensitive to noise only in a band of roughly 30–1000 kHz.

The noise fluctuations on the TFT and bias lines are identical across the array, so that the same noise is detected on each data line, since they are all sampled at the same instant. (Here it is assumed that filtering by the array is negligible – usually a fair approximation.) The noise is therefore the same on each pixel of one line, but different from line to line. Such line-correlated noise is visibly different from the random noise and can be readily distinguished in images. Indeed human vision is particularly well adapted for detecting lines, and therefore when an image contains equal amounts of line-correlated and random noise, the line noise is much more obvious. It is therefore important to minimize this visually distracting noise source.

Extrinsic noise increases with the data line capacitance and so increases with the number of pixels. The trend to increasingly large arrays makes control of the extrinsic noise more challenging. However, systems with line noise of ≈ 1000 electrons have been reported from an array with 3000 pixels along the data line, and lower values in smaller arrays [4.35, 66–68]. Figure 4.31 shows an image of the noise from a portion of such an array for which the overall noise is about 1800 electrons. The line noise becomes more noticeable as the image is viewed from further away. The fact that the noise is correlated across a line also provides a mechanism for correcting the noise even when it is present. The technique is to sample the noise in extra pixels at the edge of the array and to subtract this value from the image. This can be implemented either directly by the readout amplifier or some other part of the electronic system, or more simply by image processing software in the host computer. As we have seen before in other contexts, the subtraction of a noise signal comes with the penalty of increasing the noise by a factor $\sqrt{2}$ because of the extra sample of random noise. In this case however, it is possible to design an array with perhaps 10–20 dummy pixels to sample the line noise, in which case the subtraction can be done without significant increase in the random noise. It should be noted that the noise cancellation is only successful if the noise is the same across the array. Any additional R–C filtering of the noise voltages by the array with cause the correction to fail.

The above discussion by no means exhausts the sources of extrinsic noise. Since the array has no ground connection, the voltages of the bias and gate lines are referenced to the amplifier input and therefore ultimately to the amplifier power supplies, which must have similar low noise properties. The whole system mixes digital controls with the low noise analog inputs and it is obviously important that digital noise does not get coupled into the input of the amplifier. Finally, one should note that the whole array, measuring perhaps 40 cm across and with up to 2 km of metal lines, may act as a large antenna and pick up radiated electromagnetic signals. Shielding to eliminate these noise sources is necessary. The total noise from the above discussion is,

$$q_N^2 = q_{pix}^2 + q_D^2 + q_{amp}^2 + q_{EXT}^2 . \tag{4.44}$$

As with other high performance electronic systems, it takes a good deal of experimentation and experience to achieve the intrinsic noise performance limit.

Fig. 4.31. Example of a noise image [4.71]

In practice the reported noise performance has been improving steadily even as the array sizes and pixel number increase. Several systems have been reported with noise in the range 1000–2000 electrons [4.66–68], and there seems every reason to expect that systems will soon converge on the theoretical values of less than 1000 electrons.

In an imaging system for which the pixel noise is a small fraction of the total noise, it is obviously beneficial to have a pixel amplifier [4.57]. A gain of about 10 is sufficient to increase both the signal and noise until they are no longer limited by amplifier and bias supply noise. Pixel amplifiers have been implemented in CMOS sensors (see Sect. 4.3.6) and may be anticipated in large area array in the future.

4.4.3 X-Ray Detection

Thin amorphous silicon p-i-n photodiodes have virtually no X-ray sensitivity because there is minimal absorption of the incident beam. X-ray imaging with systems based on these devices is accomplished by placing a phosphor on the surface of the array. On the other hand, arrays designed with a thick photoconductor (see Sect. 4.3.4) are X-ray sensitive by intentional design. Either approach is effective and X-ray medical imaging systems based on both designs are being manufactured [4.68].

These two methods of X-ray imaging are compared in Fig. 4.32. The physical process for X-ray detection begins the same way in either case when a single interaction transfers some or all of the X-ray photon energy to excite a high energy electron [4.69, 70]. The mechanism is photoionization at lower energy (i.e., $< 50\,\mathrm{keV}$) and Compton scattering at higher energy. The energetic electron loses energy by creating many electron–hole pairs by successive

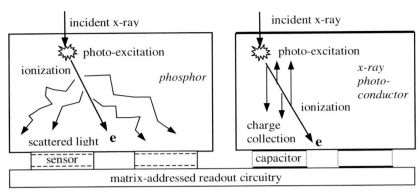

Fig. 4.32. Illustration of the phosphor/photodiode (*left*) and photoconductor (*right*) methods of X-ray detection

ionization until all its energy is dissipated. The number of pairs created, N_P, by each absorbed X-ray photon is given by,

$$N_\mathrm{P} = E_\mathrm{X}/W \, . \tag{4.45}$$

W is the average energy needed to create an electron–hole pair and is usually 2–3 times the band gap energy, and E_X is the incident X-ray energy. The ionization charge forms the latent image that is sensed by the imager.

X-ray attenuation coefficients are shown in Fig. 4.33 for some phosphor and photoconductor materials in the region of most interest for medical applications from about 10 to 100 keV. X-ray attenuation decreases with X-ray energy except for the abrupt steps at the K or L absorption edges, and also increases with atomic number. In the middle of the energy range, attenuation coefficients are roughly 10 to 100 cm^{-1} so that that the thickness of the X-ray absorber is in the range 30–1000 μm.

When a phosphor is used for detection (Fig. 4.32, left), the electron–hole pairs recombine to emit visible light which is sensed by the photodiodes on the array. The phosphor is placed in close contact with the array to allow the maximum transfer of the light. Indeed, contact imaging is vital for high sensitivity, and is one of the principal reasons why large area a-Si:H arrays are excellent X-ray imagers. In the alternative situation in which the phosphor emission is imaged onto a small detector using a lens, then the optical transfer efficiency is given by,

$$\text{lens efficiency} = \frac{1}{1 + 8f^2(1 + 1/m)^{-1}} \, , \tag{4.46}$$

where m is the demagnification, f is the f-number of the lens and the constant depends on the nature of the emitting surface. For a large phosphor screen, the typical transfer efficiency is 10^{-4}–10^{-3}, compared to $> 50\%$ for contact imaging.

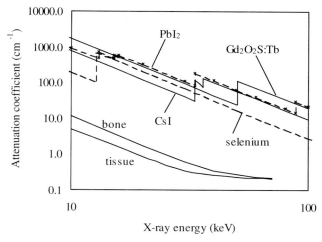

Fig. 4.33. X-ray attenuation coefficients for the detector materials described in the text, compared with the attenuation of bone and tissue

The ideal phosphor has strong X-ray attenuation, high light emission efficiency, and an emission wavelength matched to the collection efficiency of the amorphous silicon sensor. Of the numerous candidate phosphor materials, the best choices are GdO_2S_2:Tb and CsI:Tl. The emission from both of these phosphors is efficient, and is in the wavelength range 500–600 nm, which coincides with the peak of the a-Si:H quantum efficiency spectrum (see Fig. 4.2).

Since the recombination light is emitted isotropically in the phosphors, it is important that the photodiode detects the emission near where it is created, to maintain good spatial resolution in the image. The GdO_2S_2:Tb phosphor screen is fabricated as a powder in a binder and scatters light, such that that the spreading of the image is approximately equal to the thickness of the film. The detection of high energy X-rays therefore involves a compromise between the need for a thick film to absorb X-rays efficiently, and the decrease of spatial resolution in the thick film.

CsI:Tl is deposited by thermal evaporation on the surface of the array [4.35]. It was discovered long ago that under suitable deposition conditions, the deposited film has a columnar structure that causes a light piping effect and reduces the loss of spatial resolution in thick films. Thus CsI:Tl provides both higher sensitivity and better spatial resolution than GdO_2S_2:Tb, at the expense of some additional difficulty in manufacturing.

When a photoconductor is used to detect X-rays (see Fig. 4.32, right), the ionization charge is collected by the applied field. There is little lateral diffusion of the charge so that the spatial resolution of the image is maintained, regardless of the layer thickness [4.36, 37]. In principle, the photoconductor gives a higher sensitivity than the phosphors, partly because it can be made thick enough to absorb all the X-rays and partly because the light emission and subsequent detection steps are absent. However, it represents a significant materials challenge, since high charge collection in a thick film requires a high value of the $\mu\tau$ product – 1 or 2 orders of magnitude larger than for thin a-Si:H sensors. Furthermore, high charge collection efficiency is required for both electrons and holes because X-rays are absorbed throughout the photoconductor. The charge trapping length, L, is given by

$$L = \mu\tau V_B/t_F , \tag{4.47}$$

where t_F is the thickness, V_B is the bias and a uniform electric field is assumed. Ideally L should be at least twice the thickness of the film, so that the minimum bias voltage is

$$V_{MIN} > 2t_F^2/\mu\tau . \tag{4.48}$$

The required bias is even larger if the field is non-uniform due to a depletion layer. The photoconductor thickness must be roughly $1/\alpha$ to absorb X-rays, where α is the attenuation coefficient which varies with X-ray energy approximately as $E_X^{-2.5}$ (see Fig. 4.33). Hence,

$$V_{MIN} \approx E_X^5/\mu\tau , \tag{4.49}$$

so that the minimum bias voltage increases rapidly with energy [4.71].

Amorphous selenium photoconductors have been extensively investigated, and polycrystalline PbI_2 appears to be a promising new material. The selenium is doped with As and Cl to give good charge collection for both electrons and holes, and is readily evaporated onto large arrays at low temperature. However, as shown in Fig. 4.34, the ionization rate, W, for selenium is unusually small and field dependent, of the form [4.37, 39]

$$W = \text{const } F^\alpha ; \quad \alpha = -(0.65 - 0.9) ;$$
$$W \approx 50 \, \text{eV/pair} \quad \text{at} \quad 10 \, \text{V/µm} . \tag{4.50}$$

Possibly this effect is due to geminate recombination, and is related to the similarly low quantum efficiency for visible light. The films require a field of about 10 V/µm, or about 3000 V across a 300 µm layer.

PbI_2 has a high sensitivity to X-rays, but at the time of writing, the leakage current is moderately large [4.71–73]. Further development is needed to determine whether the photoconductors can approach the ideal sensitivity. When both the charge ionization rate and the charge collection are nearly ideal, then a single 50 keV X-ray should generate 5–10 000 electrons of charge, an order of magnitude more than is typically obtained with phosphors. PbO has also been shown to be an effective material [4.74].

4.4.4 The Performance of X-Ray Detectors

Detector Performance Analysis. The goal for the design of any imaging system is to optimize the performance of the detector in terms of its ability to

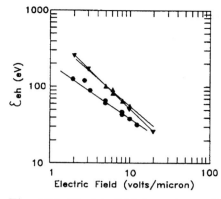

Fig. 4.34. Electric field dependence of the Se ionization rate [4.39]

faithfully reproduce the information content of the incident image. It is therefore necessary to have the appropriate analytical metrics for performance and the techniques to measure them. All aspects of the X-ray imaging system contribute in one way or another to the performance, affecting the signal, noise and spatial resolution. These can be categorized as follows [4.75–77];

1) **Signal gain.** At various stages in the detection there is a change in the number of photons or electrons that comprise the signal. For example, not all incident X-ray photons are absorbed by the detector, corresponding to an absorption gain of less than unity. On the other hand, each absorbed X-ray photon generates a large number of electron–hole pairs by ionization (i.e., a gain of 100–10 000). The detection of these charges at the pixel has some inefficiency which results in a further gain of less than unity.

2) **Image spreading.** The detection of signal generated by the phosphor or X-ray photoconductor may result in some signal spreading to neighboring pixels, and results in loss of spatial resolution of the image.

3) **Pixel sampling.** Even in the absence of image spreading, the size of the sensor and of the pixel places an upper limit on the ability of the detector to resolve fine features in the image.

4) **Additive noise.** Along with any reduction in signal-to-noise performance generated by the gain, blurring and sampling stages, there is additional noise that arises from the electronic readout, as discussed in Sect. 4.4.2.

The performance of the detector must be defined in terms of the incident signal which has its own signal-to-noise limitations. The fundamental limitation is the shot noise that arises from the counting statistics of the incident photons. N incident photons are governed by Poisson statistics and have an RMS fluctuation of $\sqrt{N}$ (and variance N). The performance of a detector is the measure of how well the incident signal-to-noise (S/N) ratio is preserved by detector. This leads to the concept of the detective quantum efficiency,

DQE, defined in its simplest form as the square of the ratio of the measured and input S/N,

$$\text{DQE} = \frac{(\text{S/N})_{\text{OUT}}^2}{(\text{S/N})_{\text{IN}}^2} \,. \tag{4.51}$$

The S/N ratio of a detector also depends on the spatial frequency of the measurement, because the signal from an incident X-ray is typically spread over many pixels, for example due to light scattering in the phosphor. The noise and spatial properties of the array are combined together in a DQE measure that is a function of spatial frequency and is the most complete expression of the imager performance. The full discussion of the DQE is beyond the scope of this book, but it is useful to summarize the main points to understand how the image performance depends on the system design and how the performance is measured.

Although the image comprises a signal with intensity $s(x)$ that varies with position, x, the analysis of the signal and noise is more conveniently accomplished in the spatial frequency domain, because of the properties of Fourier transforms, as will be illustrated for the case of image spreading by the phosphor. An incident delta function signal at position x_0 results in an output signal, $t(x - x_0)$, spread over a range of values of x, as illustrated in Fig. 4.35. (For simplicity a 1-dimensional analysis is used; the extensions to 2-dimensions are given in [4.75–77].) In a linear system, the output image, $o(x)$, corresponding to the input $s(x)$ is the convolution of $s(x)$ with $t(x)$ [4.75]

$$o(x) = \int s(x')t(x' - x)\mathrm{d}x' \,. \tag{4.52}$$

The Fourier transform of a convolution is a product of the Fourier transforms of the separate functions, such that in frequency space

$$O(u) = S(u)T(u) \,, \tag{4.53}$$

where the functions of spatial frequency, $u = 1/x$, are Fourier transforms of the corresponding real space functions. The simplification of dealing with the product of functions rather than the convolution integrals is helpful both conceptually and for analysis. $T(u)$, the Fourier transform of the point-spread, or in one dimension, the line-spread function, $t(x)$, is known as the *modulation transfer function*, the MTF. It has the value unity at spatial frequency $u = 0$, and decreases to zero at higher frequencies.

All pixel arrays sample the image at discrete sensors of finite size. The signal detected by a sensor is the convolution of the image resulting from the X-ray detection stage, $o(x)$, and a rectangle function, $r(x)$, corresponding to the size, a, of the sensor, for which the Fourier transform is a sinc function. The signal at a pixel is then given by,

$$P(u) = \text{const}.S(u)T(u)R(u) = \text{const}.\, S(u)T(u)|\text{sinc}(\pi a u)| \,, \tag{4.54}$$

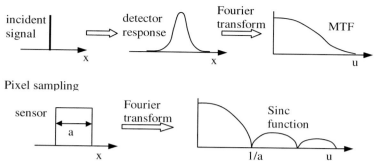

Fig. 4.35. Illustration of the analysis of image spreading and pixel sampling that determine the detector response to the incident image

where the constant includes the signal gain. The product $T(u)\mathrm{sinc}(\pi au)$ is known as the pre-sampled MTF and its measurement is discussed in the next section. The presampled MTF represents the spatial frequency response of a single sensor.

The next stage of the image chain is the sampling of the sensor signal $P(u)$ by the series of pixels that make up the array. The signal at each pixel is represented by a single quantity (the measured charge) assumed to be at the center of the pixel. The sampling is therefore represented by an periodic array of delta functions separated in real space by the pixel pitch, d, and its Fourier transform is the same function separated in spatial frequency by $1/d$.

The Nyquist frequency, $u_N = 1/2d$, represents the upper limit of the spatial frequencies that can be detected, since essentially two samples per cycle are needed to detect a modulation. If the presampled MTF extends beyond the Nyquist frequency, then an imaging artifact known as aliasing (or the Moiré effect) may result. A component of the incident image with high spatial frequency, u, beats with u_N and the difference frequency, $u - u_N$, is observed in the measured image. It is virtually impossible to distinguish aliasing from a real signal at the lower frequency.

An ideal imager has a high MTF below the Nyquist frequency to maintain the spatial resolution of the image, but a low MTF above u_N, to eliminate aliasing artifacts. A sharp cut-off in the MTF at u_N is generally not possible, so some trade-off is needed, usually with the MTF extending some way beyond u_N. If the incident image contains no spatial frequencies above u_N, then obviously no aliasing of the image can occur, although aliasing of the noise can affect the detection. The ideal imager therefore matches the pixel pitch to the MTF and to the spatial frequency content of the source.

The image noise is similarly influenced by the various gain, spreading, and sampling stages of the detector as well as the additive readout noise from the electronics. For example, the probability, g_{abs}, of absorbing an incident X-ray, is accompanied by a statistical fluctuation in the absorption with variance $g_{abs}(1-g_{abs})$ corresponding to a binary process. The spatial auto-correlation function of the signal describes the random variations of signal from site to site and the degree to which the noise is spatially correlated. Its Fourier transform is the noise power spectrum (NPS),

$$\text{NPS}(u) = \text{FT}\left[\int s(x')s(x-x')\mathrm{d}x\right] = S^2(u). \tag{4.55}$$

The zero frequency value of NPS(u) is the variance of the image under conditions of uniform illumination.

When an incident image signal is modified by a detector spreading function, the resulting NPS is given in terms of the MTF

$$\text{NPS}_{\text{OUT}}(u) = \text{NPS}_{\text{IN}}(u)|T(u)|^2. \tag{4.56}$$

In practice there is additional noise because image spreading is a stochastic process involving individual quanta rather than a continuous function, so that further statistical fluctuations are introduced.

The frequency dependence of the detector signal-to-noise performance is described by a detective quantum efficiency (DQE), related to the NPS and MTF by,

$$\text{DQE}(u) = \frac{G^2|T(u)|^2\text{NPS}_{\text{IN}}(u)}{\text{NPS}_{\text{OUT}}(u)}, \tag{4.57}$$

where G is the overall gain of the system and NPS$_{\text{OUT}}(u)$ is the NPS of the detector output signal. This expression connects together the gain, spreading and noise properties of the system into a single function that describes the imager performance. It can be seen that a high value of the measured NPS compared to the input noise corresponds to a low DQE.

The analysis of the DQE for large area detectors is fairly complete as a result of the work of Cunningham, Siewardsen and others [4.68, 75–77]. While rather involved, some general points can be made regarding the difference between phosphor and photoconductor detection, as well as the effects of electronic noise. Perhaps the most important point is that the DQE is proportional to the fraction, α_X, of X-rays that are absorbed, and this factor is the upper limit on the DQE. It is therefore important to have a high absorption.

The phosphor and photoconductor systems are distinguished by their different MTF. There is always significant image spreading when a phosphor is used, while the photoconductor devices have essentially no spreading. The absence of spreading is not always beneficial, and can reduce the DQE in two ways. First, when the fill factor, F_F, is less than unity, the absence of spreading ensures that only the fraction F_F of the absorbed flux is detectable, and consequently the upper limit on the DQE is reduced to $\alpha_X F_F$. There is no such proportional loss of DQE when there is some image spreading, because some part of the signal from all the absorbed X-rays are detected.

The second effect on the DQE is the more subtle property of noise aliasing, which places an upper limit on the DQE at the Nyquist frequency of $\approx 0.4\,\alpha_X$ [4.76]. When an X-ray hits a detector with no spreading, there is no knowledge where it hit within the pixel and some spatial resolution information is lost. However, when there is image spreading, the relative intensity of signals in adjacent pixels provides the additional spatial information. The apparently paradoxical result, that a detector with image spreading has a better performance at high spatial resolution than one with no image spreading, only applies if the gain is high enough, and comes at the price of increased sensitivity to additive noise when the MTF is low.

The DQE for a detector with some image spreading can be approximated as

$$\mathrm{DQE}(u) = \frac{\alpha_X}{1 + \beta + \dfrac{1}{gT^2(u)}\left(1 + \dfrac{q_N^2}{\alpha_X g S}\right)}, \qquad (4.58)$$

where g is the total gain (excluding the X-ray absorption), q_N is the additive noise, and S is the X-ray flux [4.68, 75–78]. The term β represents additional noise from the phosphor arising from the variance in the gain, and has a typical value of 0–0.2. The term $1/gT^2$ describes the effect of the MTF in the absence of additive noise and shows that the detector can overcome a low MTF provided that the gain is large enough that $gT^2(u) > 1$. These systems have gains of 100–1000, so that in principle the DQE can be maintained to high spatial frequency even when the MTF is rather small.

The additive electronic noise further reduces the DQE, particularly where the MTF is small. At low spatial frequency (i.e., $T(u) = 1$) the DQE falls when,

$$q_N > g\sqrt{\alpha_X S}. \qquad (4.59)$$

This result confirms the intuitive expectation that the additive noise has its greatest adverse effect at low X-ray exposure, and that a high gain helps to overcome the effects of noise. At high spatial frequencies, the condition for the noise to dominate the DQE becomes,

$$q_N > gT(u)\sqrt{\alpha_X S}, \qquad (4.60)$$

from which it is clear that the low MTF region is more readily affected by noise. In summary, the ideal DQE occurs with high X-ray absorption, high gain, an MTF with small but non-zero spreading and low additive noise. The effects of absorption and additive noise on the DQE are illustrated in Fig. 4.36.

The analysis and measurement of the DQE begs the question of what values are required. The answer is difficult to state with any precision, and certainly depends on the specific application. We have seen that the DQE decreases at low incident X-ray flux due to additive noise, and at high spatial

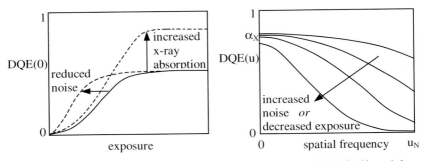

Fig. 4.36. Illustration of the form of the zero frequency DQE (*left*) and frequency dependent DQE(u) (*right*) as the X-ray absorption, exposure and additive noise are varied

frequency due to image spreading. It is essential to understand the properties of the image to decide whether the DQE is adequate for any specific situation. An image suitable for medical diagnosis must contain a sufficient incident flux to allow observation the features to be discerned. Usually this represents a combination of spatial frequency and signal-to-noise requirements. If high S/N is essential then additive noise may be relatively unimportant because the incident quantum noise is large. In situations where a very weak flux is detected, it is preferable to increase the signal by having large pixels, at the expense of spatial resolution where there may be no useful information because of the low DQE. In general the imager design must be optimized for each specific application.

Detector Performance Measurements. The measurement of the presampled MTF is commonly performed by imaging through a narrow slit with a width much smaller than the pixel size [4.78]. The slit is positioned at a small angle to the axis of the pixel array. The image along one row of pixels measures the signal that corresponds to illumination in a narrow line across one pixel. The next row of pixels is another sample of the signal, but with the illuminated line moved slightly across the pixel. Many such rows create a composite image equivalent to that of a line source being scanned across a single pixel. The result is the convolution of the phosphor line-spread function with the sensor shape, and its Fourier transform is the presampled MTF according to (4.54). An example of the MTF is shown in Fig. 4.37 for an amorphous silicon photodiode imager using a Lanex regular (GdO_2S_2:Tb) phosphor screen [4.79]. The predicted result is based on the known MTF of the phosphor, and takes into account that the pixel is not square, and has metal lines covering the surface, so that the presampled MTF is slightly different in the two orthogonal directions. The measured MTF is slightly less than the calculated value particularly at low spatial frequency where the MTF drops more rapidly than expected. Further investigation showed that the reduced MTF is due

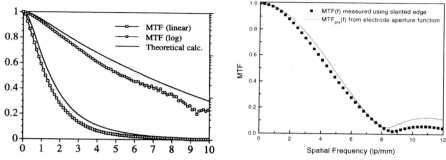

Fig. 4.37. Examples of the MTF of an a-Si:H photodiode array with a Lanex regular screen with 127 μm pixels (*left*) [4.80], and a selenium photoconductor array with 160 μm pixels (*right*) [4.39]

to X-ray fluorescence in the glass substrate arising from X-rays that are not absorbed in the phosphor [4.80]. The glass substrate contains barium, which fluoresces weakly and the spreading of the X-rays causes the small reduction in MTF. This is an illustration of the subtle effects that may influence the detector performance.

The MTF of a selenium photoconductor array is also shown in Fig. 4.37 for comparison. Because there is virtually no spreading of the image in the selenium layer, the MTF approximates to the sinc function in (4.54), for which the MTF is about 70% at the Nyquist frequency, (assuming 100% fill factor). At a spatial resolution of 3 lp/mm, the Lanex regular imager has an MTF of about 10% while the selenium imager has ≈ 70%. As the preceding discussion of DQE shows, the high MTF alone does not ensure that the selenium imager has the higher DQE, but does make the imager less sensitive to degradation of the high frequency DQE by additive electronic noise.

A one-dimensional NPS is measured by taking the Fourier transform of the signal along a row of pixels illuminated by a uniform source, with corrections to remove any fixed pattern signals, or by analyzing the response of individual pixels in a sequence of similar images. Some NPS data are shown in Fig. 4.38 for a range of X-ray exposures that are typical for fluoroscopy applications in an imager with 200 μm pixels and a CsI structured phosphor [4.35]. It is a small area imager which therefore has low data line capacitance and particularly low values of the additive electronic noise The NPS at zero frequency increases in proportion to the X-ray exposure as expected for the Poisson statistics of the source (for which the variance equals the mean), and the decrease at higher spatial frequencies generally reflects the MTF according to (4.55). The NPS of the dark signal is independent of spatial frequency, which shows that the readout noise has no spatial correlation, as is expected for the pixel and amplifier noise described in Sect. 4.4.2. The observation that the NPS does not fall off as fast as the square of the MTF in Fig. 4.29, in part due to the presence of additive noise, illustrates that the detector has

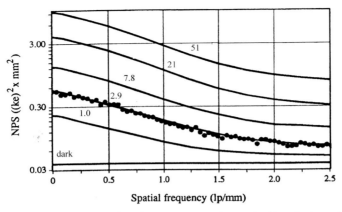

Fig. 4.38. Measured noise power spectrum in the dark and for different X-ray exposures from 1 μR to 51 μR at 75 kVp. The 192 × 192 pixel array has 200 μm pixels and a CsI phosphor [4.35]

introduced excess noise at the higher spatial frequencies. The frequency dependent DQE is usually obtained from the measured NPS and MTF, using (4.57).

Siewardsen et al. have reported a detailed DQE model for the TFT/photodiode imager with both Lanex and CsI:Tl phosphors [4.77]. The model gives an excellent fit to data from an array with 127 μm pixel size and can therefore be used to predict performance for many imaging situations. Figure 4.39 shows calculations of the zero frequency DQE for conditions that correspond to applications in radiography, fluoroscopy and mammography. The figure shows the effects of radiation dose, which generally follow the trends outlined in Fig. 4.36. The DQE reaches a constant value at large exposures that is primarily determined by the X-ray absorption, and which is larger for CsI than Lanex, and lower at higher X-ray energies. An amplifier noise of 1000 electrons is assumed and it is seen that fluoroscopy falls within the dose range where the DQE is low. Figure 4.39 shows more explicitly the effects of additive noise. The noise decreases the DQE and the magnitude of the effect depends on the X-ray energy and exposure. As expected the additive noise has the largest effect with low exposure characteristic of fluoroscopy.

Figure 4.40 shows the measured values of the spatial frequency dependence of the DQE for a diode addressed array with pixel size 196 μm [4.43]. The array uses CsI:Tl phosphor and has a sensitivity of 1150 detected electrons per incident X-ray at an X-ray energy of 70 kVp. The low frequency DQE values of 0.4–0.75 are generally consistent with the calculations in Fig. 4.4. The data show the decrease in DQE with increasing spatial frequency and the decrease with reduced exposure.

The choice of pixel size represents a compromise between sensitivity and spatial resolution, and the measurement of DQE helps achieve the optimum

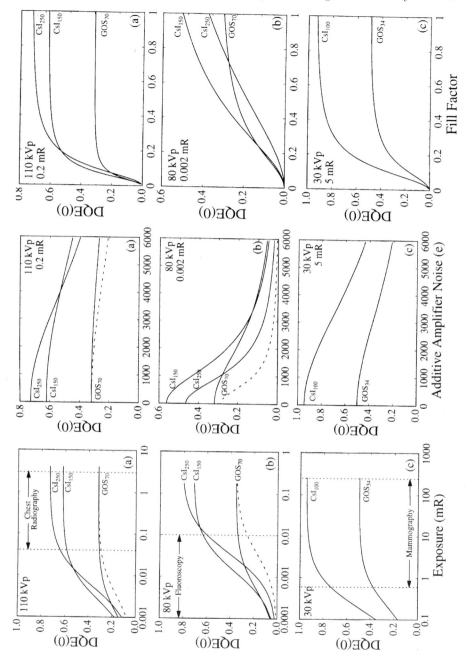

Fig. 4.39. Calculated DQE for different exposure and measurement conditions [4.77]

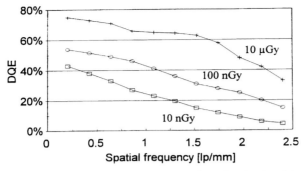

Fig. 4.40. Frequency dependent DQE of a diode addressed array

choice. At low exposure in Fig. 4.40 the DQE drops to about 10% at 2 lp/mm. A higher resolution array would have an even lower DQE because of the lower signal at each pixel. Thus little is gained by increasing the resolution because the DQE at high frequency is essentially zero. However, at high exposures a higher spatial resolution can yield additional image information.

4.5 Applications of Large Area Image Sensors

4.5.1 Medical X-Ray Imaging

Medical X-ray diagnostic imaging is expected to be the main application of a-Si:H image sensor arrays. The need to obtain images from major anatomical parts of the body – for example the chest – and the lack of a practical means to focus X-rays, necessitates the use of a large area sensing device. X-ray imaging of inanimate objects for security inspection or to confirm the internal structural integrity – known as non-destructive evaluation (NDE) – and scientific imaging applications such as X-ray crystallography, are other important applications of the technology that are discussed in Sect. 15.2. The same imagers can be configured for neutron imaging – a specialized but significant procedure for NDE.

The various techniques can be further distinguished by the need for single images or continuous video sequences. Single images are usually obtained at high resolution for careful inspection and diagnosis. The benefits of digital acquisition is that the image can be enhanced in contrast, magnified and processed in other ways to allow the optimal observation of fine detail. Video images are used to observe internal motion and are necessarily acquired in electronic form. Generally the spatial resolution requirements are more relaxed but high sensitivity is demanded. The following examples of the applications illustrate different requirements and usage of the image sensor arrays.

Radiographic Imaging. Most people have experienced an X-ray radiography procedure to diagnose a broken bone or other medical condition. Indeed, of order 1 billion X-ray medical images are taken each year. The traditional technique captures images with X-ray film, which consists of a large sheet of photographic emulsion between two phosphor screens. The film records the image of X-rays transmitted through the patient as a result of a brief X-ray exposure. After chemical development of the film, the diagnosis is performed by a radiologist who views the negative film placed on a light box. The technique has been in use for 100 years – the first diagnostic X-ray image was recorded in 1896, less than one year after X-rays were discovered by Roentgen. The procedures for the medical use of X-ray film are highly advanced as a result of the long development time.

The goal of radiographic imaging is to obtain high spatial resolution and contrast in the observation of the anatomical features under investigation, at the lowest possible dose to the patient. The principal concern with the X-ray imaging procedure is that there is no completely safe X-ray dose, and so the benefit to the patient must be matched against the risks of the exposure. An absolute requirement of any new imaging technology, such as the amorphous silicon arrays, is that the same quality of diagnosis can be achieved with an equivalent or lower dose to the patient. Fortunately, the arrays seem set to achieve this goal, and the first clinical studies appear to justify the expectations [4.81].

The attainment of high image contrast depends on the imaging conditions, the detector and the image display. The X-ray energy and dose for the procedure are chosen with care to maximize the contrast between bone and tissue. The contrast originates from the difference in their attenuation coefficients which is about a factor 3 up to 40 keV and decreases at higher energy (see Fig. 4.33). The X-ray generator emits a broad band X-ray spectrum, which varies with the accelerating voltage and can be additionally modified by filtering to remove the low energies. The X-ray energies of interest are those for which there is partial attenuation by the body with the maximum difference of transmission between bone and tissue etc. Low energy X-rays are completely attenuated by the body and are undesirable because they expose the patient without contributing to the image. High energy X-rays pass through virtually unattenuated and therefore also give no contrast. The optimum image contrast involves a choice of the X-ray energy, filtering and exposure, and differs for different parts of the body.

In a typical chest X-ray, the X-ray generator is operated at 80–120 kVp, with most of the intensity centered at about half the maximum energy. (The unit kVp refers to the generator accelerating voltage and is the upper limit on the X-ray energy spectrum.) The dose is 3–5 mR, which corresponds to an X-ray flux of $1-1.5\ 10^7$ X-ray photons/cm^2. Approximately 10–20% of the X-rays pass through the body to the detector, and the contrast in the transmitted beam is 20–50%, or 2–10% of the incident beam. The dynamic range

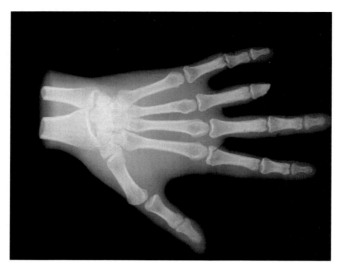

Fig. 4.41. Radiographic image

of an electronic detector should be at least 4000 (i.e., $\geq$ 12 bit digitization), so that the low contrast regions can be viewed without being significantly affected by digitization noise.

In 1998, several medical imaging companies are introducing amorphous silicon radiographic imagers with active areas up to $17 \times 17''$, which is the preferred size for chest X-rays [4.68]. The pixel size ranges from 127–160 μm (a spatial resolution of 3–4 line pairs/mm), so that the total pixel number in the imager is 7–10 million. Compared to fluoroscopic applications (see below), the radiographic dose is quite large, in order that the contrast detail is not obscured by the quantum noise of the X-ray beam. The DQE is therefore less sensitive to additive noise (see Fig. 4.37), and is largely determined by the absorption of X-rays (4.58). The pixel size is a compromise between sensitivity and spatial resolution. To give an example, an exposure of 1.3×10^7 photons/cm^2 with a pixel size of 150 μm yields 3×10^3 photons/pixel. With an average conversion gain of 500 electrons/photon, the maximum pixel charge is 0.3 pC, which is consistent with the pixel capacitance of the array. An example of a radiographic image is shown in Fig. 4.41.

Both scintillator and photoconductor systems are suitable for medical radiography and both types are being developed. Since GdO_2S_2:Tb phosphors are conventionally used for X-ray film, their use in a-Si:H imagers gives essentially the same spatial frequency performance up to the Nyquist frequency. At 4 lp/mm, a typical GdO_2S_2:Tb phosphor has a MTF of 10–20%, and so the absence of detector response at higher spatial frequencies is probably not significant, although a detailed optimization of the pixel size has yet to be reported. CsI scintillators offer improved spatial resolution, while the photoconductor arrays give a further improvement in resolution.

The means by which a radiographic image is viewed is also important to the detection of contrast detail. X-ray film is viewed in a light-box at high brightness, since this improves the perceived image by the fovea (the high resolution area of the retina). Soft displays for electronic image sensors need further development in order to match the capabilities of the detector. There are CRT displays with 2000×2500 pixels, but these only display about half a large radiographic image, and display is not very bright. Possibly, amorphous silicon active matrix LCDs will become the display of choice when their gray scale response is further improved, as they have high resolution and high brightness.

The DQE is an incomplete measure of the detector performance because it does not take into account the process of viewing the image, which depends on the complex non-linear response of the human vision system. For this reason, testing of detector performance ultimately depends on comparative observations by radiologists. New models of the human vision system are being developed which, if successful, may allow the entire system of detector, display and human observer be simulated accurately [4.82].

Mammography imaging requires a significantly higher spatial resolution than most other radiographic procedures. The purpose of mammography is to screen patients for the various precursors of breast cancer including micro-calcifications, which are very small features with low contrast. A high dose and a pixel size of 50–100 µm are needed to resolve the features. This application may be well suited to photoconductor detectors which can provide the high resolution, and the low X-ray energies used in the procedure allow for thinner photoconductor layers.

The desirable features of a digital replacement for X-ray film based on a-Si:H image sensor arrays include the following:

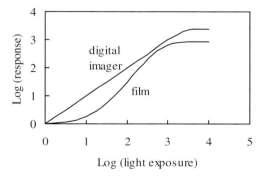

Fig. 4.42. Schematic illustration comparing of the response of film with the more linear response of a digital imager

1. The acquisition and display of the electronic image is immediate, taking at most a few seconds. This provides rapid confirmation that the correct exposure conditions were used.
2. Film has a non-linear response to exposure, which is compared to the linear response of digital imagers in Fig. 4.42. As any photographer knows, the limited dynamic range of film makes it essential to have the correct exposure, and when regions of very different optical density are of interest, it is impossible to optimize the exposure for the complete image. The digital imager, on the other hand, has a linear response over a wide dynamic range, and image processing can selectively enhance the contrast of different regions, as well as reproducing the response of film to give an image that is familiar to radiologists. Film loses sensitivity at low exposures compared to the image sensor arrays, and so the new imagers offer the ability to produce good images at lower doses, which reduces the dose to the patient.
3. The archival storage of film consumes much space and labor, and incorrect filing frequently results in the loss of an image. Electronically filed images can be carefully tracked in a data base and offer tremendous space savings.
4. Telemedicine is enabled by the ability to capture and transmit digital images.

Fluoroscopy. Fluoroscopy is the real time acquisition and display of X-ray image sequences, and is used when it is important to observe internal motion of the body. A typical application of fluoroscopy is in non-invasive surgery, such as cardiac angiography to correct heart disease. In this procedure, the cardiac artery is opened by inflating a balloon that is passed into the body along a guide. The procedure relies entirely on fluoroscopic X-ray imaging to ensure that the guide is positioned correctly. These imagers are generally mounted on large C-arms, so that they can be positioned anywhere on the patient and rotated about an axis along the patient. Often, a pair of such imagers are mounted at right angles to give alternative views, and the surgeon views the progress of the procedure on a bank of video monitors as shown in the photograph in Fig. 4.43.

The existing fluoroscopy technology uses an image intensifier tube (IIT) which is a large vacuum tube. A columnar CsI phosphor layer inside the tube absorbs the X-rays and an electron emissive photocathode causes electrons to be released from the surface, where they are accelerated and focussed by the electron optics of the IIT. The result is a small bright image on a second phosphor screen which is viewed with a conventional video camera. The IIT has excellent sensitivity due to its very high gain, but is a bulky and heavy device and suffers from some image distortions and veiling glare, as well as sensitivity to external magnetic fields.

Amorphous silicon imager technology offers a flat panel replacement to the IIT which is less intrusive to the surgeon and has less image distortion.

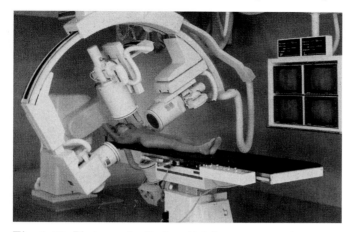

Fig. 4.43. Photograph of a hospital fluoroscopy system

Video rate image acquisition is well within the capability of the a-Si:H arrays, at least for arrays with 1000–2000 lines. The challenge for the image sensors is the exceptional sensitivity requirements. Fluoroscopic imaging does not require high contrast, but instead the emphasis is on very low dose, because the patient is exposed for an extended time. The typical exposure is 1–10 µR per frame, which is about three orders of magnitude less than the radiographic dose. Sensitivity is gained by increasing the pixel size to 200 µm or larger. Even then, an exposure of 1 µR corresponds to only about 10 X-ray photons/pixel, and the quantum limited signal-to-noise is just barely high enough to discern an image. In this regime, the DQE tends to be dominated by the electronic noise, and a good fluoroscopic imager should have high sensitivity and very low noise. A CsI phosphor of about 400–500 µm thickness, as used in IITs, has demonstrated similar performance to IITs, providing the electronic noise is no more than 1000 electrons and the remainder of the system is well optimized. Fluoroscopy is an excellent application for high fill factor array designs, and high gain X-ray photoconductors both of which have high detector sensitivity and high DQE at low exposures. Fluoroscopic imaging using a photoconductor detector has not been thoroughly studied, and may be made difficult by the build-up of trapped charge.

Image lag is an obvious concern for video imaging. A small amount of lag can actually be beneficial because the temporal averaging of the image reduces the noise, so long it is not sufficient to cause motion blurring artifacts. A more important issue is that during a fluoroscopy procedure, the dose is occasionally increased to radiographic levels to obtain a high contrast image, and then returned immediately to the low level to continue monitoring. An image lag of only 1% from the high intensity frame can overwhelm the subsequent images.

Although challenging, there is every reason to expect that a-Si:H imagers will replace the IIT for fluoroscopy because of its compact size, and fully digital operation which allows more flexibility in image processing and display. The light weight of the detector allows a considerable saving in the bulk and cost of the support structure.

Radiation Therapy. Radiation therapy is the use of large doses of radiation to destroy cancerous tumors. High energy (megavoltage) gamma rays in the range 1–50 MeV are generated by an accelerator or a radioactive source. The aim is to deliver a lethal dose to the tumor with minimal damage to the surrounding healthy tissue, and generally this is accomplished by exposing the patient from different angles with a beam that is carefully shaped to match the outline of the tumor in each direction of exposure. The procedure is one of the standard treatments for many forms of cancer, with patients receiving several doses over a period of time. An imaging device is needed to verify that the gamma beam is correctly positioned within the patient and to monitor the exposure.

The therapy imager requirements are significantly different from those used for diagnostic purposes. The therapy dose is extremely large, amounting to 10–100 R per treatment, so that radiation damage is a major concern. Most silicon integrated circuits are destroyed by total exposures of 100 kR, although special radiation hard devices can survive above 1 MR. Amorphous silicon is insensitive to high energy radiation, and both sensors and TFTs have been exposed to 10 MR of gamma rays with only small degradation of their properties [4.83]. Part of the explanation is that the devices are so thin that the probability of an interaction in the material is very low. In addition, amorphous materials may be less susceptible to radiation damage since they are already disordered.

Figure 4.44 shows that the leakage current of p-i-n photodiode only increases by a factor 5 after a rapid exposure of 2MR, and much of the excess current decays after a few days [4.83]. The TFT characteristics show similar insensitivity, with a threshold voltage shift of 1–2 V and some increase in leakage current. Neither device shows a significant increase in noise after such high exposures.

Although the radiation therapy dose is very large, image acquisition is not simple because of the high gamma-ray energy. The difference in attenuation coefficient between bone and tissue decreases such that the contrast in the image is no more than 1%. A very high dynamic range imager is therefore required to allow clear observation of the tumor. Fortunately the required spatial resolution is quite low, and pixel sizes of 500–1000 µm are sufficient to image the procedure. Figure 4.14 indicate that a dynamic range of > 100 000 is achievable from arrays with such large pixels, which is more than adequate.

The sensitivity of the detector to high energy gamma-rays is low, because of the low attenuation in the phosphor screen. The sensitivity is enhanced

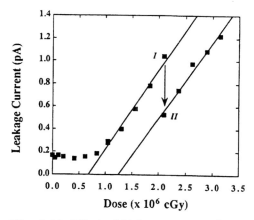

Fig. 4.44. Effect of high gamma-ray dose on the a-Si:H p-i-n photodiode leakage current. The decrease between points I and II are due to resting the sample at room temperature [4.84]

by placing a metal sheet such as copper of thickness about 1 mm on top of a phosphor such as GdO_2S_2:Tb, as illustrated in Fig. 4.45. Photons interact with the copper to create high energy electrons, which are scattered forward into the phosphor, where they cause ionization as at lower energy. The copper film increases the interaction probability and raises the overall emission intensity from the phosphor, but even with this addition the efficiency is much lower than for radiographic imaging.

Radiation therapy was the first medical imaging application proposed for a-Si:H image sensor arrays [4.84–86]. Research has confirmed that the approach is viable and gives imaging performance superior to any other digital technology. A $20'' \times 20''$ prototype imaging system operating up to 20 frames/s has been reported by Colbeth et al., who show quantum limited noise with doses from 0.6 to 25 cGy using both a 6 and 18 MV source [4.87]. Antonuk et al. discuss the merits of a dual energy imager that simultaneously acquires megavoltage and radiographic images [4.86]. This approach is enabled by stacking two flat panel imagers on top of each other in such a way that the top imager is relatively transparent to the high energy beam. The much better contrast of the radiographic image allows a more precise positioning of the megavoltage beam to the patient. This is just one example of novel imaging modalities that result from the flat panel imager structure.

4.5.2 Other Radiation Imaging Applications

Non-Destructive Evaluation. Non-destructive evaluation (NDE) involves both X-ray radiographic and fluoroscopic imaging of inanimate objects. Some examples of its use are the inspection for flaws in critical components such as airplanes, oil pipelines and nuclear power stations, monitoring for the

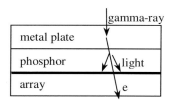

Fig. 4.45. Illustration of high energy gamma ray detection using a metal sheet and phosphor on the a-Si:H array

Fig. 4.46. X-ray image of a bullet piercing a light bulb

quality of a variety of manufactured goods and food, and security inspection systems in airports etc. The total market is about $100 M which, although significant, is considerably smaller than the medical imaging market. The traditional use of X-ray film for NDE is slowly changing to real-time digital imaging. Amorphous silicon imagers offer many of the same advantages as in medical imaging – compact size, high radiation tolerance, both radiographic and fluoroscopic modes, immediate viewing of the image and digital image enhancement, storage and transmission.

The main difference between NDE and medical imaging is that a low dose is less critical and the range of exposure conditions is much broader. Very high energies are needed to penetrate thick steel structures, and various applications require high contrast, high resolution or high speed. Figure 4.46 is an example of the latter and shows an X-ray image of a bullet piercing a light bulb and was acquired with an a-Si:H imager using a very short X-ray burst to freeze the motion.

X-Ray Crystallography. Crystallography is one example of the possible scientific applications of a-Si:H image sensors. Much research is presently directed to identifying the structure of proteins, using X-ray diffraction as the main tool. Proteins are very large molecules containing up to 100 000 atoms which form rather delicate crystals with large unit cell size. The diffraction pattern is correspondingly complex and contains a similarly large number of unique spots. Fairly low energy X-rays (5–20 keV) are employed and the usual procedure is to acquire a set of several hundred diffraction patterns by rotating the crystal in roughly 1 degree increments about an axis perpendicular to the beam. The inversion of the data set into a crystal structure is a formidable computational problem. Digital acquisition of the images eliminates the equally formidable task of obtaining all the diffraction spot coordinates by digitizing film.

A large digital imager is desirable because measurement far from the crystal reduces the effects of scattered X-rays and the blurring effect of the finite size beam. Relatively rapid image acquisition is desirable when measurements are done at a synchrotron sources where the flux is high enough to acquire an image in a few seconds. On the other hand laboratory X-ray sources produce very weak diffracted intensity and in this case, long integration times are needed. High sensitivity is important because the protein crystals are easily damaged by X-rays, and therefore the exposure must be minimized. Finally, the imager needs to have a high dynamic range, because important structural information is contained in the wide range of intensities of the Bragg peaks.

Amorphous silicon image sensors satisfy most of the requirements, with sensitivity being the most challenging [4.88–90]. As with fluoroscopy, the low flux places a particular premium on high sensitivity and low electronic noise. It is of interest to compare the amorphous silicon sensors with the charge coupled devices (CCD) that represent the present state-of-the-art in detectors. A cooled CCD has an electronic noise of < 10 electrons, which is 100 times lower than the amorphous silicon arrays. However, CCDs have too small an active area for the application. We noted in Sect. 4.3.1 that imaging from a large phosphor with a lens has very low optical transfer efficiency, which more than offsets the lower noise. The solution applied to crystallography and similar applications is to use multiple CCD devices and to couple the light from the phosphor using a tapered fiber-optic plate. Naday et al. describe an imager using 9 CCDs and a taper which de-magnifies the image by a factor 3.5, yielding a sensitive area of 15×15 cm [4.90]. Even so, they estimate that an a-Si:H imager with a phosphor converter would have a comparable performance [4.88]. Although the fiber-optic taper is more efficient that a lens, contact imaging with the phosphor has a larger optical efficiency by about a factor 20, and the larger area of the a-Si:H device yields a further performance gain which together compensate for the lower noise of the CCD.

Low energy applications such as protein crystallography are well suited to image sensors using X-ray photoconductor detectors, because thinner layer and lower fields are required. Either selenium of PbI_2 could provide a crystallography detector with better performance that the CCD devices. Other scientific applications that have been considered for the large area imagers include gamma ray space astronomy and recording DNA fragments in gel electrophoresis.

Neutron Imaging. Neutron imaging is a highly specialized branch of non-destructive testing, but illustrates the range of possible radiation imaging applications of a-Si:H sensor arrays. Hydrogen has almost no absorption of X-rays, but scatters neutrons effectively. Thus, neutron imaging can reveal features not seen in an X-ray image, and is particularly useful for detecting corrosion which usually has water associated with it [4.91]. The lack of a convenient source of neutron clearly restricts the applications, but the technique is used on military aircraft in locations that have a nuclear reactor. Detection of neutrons is actually quite easy because of the existence of some specific neutron reactions. Neutrons react with ^{6}Li and one of the fragments is a high energy alpha particle. Phosphors are readily made containing ^{6}Li, so that the alpha particles cause ionization and light emission. The gain is actually higher than for radiographic X-rays, since one absorbed neutron can yield about 175 000 visible photons. This gain is sufficiently larger than the readout noise that the imager is able to detect and count individual neutrons, which is a desirable imaging mode since the neutron flux is very low. The second detection method uses a similar nuclear reaction with gadolinium, and therefore the standard GdO_2S_2:Tb phosphor is sensitive to neutrons.

4.5.3 Document Scanning

Document scanning plays an increasing role in office management systems. Most scanners use linear sensor arrays and either the paper or the detector is moved to image the page. Two-dimensional devices capture the complete page without the need for mechanical motion. Since imaging is done in visible light, the image can be focussed onto a small detector and so the reason for a large area device is less compelling than with X-ray imaging. However, the large area arrays do offer some advantages and some novel system designs relating to high speed scanners, compact scanners and book scanning.

An a-Si:H image sensor allows a contact imaging approach to be used when the array is as large as the paper to be imaged [4.92–95]. The paper is placed in direct contact with the imager and illuminated from behind, as illustrated in Fig. 4.47. The illumination source is immediately below the array and the light passes through the gaps in the pixel between the sensor and the address lines, where it is reflected off the paper and imaged by the sensor. The sensor is protected from direct illumination by the opaque bottom metal contact. The gap between the paper and the sensor must be less than about half the pixel size enough to prevent light scattering to adjacent pixels, but large enough to allow the light to reach the front surface of the sensor. In practice, the light scattering within the paper is usually sufficient to ensure

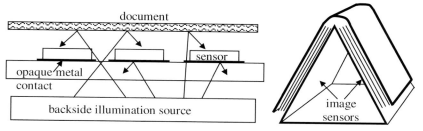

Fig. 4.47. Contact imaging of a document using back illumination (*left*). The same device configured as a compact book scanner with no moving parts

that the sensor is illuminated. This approach to document imaging has been successfully demonstrated and provides a compact imaging system with no moving parts.

Contact imaging enables other scanner designs that cannot be made with a conventional lens coupled system. Figure 4.47 illustrates a design for a book scanner that comprises two large area arrays forming a wedge-shaped structure. The book is placed over the wedge and is imaged from both surfaces. The illumination source is behind the two arrays.

Document scanning generally requires a higher resolution that X-ray imaging to avoid blurring the edges of letters and to prevent lines from showing jagged edges. FAX machines have a low resolution of 200 spot/inch (spi), while high quality full page document scanners typically have resolution of 300–600 spi. Wu et al. have implemented a design to double the spatial resolution of a a-Si:H sensor array with pixel size 127 μm (200 spi) [4.94]. The image is translated across the array by the distance of half a pixel in both directions, so that four sub-sampled images are captured and can be reassembled into a higher resolution image. This approach takes advantage of a low fill factor, which should be about 25% to image one quarter of the pixel at each sample. A slightly larger fill factor provides some oversampling which helps minimize aliasing effects. There are several ways in which the small shifts of the image can be accomplished. Wu et al. describe a one-to-one magnification lens-coupled system with a rotating glass wedge at the projection lens [4.94]. The wedge is rotated by 90° to move the image to the correct position for each sub-sample and the angle of the wedge determines the shift of the image. Measurements confirmed the increase the spatial resolution and that aliasing effects are eliminated.

The a-Si:H sensor arrays are well suited to color imaging because the spectral response of a-Si:H spans the visible spectrum (see Fig. 4.2). The quantum efficiency is greater that 50% at 400 nm in a well designed sensor, increases to 80–90% at the maximum near 550 nm, and decreases strongly aboveabout 700 nm, due to the energy of the band gap. In principle, color imaging may be performed by placing a color filter at each pixel, in a similar manner to the approach used for liquid crystal displays. Usually it is easier to capture three consecutive images with red, green and blue illumination and superimpose the three color separations [4.93]. Provided that the color filters are chosen correctly, excellent color images are obtained. Chapter 7 describes how to fabricate color-selective a-Si:H photodiodes that could be integrated into matrix-addressed arrays.

4.6 Future Developments

At the beginning of 1999, X-ray medical imaging systems are just being introduced onto the medical imaging market. Most of these devices are $\approx 17''$ radiological imagers with pixel sizes of 127–160 µm, but smaller fluoroscopy systems are also close to product introduction. While these systems meet the requirements of the technology in terms of size and imaging performance, the technology is capable of significant improvements in sensitivity and resolution, and numerous other innovations can be expected to reduce the cost and improve the device yield. Most of the new systems use a-Si:H TFTs and photodiodes, with Lanex or CsI scintillators for X-ray detection, or a selenium photoconductor with a-Si:H TFTs. Continued experience with the technology will determine the relative roles for the scintillator and photoconductor approaches to detection. Some of the expected developments in the technology are as follows:

Array Size. LCD manufactures are introducing fourth generation process lines which use substrates of 80×80 cm or larger. The technology is therefore available to make $17'' \times 17''$ imagers from a single plate, rather than from 2 or 4 tiled arrays as in most large imagers made at this time. There seems little reason for any further increase in array size for medical imaging, and so a very large substrate is only useful when more than one array can be made on a substrate. Tiling of small arrays to make a large imager may be a preferred approach, since smaller arrays usually have significantly higher yield. The added expense of attaching readout electronics to both sides of the array might be worth the lower noise that results from dividing the data line in half.

Sensitivity. Many of the approaches discussed earlier to improve the array sensitivity can be expected to reach manufacturing over the next few years. The high fill factor designs are particularly applicable to high resolution arrays and PbI_2 or other new materials may improve the sensitivity of photoconductor arrays.

Noise Reduction. The performance of readout amplifiers can be expected to improve further with the anticipated developments of CMOS fabrication technology, and the experience of the chip design. Further improvement in noise will result from array designs that lower the data line capacitance for which many approaches are possible. The TFT parasitic capacitance may be reduced by using a fully self-aligned device, and new developments in low dielectric constant materials as well as reduced process feature sizes can also be expected to trim the capacitance. An alternative approach is to introduce pixel amplifiers, since a gain of no more than 10 would completely remove the need for improved readout amplifiers or lower capacitance. Once the noise level is reduced to that of the kTC noise of the pixel, further improvements will involve reducing the capacitance of the sensors or developing a novel sampling scheme.

Electronics and System Integration. LCD technology is leading the way towards integrated drivers that can be applied to the image sensors. The vision is of a complete imager with gate drivers, readout amplifiers and control logic on the glass substrate. Drivers and control logic can certainly be made from polysilicon TFTs, and at least for radiography applications, they might even be made from a-Si:H. It is unclear whether a very low noise readout amplifier could be fabricated from polysilicon, but if the arrays have pixel amplifiers then extremes of amplifier performance are not needed. Other approaches to integration include chip-on-glass techniques to bond silicon ICs for drive and readout directly on the glass surface.

Processing. The feature sizes for the array are expected to decrease as the technology for large area processing improves. Such improvements will increase imager performance even without any fundamental change in design, by lowering the capacitance and increasing the sensor fill factor.

Applications. The main applications of a-Si:H image sensors have already been identified and it is clear that the medical imaging devices will dominate the field for the next few years. Document scanning applications are presently limited by the high cost of the image sensors, but these applications may expand rapidly as the manufacturing business grows. The market for large area arrays for general scientific applications is probably small compared to medical imaging, but could offer some interesting opportunities in astronomy, DNA sequencing etc.

References

4.1 P. G. LeComber, W. E. Spear and A. Ghaith, Electron. Lett. **15**, 179 (1979).
4.2 A. J. Snell, P. G. LeComber, K. D. Mackenzie, W. E. Spear and A. Doghmane, J. Non-Cryst. Solids **59&60**, 1187 (1983).
4.3 M. Matsumura, H. Hayama, Y. Nara and K. Ishibashi, IEEE Elect. Dev. Lett. **1**, 182 (1980).

4.4 F. Okumura and S. Kaneko MRS Symp. Proc. **33**, 275 (1984).

4.5 H. Ito, Y. Nishihara, M. Nobue, M. Fuse, T. Nakamura, T. Ozawa, S. Tomiyama, R. Weisfield, H. Tuan and M. Thompson, Proc. IEDM **85**, 436 (1985).

4.6 R. A. Street, S. Nelson, L. Antonuk and V. Perez Mendez, MRS Symp. **192**, 441 (1990).

4.7 G. Juska and K. Arlauskas, Phys. Stat. Sol. (a), **59**, 389 (1980); K. Tsuji, Y. Takasaki, T. Hirai and K. Taketoshi, J. Non-Cryst. Solids **114**, 94 (1989).

4.8 R. A. Street, Hydrogenated amorphous silicon, Cambridge University Press (1991).

4.9 S. Guha, Current Opinion in Solid State and Material Science **2**, 425 (1997).

4.10 R. A. Street, I. Fujieda, R. L. Weisfield, S. Nelson and P. Nylen, MRS Symp. Proc. **258**, 1145 (1992).

4.11 R. A. Street, Appl. Phys. Lett. **57**, 1334 (1990).

4.12 R. A. Street, Philos. Mag. B **63**, 1343 (1991).

4.13 E. A. Schiff, R. A. Street and R. W. Weisfield, J. Non-Cryst. Solids **198–200**, 1155 (1996).

4.14 R. A. Street and M. Hack, MRS Symp. Proc. **219**, 135 (1991).

4.15 C. van Berkel, M. J. Powell and I. D. French, MRS Symp. Proc. **219**, 369 (1991).

4.16 S. M. Sze, Physics of semiconductor devices, Wiley (1981).

4.17 M. J. Thompson, N. M. Johnson, R. J. Nemanich and C. C. Tsai, Appl. Phys. Lett. **39**, 274 (1981).

4.18 D. L. Staebler and C. R. Wronski, Appl. Phys. Lett. **31**, 292 (1977).

4.19 V. Perez-Mendez, J. Morel, S. N. Kaplan and R. A. Street, Nucl. Instr. and Methods **252**, 478 (1986).

4.20 V. Perez-Mendez, G. Cho, J. Drewery, T. Jing, N. Kaplan, S. Qureshi, D. Wildermuth, I. Fujieda and R. A. Street, J. Non-Cryst. Solids **137& 138** 1291 (1991).

4.21 A. Ilie, T. Pochet, F. Foulon and B. Equer, MRS Symp. Proc. **336**, 121 (1994).

4.22 S. Qureshi, V. Perez-Mendez, S. N. Kaplan, I. Fujieda, G. Cho and R. A. Street, MRS Symp. Proc. **149**, 649 (1989).

4.23 I. Fujieda, G. Cho, M. Conti, J. Drewery, S. N. Kaplan, V. Perez-Mendez and S. Qureshi, IEEE Trans. Nucl. Sci. **37**, 124 (1990).

4.24 R. L. Weisfield, M. A. Hartney, R. A. Street and R. B. Apte, The Physics of Medical Imaging, Proc. SPIE **3336**, 444 (1998).

4.25 S. D. Brotherton, Semicond. Sci. Technol. **10**, 721 (1995).

4.26 F. Lemmi and R. A. Street, MRS Symp. Proc. **507**, 79 (1998).

4.27 J. B. Boyce, P. Mei, R. T. Fulks and J. Ho, Phys. Stat. Sol. (a), **166**, 729 (1998).

4.28 J. Farrell, M. Westcott, A. Van Calster, J. De Baets, I. De Rycke, J. Capon, H. De Smet, J. Doutreloigne and J. Vanfleteren, Microelectronic Engineering **19**, 187 (1992).

4.29 R. A. Street and L. E Antonuk, IEEE Trans., Circuits and devices **9**, 39 (1993).

4.30 R. A. Street, R. L. Weisfield, R. B. Apte, S. E. Ready, A. Moore, M. Nguyen, W. B. Jackson and P. Nylen, Thin Solid Films **296**, 172 (1997).

4.31 L. E. Antonuk, Y. El-Mohri, J. H. Siewerdsen, J. Yorkston, W. Huang, V. E. Scarpine and R. A. Street, Medical Physics **24**, 51 (1997);
I. Fujieda, S. Nelson, R. A. Street and R. L. Weisfield, IEEE Trans, Nucl. Sci. **39**, 1056 (1992).

4.32 J. T. Rahn, F. Lemmi, R. L. Weisfield, R. Lujan, P. Mei, J.-P. Lu, J. Ho, S. E. Ready, R. B. Apte, P. Nylen, J. Boyce and R. A. Street, The Physics of Medical Imaging, Proc. SPIE **3659**, 510 (1999).

4.33 M. J. Powell, C. Glasse, I. D. French, A. R. Franklin, J. R. Hughes and J. E. Curran, MRS Symp. Proc. **467**, 863 (1997).

4.34 L. E. Antonuk, J. Boudry, W. Huang, D. L. McShan, E. J. Morton, J. Yorkston, M. J. Longo, and R. A. Street, Medical Physics **19**, 1455 (1992).

4.35 U. Scheibel, N. Conrads, N. Jung, M. Weibrecht, H Wieczorek, T. Zaengel, M. J. Powell, I. D. French, and C. Glasse, Medical Imaging, SPIE Proc. **2163**, 129, (1994).

4.36 W. Zhao, and J. A. Rowlands, Medical Imaging VI: Instrumentation, SPIE Proc. **1651**, 134 (1992).

4.37 D. L. Lee, L. K. Cheung and L. S. Jeromin, The Physics of Medical Imaging, Proc. SPIE **2432**, 237 (1995).

4.38 W. Zhao, J. A. Rowlands, S. Germann, D. Waechter and Z. Huang, The Physics of Medical Imaging, Proc. SPIE **2432**, 250 (1995).

4.39 W. Zhao, I. Blevis, S. Germann, J. A. Rowlands, D. Waechter and Z. Huang, The Physics of Medical Imaging, Proc. SPIE **2708**, 523 (1996).

4.40 D. L. Lee, L. K. Cheung, L. S. Jeromin, E. F. Palecki and B. Rodericks, The Physics of Medical Imaging, Proc. SPIE **3032**, 88 (1997).

4.41 J. A. Rowlands, W. Zhao, I. Blevis, G. Pang, W. G. Ji, S. Germann, S. O. Kasap, D. Waechter and Z. Huang, The Physics of Medical Imaging, Proc. SPIE **3032**, 97, (1997).

4.42 T. Graeve, E. W. Huang, S. M. Alexander and Y. Li, SPIE Proc. **2415**, 177 (1995).

4.43 J. Chabbal, C. Chaussat, T. Ducourant, L. Fritsch, J. Michailos, V. Spinnler, G. Vieux, M. Arques, M. Hoheisel, H. Horbaschek, R. Schulz and M. Spahn, The Physics of Medical Imaging, Proc. SPIE **2708**, 499, (1996).

4.44 M. Hoheisel, M. Arques, J. Chabbal, C. Chaussat, T. Ducourant, G. Hahm, H. Horbaschek, R. Schulz and M. Spahn, J. Non-Cryst Solids **227–230**, 1300 (1998).

4.45 G. Cho, J. S. Drewery, W. S. Hong, T. Jing, H. Lee, S. N. Kaplan, A, Mireshghi, V. Perez-Mendez, and D. Wildermuth, MRS Symp. Proc. **297**, 969 (1993).

4.46 T. Graeve, Y. Li, A. Fabans and W. Huang, SPIE Proc. **2708**, 494 (1996).

4.47 M. Bohm, F. Blecher, A. Eckhardt, K. Seibel, B. Schneider, J. Stertzel, S. Benthein, H. Keller, T. Lule, P. Rieve, M. Sommer, B. Van Uffel, F. Librecht, R. C. Lind, L. Humm, V. Efron and E. Roth, MRS Symp. Proc. **507**, 327 (1998).

4.48 H. Fischer, J. Schulz, J. Giehl, M. Bohm and J. P. M. Schmitt, MRS Symp. Proc. **258**, 1139 (1992).

4.49 L. E. Antonuk, J. Boudry, Y. El-Mohri, W. Huang, J. H. Siewerdsen, J. Yorkston, and R. A. Street, SPIE Conference Proceedings **2432**, 217 (1995).

4.50 J. Wu, R. A. Street, R. Weisfield, S. Nelson, M. Young, M. Nguyen, and P. Nylen, Proc. SPIE **2172**, 144 (1994).

4.51 R. Apte, N. Nickel, R. A. Street, R. Weisfield, X.-D. Wu, S. Ready, M. Nguyen, P. Nylen, MRS Symp. Proc. **420**, 177 (1996).

4.52 R. B. Apte, R. A. Street, S. E. Ready, D. A. Jared, A. M. Moore, R. L. Weisfield, T. A. Rodericks and T. A. Granberg, Proc. SPIE **3301**, 2 (1998).

4.53 S. A. Kleinfelder, W. C. Carithers, R. P. Ely, C. Haber, F. Kirsten and H. G. Spieler, IEEE Trans. Nucl. Sci. **35**, 171 (1988).

4.54 L. E. Antonuk, Y. El-Mohri, W. Huang, K. W. Lee, M. Maolinbay, V. E. Scarpine, J. H. Siewerdsen, M. Verma, J. Yorkston and R. A. Street, The Physics of Medical Imaging, Proc. SPIE **3032**, 2 (1997).

4.55 C. A. Harper (Ed) Electronic Packaging and Interconnection Handbook, McGraw Hill (1996).

4.56 I-W. Wu, A. G. Lewis, T. Y. Huang and A. Chiang, Digest of SID Conference **307** (1990).

4.57 G. Cho, S. Qureshi, J. S. Drewery, T. Jing, S. N. Kaplan, H. Lee, A, Mireshghi, V. Perez-Mendez, and D. Wildermuth, Proc IEEE Nucl. Sci. Symp 1991.

4.58 V. Radeka, Ann. Rev. Nucl. Part. Sci. **38**, 217 (1988).

4.59 J. M. Boudry, and L. E. Antonuk, IEEE Trans, Nucl. Sci. **41**, 703 (1994).

4.60 J. M. Boudry, and L. E. Antonuk, Med. Phys. **23**, 743, (1996).

4.61 G. Cho, J. S. Drewery, I. Fujieda, T. Jing, S. N. Kaplan, V. Perez-Mendez, S. Qureshi, D. Wildermuth and R. A. Street, MRS Symp. Proc. **192**, 393 (1990).

4.62 I. Fujieda, R. A. Street, R. L. Weisfield, S. Nelson, P. Nylen, V. Perez-Mendez, and G. Cho, Jpn. J. Appl. Phys. **32**, 198 (1993).

4.63 R. L. Weisfield, R. A. Street, R. B. Apte, A. Moore, The Physics of Medical Imaging, Proc. SPIE **3032**, 14 (1997).

4.64 A. van der Ziel, "Noise in solid state devices and circuits", Wiley (1986).

4.65 J. C. Santiard and F. Faccio, Nucl. Inst. And Methods A**380**, 350 (1996).

4.66 R. A. Street, R. B. Apte, T. Granberg, P. Mei, S. E. Ready, K. S. Shah and R. L. Weisfield, J. Non-Cryst. Solids **227–230**, 1306 (1998).

4.67 N. Jung, P. L. Alving, F. Busse, N. Conrads, H. Meulenbrugge, W. Rutten, U. Schiebel, M. Weibrecht and H. Wieczorek, The Physics of Medical Imaging, Proc. SPIE **3336**, 396 (1998).

4.68 Numerous papers can be found in, The Physics of Medical Imaging, Proc. SPIE **3336**, in press.

4.69 B. Hasagawa, The Physics of Medical X-ray Imaging, Medical Physics Publishing (1991).

4.70 H. E. Johns and J. R. Cunningham, The Physics of Radiology, (Charles C. Thomas) (1983).

4.71 R. A. Street, R. B. Apte, S. E. Ready, R. L. Weisfield and P. Nylen, MRS Symp. Proc. **487**, 399 (1998).

4.72 K. S. Shah, F. Olschner, L. P. Moy, P. Bennett, M. Misra, J. Zhang, M. R. Squillante, J. C. Lund, Nucl. Instr. and Methods, A **380**, 266 (1996).

4.73 K. S. Shah, P. Bennett, M. Klugerman, L. P. Moy, G. Entine, D. Ouimette and R. Aikens, The Physics of Medical Imaging, Proc. SPIE **3032**, 395 (1997).

4.74 A. Brauers, N. Conrads, G. Frings, U. Schiebel, M. J. Powell and C. Glasse, MRS Symp. Proc. in press.

4.75 I. A. Cunningham, M. S. Westmore and A. Fenster, The Physics of Medical Imaging, Proc. SPIE, **2432**, 143 (1995).

4.76 I. A. Cunningham, The Physics of Medical Imaging, Proc. SPIE **3032**, 22 (1997).

4.77 J. H. Siewerdsen, L. E. Antonuk, Y. El-Mohri, J. Yorkston, W. Huang, J. M. Boudry, and I. A. Cunningham, Medical Physics **24**, 71 (1997).

4.78 H. Fujieta, IEEE Trans. Med. Im. **11**, 34 (1992).

4.79 J. Yorkston, L. E. Antonuk, N. Seraji, W. Huang, J. Siewerdsen, and Y. El-Mohri, Proc. SPIE **2432**, 260 (1995).

4.80 I. Yorkston, L. E. Antonuk, Y. El-Mohri, K-W Jee, W. Huang, J. H. Siewerdsen and D. P. Trauernicht, The Physics of Medical Imaging, Proc. SPIE **3336**, 556 (1998).

4.81 G. S. Shaber, D. L. Lee, J. Bell, G. Powell and A. D. A. Maidment, The Physics of Medical Imaging, Proc. SPIE **3336**, 463 (1998).

4.82 W. B. Jackson, M. R. Said, D. A. Jared, J. O. Larimer, J. L. Gille, J. Lubin, Medical Imaging: Image Perception, SPIE Conf. Proc. **3606**, 64 (1997).

4.83 J. M. Boudry and L. E. Antonuk, IEEE Trans. Nucl. Sci. **41**, 703 (1994), J. M. Boudry and L. E. Antonuk, Med. Phys. **23**, 743 (1996).

4.84 L. E. Antonuk, J. Boudry, W. Huang, D. L. McShan, E. J. Morton, J. Yorkston, M. L. Longo and R. A. Street, Med. Phys. **19**, 1455 (1992).

4.85 L. E. Antonuk, J. Yorkston, J. Boudry, M. J. Longo and R. A. Street, Nucl. Instr. and Methods A **310**, 460 (1991).

4.86 L. E. Antonuk, J. Boudry, W. Huang, K. L. Lam, E. J. Morton, R. K. Ten Haken, J. Yorkston and N. H. Clinthorne, IEEE Trans. Med. Imaging **13**, 482 (1994).

4.87 R. E. Colbeth, M. J. Allen, D. J. Day, D. L. Gilblom, M. E. Klausmeier-Brown, J. Pavkovich, E. J. Seppi and E. G. Shapiro, The Physics of Medical Imaging, Proc. SPIE **3032**, 42 (1997).

4.88 S. W. Ross, I. Naday, M. Kanyo, M. L. Westbrook, E. M. Westbrook, W. C. Phillips, M. J. Stanton and R. A. Street, Charge Coupled Devices and Solid State Optical Sensors IV, Proc. SPIE **2415**, 189 (1995).

4.89 S. Ross, G. Zentai, K. S. Shah, R. W. Alkire, I Naday and E. M. Westbrook, Nucl. Inst. and Methods **A399**, 38 (1997).

4.90 I. Naday, S. Ross, M. Kanyo, M. L. Westbrook, E. M. Westbrook, W. C. Phillips, M. J. Stanton, and D. M. O'Mara, SPIE Proc. **2415**, 236 (1995).

4.91 E. W. McFarland and R. C. Lanza, SPIE Conf. Proc. **1737**, 101 (1992).

4.92 K. Kobayashi, S. Makida, Y. Sato and T. Hamano, Charge Coupled Devices and Solid State Optical Sensors III, Proc. SPIE **1900**, 41 (1993).

4.93 R. A. Street, X. D. Wu, R. Weisfield, S. Nelson and P. Nylen, MRS Symp. Proc. **336**, 873 (1994).

4.94 X. D. Wu, R. A. Street, D. Biegelsen, L. Swartz and D. Jared, Proc OSA symposium, in press.

4.95 M. J. Powell, I. D. French, J. R. Hughes, N. C. Bird, O. S. Davies, C. Glasse and J. E. Curren, MRS Symp. Proc. **258**, 1127 (1992).

5 Novel Processing Technology for Macroelectronics

S. Wagner[1], H. Gleskova[1], J.C. Sturm[1], and Z. Suo[2]

[1] Department of Electrical Engineering and Center for Photonics and Optoelectronic Materials, Princeton University, Princeton, NJ 08544, USA
E-mail: wagner@princeton.edu

[2] Department of Mechanical and Aerospace Engineering, Princeton University, Princeton, NJ 08544, USA

Abstract. Large-area electronic circuits on thin foil substrates can be made with techniques adapted from conventional printing. Conventional printing can provide a resolution and an overlay registration of 10 µm and ±5 µm, respectively, which will allow making thin-film transistors (TFTs) at densities above 10 000 per square centimeter. An early example is the use of laser-printed toner etch masks for the fabrication of amorphous silicon TFTs. Patterned devices can be made by direct printing, as demonstrated by the jet printing of the active polymer for organic light-emitting diodes (OLEDs). Paper-thin foils of glass, steel, and polyimide can serve as substrates for making TFTs with characteristics comparable to those made on glass plates. Materials options for thin foil substrates are described. A study of the mechanics of films on stiff and compliant foil substrates shows that particularly rugged and flexible device structures can be made when the foils are very thin. Integrating OLEDs with thin-film transistors on steel foil substrates provides an early example of 3-D integrated components for macroelectronics.

5.1 Introduction

Following many years of experimentation, development and manufacturing, the active-matrix liquid crystal display (AM-LCD) has become a very successful product in laptop computers, workstations and personal computers (see Chap. 2). Its success seems assured in view of the favorable reaction of early users, who experience less eye fatigue when working with AM-LCD than with cathode ray tube (CRT) screens. For the wide-spread application of large-area electronics such as digital wallpaper, smart materials, and intelligent barcodes, its cost must be reduced to well below the present value of approximately \$5 per square inch (\$8 000/m^2). This need for cost reduction has been recognized for some time. In the Giant Electronics program a group of Japanese laboratories sought to introduce printing techniques to the fabrication of AM-LCDs [5.1]. A group at the Philips Research Laboratories has been developing a technology for producing organic electronics by using a combination of photochemical and wet chemical reactions in spun-on layers on plastic substrates [5.2]. In this chapter we explore the application of printing techniques and thin foil substrates to large-area electronic, or macroelectronic, technology.

We begin with a brief discussion of the per-area cost of macroelectronics and make a case for the advantages of direct printing. The practicality of printing depends on the resolution and the overlay registration achievable by conventional printing techniques. These two parameters are discussed next. Then we describe the fabrication of amorphous silicon thin-film transistors with a process that uses laser printed toner masks instead of photoresist. This process illustrates the reduction in the number of steps achievable by printing. Our goal, the direct printing of patterned device materials, is illustrated by the jetting of polymer solutions to produce organic light-emitting diodes. Printing equipment often relies on the use of flexible substrates, which can be of particular advantage to macroelectronics because they are lightweight and can be very rugged. In this spirit we continue with a discussion of substrate materials, and of the mechanics of thin films on stiff and on compliant substrates. The practicality of a compliant substrate is illustrated with transistors on polyimide foil. Finally, we show that TFTs and OLEDs can be integrated to create a thin-film display on a foil substrate.

Macroelectronic products, for example display banners, digital wallpaper, or smart greeting cards in many cases must be large but need not necessarily have a high device density. Hence it is useful to gauge the cost of electronics per unit area rather than by cost of function.

We do this in Table 5.1. Inspection of the cost of integrated circuits (ICs) and displays shows that AM-LCDs cost an order of magnitude less than silicon ICs, but an order of magnitude more than present-day large-area photovoltaic (PV) modules. Thin-film PV modules are complex multilayer devices (Chap. 6). For example, the triple-junction module made by United Solar Systems Corporation contains 12 layers, some of which are only 10 nm thick [5.3]. These modules are integrated monolithically over large areas. In the Solarex module, a $0.80\,\mathrm{m}^2$ large ten-layer device structure is interconnected using laser scribing [5.4]. Experience with PV modules shows that complex, multilayer, active electronics can be made at a cost well below \$ 1 000 per square meter. However, integrated PV modules need few patterning steps, and the individual diodes still measure many square centimeters. What will it take to fabricate inexpensive, large-area transistor electronics with a useful device density?

Table 5.1. The cost of integrated circuits and information displays

Technology	Cost ($ $\cdot$m^{-2})
Silicon integrated circuit (microprocessor – memory)	500 000–50 000
Active-matrix liquid crystal display	10 000–6 000
Cathode ray tube	3 000–1 000
Amorphous silicon solar module (present → target)	400 → 100
Mail-order catalog (four-color print)	0.1

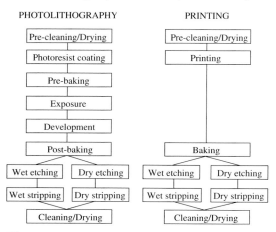

Fig. 5.1. *Left*: Steps in typical photolithography. *Right*: Steps in a process by which devices are patterned using printed etch masks. This figure illustrates the reduction in the number of process steps made possible by printing (from [5.5])

If we take the development of thin-film PV as a guide, we see that the introduction of large-area electronics relies on two coupled developments. One is the reduction of materials input, and the other is the reduction of the number of process steps. Using thin films, thin substrates and thin encapsulation reduces materials input. Reducing the number of process steps lowers the cost of equipment, and the consumption of the ancillary chemicals that are used in circuit processing. The number of process steps could be reduced greatly by the direct printing of the active materials as illustrated in Fig. 5.1 [5.5]. In drawing an analogy between the printing of books and the fabrication of integrated circuits this reduction becomes evident. If books were made like ICs, the paper would be first blackened with ink, and then the ink would be removed selectively to leave letters standing on a white background. Thus the use of laser printed masking layers, instead of photoresist, for etch or lift-off provides a good example for the transition from IC fabrication to book-printing techniques.

In IC fabrication the application or modification of the active material is separate from its patterning, because the material properties and the material pattern are optimized best when done in separate steps. If one wishes to print active circuits directly, one must devise materials that can be applied and patterned in a single step. The materials needed for the printing of active circuits include metallic conductors, insulators, semiconductors for transistors and light emitters, piezoelectric materials, etc. This approach to the printing of active circuits explores the territory that lies between ICs and printed-wire boards. In effect, large-area printed electronics will be active circuits monolithically integrated with their packaging.

Completed thin-film circuits are at most a few micrometers thick. Therefore the substrate and encapsulation constitute the bulk of the finished product. Reduction of their weight and thickness becomes important. When the substrate is reduced to a thickness where it becomes flexible, it also becomes usable in continuous, roll-to-roll paper-like production. The finished circuit then is a flexible foil, and using equally thin encapsulation will preserve this flexibility. We shall see that rugged thin-film circuits are a natural consequence of the mechanics of thin foil substrates.

Large-area transistor circuits may be integrated with light valves or emitters, light sensors, piezoelectric actuators, and many other opto-electronic or microelectromechanical devices. At the end of this chapter we illustrate the beginning of this integration with a TFT/organic light-emitting diode (OLED) pixel.

When large-area circuits are made by programmable printing, for example xerography or ink jetting, they can serve as vehicles for rapid prototyping [5.6]. It is conceivable that in an electronic printing shop single copies or short runs of circuits will be programmed and printed on short order. The laser printing of toner etch masks, which are designed on the computer that drives the laser printer, and the jet printing of polymers for organic light-emitting diodes illustrate this emerging capability.

5.2 Resolution and Registration: The Density of Functions Achievable by Printing

Programmable advertising banners may have square-centimeter size pixels and cover an area of many square meters. Intelligent barcodes, on the other hand, may cover only a few square centimeters but need high resolution. The development of microelectronics has shown that the search for high pattern density is one of the main drivers of IC technology. Therefore, it is instructive to estimate the density of active devices that could be produced by using conventional printing techniques.

Let us look at the density of amorphous-silicon thin-film transistors achievable by printing. The smallest size of a TFT will be set by two parameters. One is the smallest size of a pattern that can be defined by additive printing, which in IC technology is specified as the design rule. We give it the symbol λ. The second parameter is the accuracy of overlay, or registration, of subsequent patterns, which is quantified as the overlay alignment error $\pm\delta$. We illustrate the effect of these two parameters on the achievable density of thin-film transistors with the layout illustrated in Fig. 5.2, of a conventional inverted-staggered a-Si TFT without back channel passivation. Here we assume that any subsequent layer can be registered to any preceding layer with an accuracy of $\pm\delta$. Figure 5.2 shows that the overall length of the TFT is $(\lambda + 8\delta)$, its overall width $(W + 6\delta)$. We assume an inactive fringe of width δ. Thus the area occupied by the device is $(\lambda + 8\delta)(W + 6\delta)$.

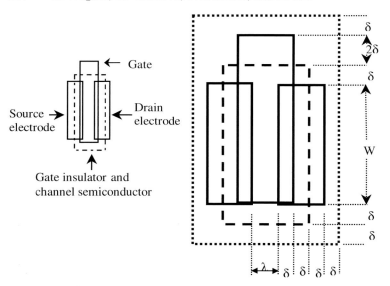

Fig. 5.2. *Left*: Layout of a thin-film transistor in the inverted staggered configuration. *Right*: The effects of resolution λ and overlay registration δ on the size of the TFT. λ sets the channel length L. The effect of the channel width W is discussed in the text

The printing industry defines resolution differently from the semiconductor industry, because it gauges by visual impact rather than device geometry and electrical function. However, the term resolution in spots per inch (spi) employed in digital imaging by the printing industry [5.7] is coming close to its use in microelectronics. For our first-order estimate we will assume that to both industries "resolution" means the number of resolvable, unbroken, line pairs per unit length. Resolution of present-day high-quality, large volume color printing is exemplified by the offset printing for the National Geographic Magazine, which is $\approx 2\,500$ spi. Experts agree that current printing technology can achieve a resolvable line separation of $\lambda = 10\mu m$. This value will be our TFT gate length L. The accuracy of overlay registration δ achievable with present-day printing technology is $\pm 5\mu m$. This value will be our alignment tolerance [5.8].

The value of the W/L ratio will depend on the TFT application. It will be small in switches, and large in power transistors for driving OLEDs. Let us assume that we are building a logic circuit made of switches that need an ON current of $I_{ON} = 5\,\mu A$. The drain current in saturation is given by,

$$I_D(\text{sat}) = (W/L)\mu_e \left(\varepsilon_{\text{ins}}\varepsilon_0/d_{\text{ins}}\right)(V_g - V_{\text{th}})^2/2 \,. \tag{5.1}$$

For the typical values of the electron mobility μ_e of $1\,\text{cm}^2\text{V}^{-1}\text{s}^{-1}$, insulator dielectric constant ε_{ins} of 7.5, insulator thickness d_{ins} of 320 nm, and $(V_g - V_{\text{th}}) = V_{S/D} = 10\,\text{V}$, $I_D(\text{sat})$ becomes $(W/L)\,\mu A$. The equivalent source–

drain resistance is $R_{SD} = 10\,(L/W)$ MΩ. A TFT that delivers $I_{ON} = 5\,\mu\text{A}$ will need $W/\lambda = 5$. The area of each TFT will be $4\,000\,\mu\text{m}^2$ (Fig. 5.2), and the packing density will be 25 000 transistors per cm^2.

Note that this packing density of a-Si TFTs is less than 1/1 000 that of IC MOSFETs, and that the speed of a printed a-Si:H TFT will be 1/10 000 the MOSFET speed. Obviously, printed TFTs will not make sense as competitors of IC MOSFETs. But at a packing density of 25 000 cm^{-2}, printed TFTs will be acceptable for many applications that need large area, or integrated packaging. Building-sized displays may require only a few TFTs per square centimeter, and a printed-TFT smart card with a surface of $\approx 50\,\text{cm}^2$ will hold as many as 10^6 printed TFTs on one level.

The physical limits of the resolution λ and the overlay registration $\pm\delta$ depend on the tools and materials that are used for printing. First, the ink must have sufficiently fine grain. This condition is easy to meet in principle by using molecule-based inks (solutions) or inks containing fine particles like polystyrene latex, which is available commercially in sizes down to $\approx 10\,\text{nm}$. Second, the tools for printing and alignment must have sufficient resolution. When light is used for both printing (as in electrophotography) and alignment, the physical limits for both λ and $\pm\delta$ are $\leq 1\,\mu\text{m}$. Thus, in principle, both ink and tools can easily meet our requirements of 10 and $\pm 5\,\mu\text{m}$, respectively.

Commercial printing is carried out in two steps. In the pre-press step, text or image is transferred to a printing plate. In a second step this plate is used to do the printing [5.7]. The resolution of plate-making and printing techniques can be as high as $2\,000$–$4\,000$ spi (12.7–6.4 μm). By using either three-point mechanical or optical alignment the overlay registration can be brought to $\pm 5\,\mu\text{m}$ [5.9]. Thus the know-how and the components for the printing of active electronics at useful resolution and registration do exist.

The ratio of resolution to registration accuracy appears to be the same at all scales ranging from microelectronics to large-scale printing [5.10]. The IC resolution is $\approx 0.25\,\mu\text{m}$ and the overlay tolerance $\leq 0.1\,\mu\text{m}$. High-speed gravure presses print a linewidth of $\approx 150\,\mu\text{m}$ with a registration of $\approx 50\,\mu\text{m}$. The printing speeds are very different, though. While a modern IC plant produces approximately $10\,000\,\text{m}^2$ of ICs per year, a modern gravure press prints on this surface in about 5 min. Indeed the speed of today's printing presses is so high that large-area electronics may first be printed on plate-making equipment rather than printing presses. Note that a ratio of $\lambda/\delta = 3$ is easily compatible with the assumptions we used in calculating the packing density of printed TFTs.

The physical limits of several printing techniques are considerably finer than the resolution and registration of conventional printing equipment. Laser writing can produce a resolution of the order of $1\,\mu\text{m}$. Nanoimprinting has demonstrated a resolution in the tens of nanometer range [5.11]. Therefore, the density of directly printed devices can be raised orders of magnitude above

≈ 10 000 per square centimeter. We can anticipate continuous improvements of resolution and registration once a direct-printing industry for electronics has come into existence.

5.3 Printed Toner Masks for Etching and Liftoff

The fabrication of integrated circuits alternates between deposition or modification of a layer of material and the patterning of this layer. The established patterning process is photolithography, which typically consists of the nine steps shown in Fig. 5.1. The patterned photoresist serves as the etch mask for the active material that it covers. This step is shown as wet etching or dry etching in Fig. 5.1. Patterned photoresist also can serve for patterning overlying active material in a process called Liftoff. Both processes can be shortened by three steps if the etch or liftoff masks are printed directly. A typical patterning sequence using a printed mask is shown on the right-hand side of Fig. 5.1. We now describe two practical processes that use laser-printed toner as masks for etch and liftoff. The first process relies on a combination of mask printing and of toner transfer via an auxiliary paper substrate. The second exclusively uses masks that are printed directly on the substrate. The mask material is xerographic toner, which is applied with a 600 dpi laser printer. The two processes represent chronological developments and are not necessarily tied to either glass or steel substrates.

5.3.1 Toner Masks via Paper Transfer: TFTs on Glass Foil

TFTs from a-Si:H have been fabricated on 50 µm thick alkali-free glass foil (Schott AF45), with the glass foil providing the flexibility required by an office-type laser printer. The bottom-gate, back-channel-etch TFTs need four mask levels, or patterning steps. These steps include direct laser printing of the first mask [5.6] and toner transfer for the higher levels. The channel length L and width W of the TFTs are 100 µm and 500 µm, respectively. The gate electrode is 300 µm long. The complete TFT process is shown in Fig. 5.3 [5.5]. The first step is to run the glass foil through the laser printer to print a negative toner gate mask. Then an ≈ 100 nm thick Cr layer is thermally evaporated and the Cr-on-toner is lifted off in toluene. The silicon stack is deposited in a three-chamber plasma-enhanced chemical vapor deposition system: ≈ 410 nm of SiN_x, ≈ 160 nm of undoped a-Si:H and ≈ 50 nm of (n^+) a-Si:H. An ≈ 100 nm thick Cr layer for source/drain contacts is thermally evaporated. Then a positive source/drain pattern is printed using transfer paper [5.12, 14]. The Cr layer is wet etched, the (n^+) a-Si:H layer dry etched in CF_4, and the toner is removed. Next, a positive toner mask is printed to define the TFT island, again using transfer paper. The undoped a-Si:H layer is dry etched in CF_4. Without stripping the island mask, the final toner mask for opening the gate electrode pad is printed, again using transfer paper. The

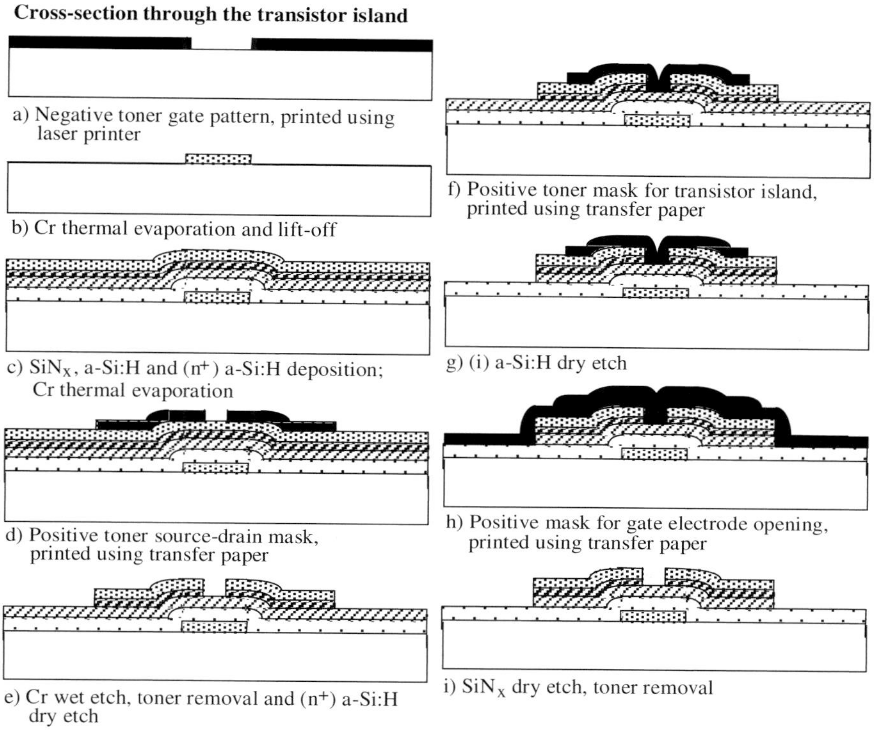

Cross-section through the transistor island

a) Negative toner gate pattern, printed using laser printer

b) Cr thermal evaporation and lift-off

c) SiN_x, a-Si:H and (n^+) a-Si:H deposition; Cr thermal evaporation

d) Positive toner source-drain mask, printed using transfer paper

e) Cr wet etch, toner removal and (n^+) a-Si:H dry etch

f) Positive toner mask for transistor island, printed using transfer paper

g) (i) a-Si:H dry etch

h) Positive mask for gate electrode opening, printed using transfer paper

i) SiN_x dry etch, toner removal

Fig. 5.3a–i. Process for making TFTs using toner masks. Steps (**a**) and (**b**) show lift-off. The other toner steps are etch masks applied by using transfer paper. Function and composition of the layers is evident from the captions (from [5.5])

SiN_x is dry etched in CF_4 and the toner is removed. Finally the TFTs are annealed in forming gas at 200°C to anneal out the radiation damage caused by the plasma during dry etching.

The photograph of a TFT made by this process, and the transfer characteristics of this transistor are shown in Fig. 5.4 [5.5]. These characteristics are identical to those of TFTs made on glass plate substrates by conventional photolithography. The channel of this transistor lies at the bottom center of the photograph, which is dominated by the contact pads, which were made large to ease the probing of curved substrates.

The size distribution of toner particles for commercial laser printers is centered at 7–10 µm [5.15]. This large particle size produces irregular edges, as can be seen in Fig. 5.5a [5.16]. The toner particles also spread into open areas, and they do not coat the printed areas perfectly. This irregularity is a consequence of the mutual repulsion of toner particles, of variations in the surface charge density of the toner particles, and of the attenuation by the glass substrate of the electric field gradient that directs the toner particles.

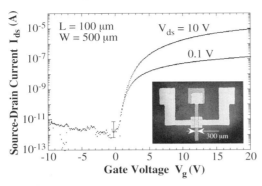

Fig. 5.4. Transfer characteristics of an a-Si:H TFT made by the process of Fig. 5.3 on glass foil. The inset is a photograph of the TFT structure. Source, gate, and drain contact pads lie near the upper edge of the photograph. The 100 μm long channel lies in the center of the TFT island, which is close to the lower edge. The gate electrode is 300 μm wide (from [5.5])

Fig. 5.5. An as-printed toner pattern before (*left*) and after (*right*) post-printer fusing. The vertical line is ≈ 100 μm wide ([5.16])

The laser printer has a fusing stage in which the toner particles are sintered to the substrate and to each other. The bulk of a typical toner particle is a polystyrene-based copolymer with a glass temperature near 100°C. During the fusing step in the printer the temperature may reach 170°C for a fraction of a second, but this treatment does not suffice to melt the toner particles completely. Printed areas therefore remain porous to etchants. A post-printer anneal for 1 h at 120°C melts the toner to the extent that it forms a contiguous, impermeable layer, and that small holes in the printed layer close. The pattern of Fig. 5.5b was fused at 160°C for 30 min. Figure 5.6 [5.17] shows surface profiler traces of a toner layer out of the laser printer, and after the post-printer anneal, respectively.

5.3.2 All Masks Printed Directly: TFTs on Steel Foil

If provisions are made for reproducible alignment of the substrate in the laser printer, all toner masks can be printed directly onto the substrate [5.18]. To do so, the substrate is aligned mechanically for each patterning step by

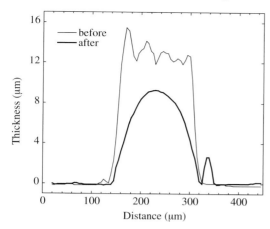

Fig. 5.6. Surface profiler traces across a toner line as produced by the printer, and after the post-print anneal. Note the reduction in cross section caused by compaction of the toner (from [5.17])

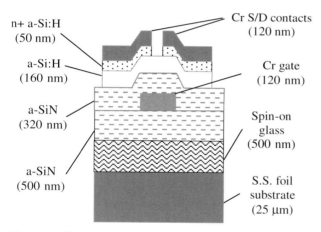

Fig. 5.7. Cross-section of a TFT on steel foil, showing materials and typical layer thicknesses (from [5.18])

matching registration marks between the substrate and a rigid carrier. The carrier is fed into the laser printer and is aligned in the printer just like a sheet of paper. The schematic cross section of a TFT on steel foil is shown in Fig. 5.7 [5.18] The performance of TFTs made by printing all toner masks directly is comparable to that of TFTs made on similar steel foil but using photolithography, as is evident from Fig. 5.8 [5.18].

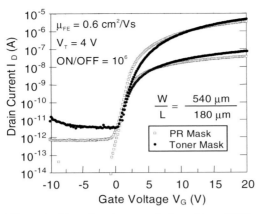

Fig. 5.8. Transfer characteristics of two TFTs made on 25 μm thick steel foil. One TFT was made using conventional photolithography, the other by printing all mask layers directly in a laser printer (from [5.18])

5.4 Printing Active Materials: Jetting Doped Polymers for Organic Light Emitting Devices

Direct printing of active materials is expected to produce large-area electronics at very low cost. An early example of direct printing is the printing of the active polymers for organic light-emitting diodes [5.19]. One group of organic materials for OLEDs consists of polymer hosts that permit charge transport and are doped with color centers. The color centers are organic dyes that emit light upon electron–hole recombination. The choice of dopant dye determines the color of emitted light. In the case of organic device materials the direct printing provides a particular advantage: It is the first simple process for the patterning of monolithically integrated three-color OLEDs.

An ink-jet printer was used with a resolution of 640 dots per inch. The piezoelectric printhead squirts ink droplets from a nozzle with a 65 μm opening. Four ink cartridges and four nozzles enable the printer to print four different colors simultaneously. As the printer head scans the page and the piezoelectric materials are pulsed, ink is squirted from the nozzles onto the page. The only modification to the ink-jet printer for printing OLEDs was to replace the original inks in the cartridges with polymer solutions.

The hole-transport polymer poly(N-vinylcarbazole) (PVK), and one of the fluorescent dyes coumarin 6 (C6) for green, coumarin 47 (C47) for blue, and nile red were dissolved into chloroform solution, which was then deposited by jetting, or by spin coating for comparison. No electron transport material was used in this initial work. After deposition the chloroform evaporates leaving a doped polymer material. Typical concentrations of PVK dissolved in chloroform were 10 g/l, and dye dissolved in chloroform were 0.1 g/l, yielding on the order of 1% dye in the PVK. Chloroform solutions containing varying

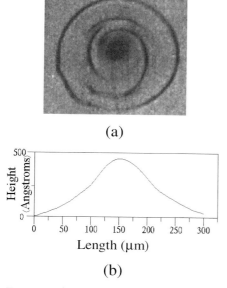

Fig. 5.9. (a) Photograph of a jet-printed dot; (b) surface profile of a jet-printed dot 300 μm wide (from [5.19])

amounts of PVK and luminescent materials were prepared by stirring and were passed through 0.45 μm filters. Thin films were jetted onto 175 μm thick flexible polyester coated with indium tin oxide (ITO). Before deposition, the ITO was treated with an oxygen plasma to modify its surface properties for making ohmic contact to the OLEDs.

The optical micrograph of a jet-printed dot and a surface profile of the shape of a typical dot are shown in Fig. 5.9 [5.19]. The thickness of the dots ranged from 40 to 70 nm and the dot widths ranged from 150 to 300 μm. The larger-diameter dots tended to be thicker, indicating that the thickness variation is due to the total amount of deposited solution. One can see some structure within the dot, which may be evidence of mass transport due to segregation as the solvent evaporated. To demonstrate the ability of the jet-printing technique to deposit patterns, PVK doped with C6 or nile red was deposited in a jet-printed test pattern. This sample was then illuminated using ultraviolet light, which excited green (or red) emission from the patterned polymer.

Figure 5.10 shows the photoluminescence spectra of three individual jet-printed thin films and of spin-coated films made from the same solution, each with a different dye [5.19]. The photoluminescence was measured with an excitation wavelength of 380 nm for the C47 doped film and 440 and 520 nm for C6 and nile red doped films, respectively. No significant difference in shape or magnitude is seen between the films prepared by jet-printing vs.

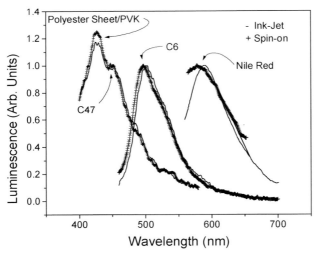

Fig. 5.10. Photoluminescence spectra of PVK doped with coumarin 47, coumarin 6, or nile red. The spectra of jet-printed and spin-coated films are shown for each band. The luminescence spectra are identical for the two techniques of application (from [5.19])

spin coating. It should be noted that the peak of 420 nm in the C47 spectra is due to a combination of the polyester substrate and the PVK host, not the deposited film. The dye luminescence appears as a peak at 450 nm on the shoulder of the 420 nm peak.

It is difficult to fabricate devices directly on top of the polymer dots made by jet-printing because of the difficulty in aligning a shadow mask for metal cathode formation directly over a polymer dot. Therefore, to fabricate test devices the ink-jet printer was operated in a mode to create a continuous film of polymer rather than discrete dots. After jet-printing, the samples were loaded into a vacuum chamber with a base pressure of $< 10^{-7}$ torr for the metallization step of device fabrication. Typically, at least ≈ 90 min were allowed between loading of samples and metal evaporation, and no further heating of samples was done. Top metal cathodes were deposited through a shadow mask to form an array of 250 μm diameter devices on the polymer film. Metal alloys such as Mg:Ag (10:1) were deposited by co-evaporation from two separate sources, followed by the deposition of Ag as a protective layer. The ITO on the polyester sheet served as the anode. The devices were then measured in air without any protective coating.

Figure 5.11 shows the I–V curves of typical devices made by this procedure and a control device fabricated on a film spin-coated from the same polymer solution [5.19]. The organic film thickness of the control device was about 50 nm. Current densities of ≈ 2 mA cm^{-2} are achieved at a voltage of ≈ 7 V for the control device, and ≈ 7–11 V for the jet printed devices. The

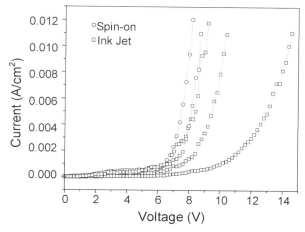

Fig. 5.11. Current-voltage characteristics for one spin-coated and four jetted OLEDs. The turn-on voltage of diodes made by jetting can come close to that of the spin-coated diode (from [5.19])

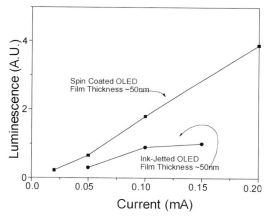

Fig. 5.12. Luminescence versus current characteristics of a spin-coated and a jetted OLED. The efficiency of the jetted OLED is lower than that of the more-developed spin-coated OLED (from [5.19])

luminescence versus current characteristics of the two OLEDs show that the efficiency of this early jet-printed device is about a factor of two lower than that of the spin coated device (Fig. 5.12 [5.19]).

A very promising modification of the approach described above is the local modification of a large-area spin-coated layer of host material by the jetting of solutions of the fluorescent-dye light-emitting dopants [5.20]. The jetted dye diffuses into the host (PVK) and forms the OLED material. This approach allows easy tuning of the UV fluorescence spectra of the film and

Fig. 5.13. Photograph of a page-size glass foil of 80 μm thickness printed with a transistor pattern of toner (from [5.6])

the spectra of the OLEDs made from the film, while retaining the planarity and film thickness advantages of spin-coating.

5.5 Substrates and Encapsulation for Macroelectronic Circuits

Substrates and encapsulation function as the mechanical support and protection of the thin film circuits. For use in AM-LCDs both must be transparent, while in an OLED display only the encapsulation may need to be transparent. The choice of substrate materials broadens considerably when the process temperature is reduced to below 200°C, and then ranges from the conventional silicate glasses over metals to organic polymers. A concurrent trend to thinner substrates and encapsulation seeks to reduce weight, and for very thin substrates, to increase flexibility and ruggedness. Tables 5.2 and 5.3 list properties of twelve materials that may be used as substrates or encapsulants. Note that the melting, strain or glass temperature, the thermal conductivity, and the coefficient of thermal expansion do vary widely between the materials of the table.

Glass substrates for active TFT circuits are best known and characterized [5.21]. One of the main advantages of borosilicate glasses is that their coefficients of thermal expansion are similar to those of the silicon TFT materials. Glass foil substrates that can be fed through printers with curved paper paths are available with thickness down to 50 μm from DESAG, a Schott subsidiary. The DESAG glass foil is made of borosilicate. This glass needs no

Table 5.2. Optical and electrical properties of materials for substrates and encapsulation[a]

Material	Brand name	Density (g cm^{-3})	Appearance and range of optical transparency (μm)	Refract. index at 0.55–0.6 μm	Dielectric constant at 0.1–1 MHz
Cryst. silicon		2.33	Opaque	3.4	11.8
Soda lime glass	Corning 0211	2.53	Clear 0.4–2.2	1.52	6.7
Borosilicate glass	Corning 7059 Schott AF 45	2.76	Clear 0.35–>2.5	1.53	5.9
Borosilicate glass	Corning 1737	2.54			
Invar		8.0	Opaque		
Stainless steel	AISI 304	7.93	Opaque		
Aluminum		2.70	Opaque		
Perfluorocyclobutane aromatic ether polymer	DowXU-35033.00		Milky white <0.4–>0.6	1.50	2.45
Polyimide	DuPont Kapton E	1.46	Amber		3.4 50% RH
Polyethersulfone		1.37	Pale amber	1.65	3.7
Polyarylate	Kaneka F-1100	1.20	Clear	1.60	2.7
Polycarbonate	Lexan	1.2	Clear	1.58	2.9

[a] For references and comments see footnotes to Table 5.3

further passivation prior to a standard a-Si:H TFT process. Foils of soda lime glass with thickness down to 50 μm are available from Corning, or down to 30 μm from DESAG. Soda lime glass must be passivated with a deposited SiO_2 or SiN_x barrier to prevent Na diffusion. Fusing or HF-etching the edges reduces breakage, because cracks propagate easily from the saw-damaged edges. When supported uniformly, glass foil is surprisingly resistant against impact by blunt objects, but it does shatter upon impact of sharp objects. The Corning glass foils tend to have a wavy surface that gives it anisotropic flexibility but does not affect the a-Si:H fabrication and laser printing. The flexibility of an 80 μm thick page-size glass foil is demonstrated in Fig. 5.13, which shows a laser-printed toner pattern [5.6]. A transistor made on 50 μm

Table 5.3. Mechanical and thermal properties of materials for substrates and encapsulation

Material (same brands as in Table 5.2)	Young's modulus Y (GPa)	T_{melt}, T_{strain}, or T_{glass} (°C)	Coefficient of thermal expansion at 20°C (10^{-6} K^{-1})	Thermal conductivity (Wm^{-1}K^{-1})	(Primary lit. references) Comments Chemical composition
Cryst. Silicon	190	1415 T_m	2.4	165	[a,b] For comparison.
Soda lime glass	74	508 T_s	7.4 at 300°C	0.12	[c] KNaZn borosilicate T_{strain} at 10^{14} Poise
Borosilicate glass	68	593 T_s	4.7	≈ 1	[d,e] BaAl borosilicate T_{strain} at $10^{14.6}$ P Shrinks 10^{-4} at 450°C 8 h
Borosilicate glass		670 T_s	3.7	≈ 1	[f] Shrinks 3×10^{-4} at 450°C 8 h
Invar	140	1430 T_m	1.2	13	[g] 64 Fe, 36 Ni (wt.%)
Stainless steel	190	≈ 1400 T_m	18	16	[g] 18 Cr, 10 Ni (wt.%)
Aluminum	70	660 T_m	24	237	[g] Y for hard Al
Perfluorocyclo-butane aromatic ether polymer	2.3	400 T_g	90		[h] Experimental grade. Contains 0.021wt.% water. Wt. loss 0.4%/h air 350°C
Polyimide	5.2	≥ 250 T_g (est.)	12 50–200°C	≥ 0.1	[i,g] CHE 9.10^{-6}/%RH. Shrinks 0.3×10^{-3} at 200°C
Polyether-sulfone	2.4	200 T_g	55	0.15	[g]
Polyarylate	≥ 2 (est.)	215 T_g	51		[j] Shrinks 2×10^{-3} at 180°C 3 h
Polycarbonate	2.3	160 T_g	70	0.2	[g,k] Degasses > 160°C

[a] Bell Labs Quick Reference Manual (May 1975).
[b] AIP Handbook, McGraw Hill, New York (1972).
[c] Corning Product Information PI-0211-87.
[d] N.P. Bansal, R.H. Doremus, Handbook of Glass Properties. Academic (1986).
[e] Schott/DESAG spec. sheet AF 45 and D 263.
[f] D.M. Moffatt, Mat. Res. Soc. Symp. Proc. **377** (1995).
[g] Goodfellow Catalog (1996/97).
[h] D. Perettie, L. Bratton, J. Bremmer, D. Babb, Liq. Cryst. Mater., Devices and Applications II, SPIE **1911** 15 (1993).
[i] J.A. Kreuz, S.N. Milligan, R.F. Sutton: DuPont Films Technical Paper 3/94, Reorder No. H-54504 Properties, Academic (1986).
[j] Kaneka spec. sheet.
[k] S.M. Gates: Mat. Res. Symp. Proc. **467**, 843 (1997)

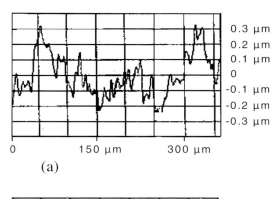

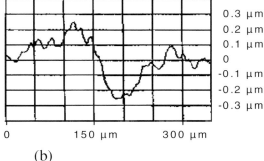

Fig. 5.14a,b. Surface profiles of a steel foil (**a**) as received and (**b**) after planarization with spin-on-glass. SOG removes the short-wavelength roughness that reduces transistor yield. The full vertical scales measures 0.8 µm, and the traces are 400 µm wide (from [5.24])

thick Schott AF45 glass foil and its transfer characteristics are shown in Fig. 5.4 [5.5].

Stainless steel foil with thickness between nominally 125 and 200 µm (5 and 8 mils) has been used for many years as the substrate for amorphous silicon solar cells made in a roll-to-roll process [5.3]. It is also possible to make TFTs on steel foil [5.22], even with less than 10 µm thickness [5.23]. At low thickness the handling of the foil – if processed in free-standing form – dominates TFT fabrication. Steel foil must be coated with an insulator prior to TFT fabrication. A convenient insulator is the silicon nitride that is used as the a-Si:H TFT gate dielectric. Steel foils are made with a wide variation of surface finish, which depends on the state of the rolling equipment. Some surface finish will give high transistor yield with just the SiN_x barrier layer alone. Other surfaces have short-wavelength roughness that needs planarization, which can be done by using the sol-technique of spin-on glass (SOG) [5.24]. The most viscous spin-on-glass precursor results in $\approx$ 500 nm thick oxide layers. A surface profile of an as-received 75 µm thick AISI 304 stain-

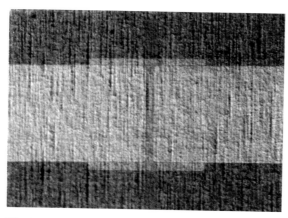

Fig. 5.15. Photograph of a working TFT made on non-planarized 15 μm thick steel foil. The surface finish of the as-received foil was adequate for high TFT yield. Source and drain contacts lie to the left and right of the channel at the center of the TFT island. The channel is 48 μm long (from [5.23])

less steel foil is shown in Fig. 5.14a [5.24]. The TFT yield on this substrate was ≈ 50%. Figure 5.14b illustrates the effect of SOG planarization, which suppresses the short-wavelength asperities [5.24]. The TFT yield on this substrate approaches 100%. Figure 5.15, which is a photograph of a functioning TFT on a 15 μm stainless steel foil, illustrates the ruggedness of a-Si:H TFT technology [5.23]. The gap L between the source/drain electrodes on this photograph is 48 μm. Despite its rough appearance, the surface of this foil was sufficiently smooth on a microscale that it needed no SOG planarizing layer.

Foil substrates lend themselves to roll-to-roll fabrication and curved process paths. They also provide flexibility and ruggedness during use. Figure 5.16 illustrates the flexibility of TFTs fabricated on a 25 μm thick stainless steel foil [5.25]. The transistors were tested after successive bending to decreasing radii of curvature. They failed by delamination of the spin-on glass from the steel under concave (facing in) bending to 2.5 mm radius, and under convex (facing out) bending to 1.5 mm radius.

This result can be understood using a first-order analysis of the strain of a continuous thin film of TFT material deposited on a steel foil. Film and substrate have nearly identical Young's moduli of ≈ 200 GPa (Table 5.3). The bending is dominated by the substrate, which is much thicker than the ≈ 1 μm thick transistor film. Bending the film/foil couple, if free of strain when flat, to a cylinder as shown in Fig. 5.17 induces a strain ε in the surface [5.18].

For a film thickness d_{film} much smaller than the substrate thickness $d_{\text{substrate}}$, the strain-free neutral plane lies in the center of the foil,

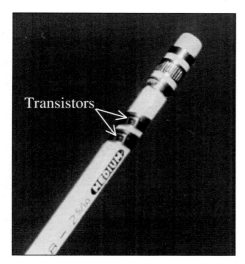

Fig. 5.16. The TFT characteristics are not affected by wrapping the TFT-on-foil around a pencil. The steel foil substrate is 25 μm thick. Note the TFT islands (from [5.25])

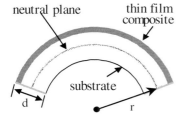

Fig. 5.17. The principal parameters of a film/substrate couple bent into cylindrical shape. The neutral plane is free of strain (from [5.18])

$\approx d_{\text{substrate}}/2$ away from the film [5.18]. The strain ε in the film induced at the bending radius r is given by:

$$\varepsilon \cong d_{\text{substrate}}/2r \,. \tag{5.2}$$

Assuming that the transistors fail when a certain level of strain $\varepsilon_{\max}$ is reached, it can be seen that the minimum allowable radius of bending $r_{\min}$ scales linearly with foil thickness $d_{\text{substrate}}$

$$r_{\min} = d_{\text{substrate}}/2\varepsilon_{\max} \,. \tag{5.3}$$

When the sample is held flat during film growth, it usually is under strain after fabrication, before intentional bending. Transistors grown on steel (Figs. 5.7, 5.15, 5.16) are under compressive strain as a result of differential thermal contraction between the steel ($\alpha_{\text{substrate}} = 18 \times 10^{-6} \,\text{K}^{-1}$) and the silicon layers grown at $\approx 300°C$ (α_{film} of a-Si:H $\cong 4\times 10^{-6}$ K^{-1} [5.26]). This contraction produces a mismatch strain ε_M, which is given by,

$$\varepsilon_M = (\alpha_{film} - \alpha_{substrate})\Delta T, \tag{5.4}$$

where ΔT is the difference between the temperature of growth and room temperature. The stress in the film σ_{film} produced by the mismatch strain ε_M is given by,

$$\sigma_{film} = \varepsilon_M Y^*_{film}. \tag{5.5}$$

Here $Y^*_{film} = Y_{film}/(1 - \nu_{film})$ is the biaxial elastic modulus of the film, with Y_{film} being Young's modulus of the film and ν_{film} its Poisson ratio. This stress causes the substrate to bend to a radius of curvature R, which is given by the Stoney formula [5.27]

$$R = Y^*_{substrate} d^2_{substrate}/6\sigma_{film} d_{film}. \tag{5.6}$$

When the in-plane stiffness of film and substrate becomes comparable, $Y_{film}d_{film} \approx Y_{substrate}d_{substrate}$, the substrate may deform considerably, which in turn reduces the stress in the film. Thus a substrate becomes compliant in two circumstances. In one, a high-modulus substrate material is made very thin (e.g., steel foils a few μm thick). In the other a low-modulus substrate material is chosen (typical plastic substrates have Young's moduli of ≈ 2 GPa, which is 1% of those of the TFT materials). When the film/substrate couple is held flat in a frame, the stress in the film is given by,

$$\sigma_{film} = \varepsilon_M Y^*_{film}/(1 + Y^*_{film}d_{film}/Y^*_{substrate}d_{substrate}). \tag{5.7}$$

It can be seen that for $Y_{film}d_{film} = Y_{substrate}d_{substrate}$ the stress produced in a device film deposited on a compliant substrate is reduced by a factor of 2 below that on a stiff substrate.

When a film is deposited on a compliant substrate that is held flat in a frame, and the structure then is released from the frame, the structure bends. A mismatch strain ε_M, which may include thermal and intrinsic components, produces a radius of curvature R that is given by [5.31],

$$R = \frac{(\overline{Y_{substrate}}d^2_{substrate} - \overline{Y_{film}}d^2_{film})^2}{6\varepsilon_M(1+\nu)\overline{Y_{film}Y_{substrate}}d_{film}d_{substrate}(d_{film} + d_{substrate})} + 4(d_{film} + d_{substrate})/6\varepsilon_M(1+\nu). \tag{5.8}$$

Here we assume that $\nu = \nu_{film} = \nu_{substrate}$. The normalized radius of curvature as a function of the film/substrate thickness ratio $d_{film}/d_{substrate}$ is plotted in Fig. 5.18 for two different ratios of Young's moduli, $Y_{film}/Y_{substrate}$ [5.31]. A ratio of $Y_{film}/Y_{substrate} = 1$ corresponds to steel or glass substrates, and a ratio of 100 to plastic substrates. Note that for very small and very large $d_{film}/d_{substrate}$ ratios the substrate or the film dominate, and the radius of curvature R assumed by the structure after processing is large, reflecting

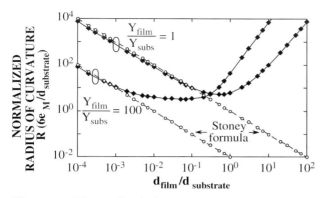

Fig. 5.18. When a film/substrate couple is released after deposition, it will assume a radius of curvature R that depends on the mismatch strain ε_M and on the elastic moduli of film and substrate. The Stoney equation is seen to be valid for $d_{\text{substrate}} \gg d_{\text{film}}$. Results of the complete theory are shown by diamonds (from [5.31])

a flat sample. The radius of curvature becomes smallest for thickness ratios near unity. Figure 5.18 also shows the results of the Stoney equation, which is seen to be a good approximation only if $d_{\text{substrate}} \gg d_{\text{film}}$.

We have seen that the stress induced in a film deposited on a compliant substrate can be reduced by a factor of 2 below that on a stiff substrate. Returning to our initial discussion of the bending of TFTs on steel foil, we now proceed to analyze the forced bending of a compliant substrate after a device film has been deposited on it. We shall see that this bending may induce much less strain in the film than on a stiff substrate. This consequence becomes qualitatively obvious from a re-inspection of Fig. 5.17. When the sheet is bent, the outside surface is in tension, and the inside surface is in compression. If the film/substrate foil has uniform elastic constants, the neutral plane coincides with the mid-plane of the sheet, as in Fig. 5.17. But if the structure consists of a stiff layer on top on a compliant substrate, the neutral plane is shifted from mid-plane toward the stiff layer. Consequently, for a given curvature R, the strain in the outside (film) surface is reduced, and is given by,

$$\varepsilon_{\text{top}} = \left(\frac{d_{\text{f}} + d_{\text{s}}}{2R} \right) \frac{(1 + 2\eta + \gamma\eta^2)}{(1 + \eta)(1 + \gamma\eta)} . \tag{5.9}$$

Here R is the radius of curvature, $\eta = d_{\text{f}}/d_{\text{s}}$ and $\gamma = Y_{\text{f}}/Y_{\text{s}}$. This situation is illustrated by Fig. 5.19, where the normalized strain in the film is plotted vs. film/substrate thickness ratio for the two ratios of Young's moduli of 1 (stiff substrate) and 100 (compliant substrate). Note that a compliant substrate can reduce the normalized strain by as much as a factor of five. In this way thin-film circuits on compliant substrates become particularly insensitive to bending. When favorable combinations of Y and d are used, the structure

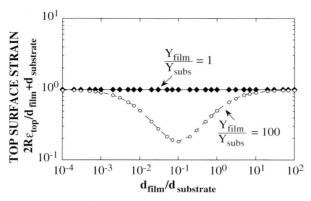

Fig. 5.19. In-plane strain ε_{top} induced in the top surface by bending an initially flat and strain-free film/substrate couple. The strain ε_{top} is reciprocal to the radius of curvature R. For a given value of R, choosing a compliant substrate and the appropriate film/substrate thickness ratio can reduce the strain by a factor of up to five

may allow extremely small radii of curvature. It even might be folded like a map.

It is noted above that TFTs made on steel foil peel off at a radius of curvature that is smaller for convex than for concave bending. This behavior is a consequence of the way the thermal strain ε_M and the strain induced by bending ε_{top} add [5.18]. For our sample this strain ε_M is $\approx 4.5 \times 10^{-3}$. (We assume that the sample is free of any intrinsic strain that often is established during growth.) Bending to a 1.5 mm radius adds a strain of $\pm 8.5 \times 10^{-3}$. The sum of bending and thermal strain in the convex (facing out) surface is $+4.5 \times 10^{-3}$, and on the concave (facing in) surface -13×10^{-3}. (We define tensile strain as positive, and compressive strain as negative). It is easy to see why the film peels off the concave surface at a larger radius than from the convex surface. Note that the peeling occurs at the large strain of close to 0.5%.

5.6 Plastic Substrate Foil: TFT on Polyimide

Thin film transistors made on thin foils of plastic serve as a good illustration of the compliant-substrate situation discussed in the preceding section. The glass transition temperatures of most organic polymers lie below 200°C, much lower than those of inorganic glass substrates (Tables 5.2 and 5.3). Therefore, the standard a-Si:H TFT processes cannot be used on plastics because they require temperatures of up to 350°C. We have fabricated a-Si:H TFTs on 51 μm (2 mil) thick polyimide foil after re-optimizing the TFT process for a maximum temperature of 150°C [5.28]. The commercially available polyimide grade Kapton® E was selected because it has a low coefficient of thermal

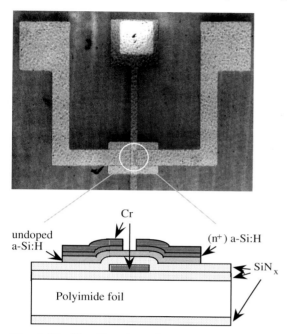

Fig. 5.20. *Top*: Top view of a TFT on a polyimide substrate. The channel is 15 μm long. *Bottom*: Cross sectional view (from [5.28])

expansion of $12 \times 10^{-6}/°C$, and its coefficient of humidity expansion is $9 \times 10^{-6}/\%$ Relative Humidity.

The transfer of such low-temperature technology from glass to plastic substrates brings new processing and mechanical issues to a-Si:H TFT fabrication. A widely usable low-temperature TFT needs to be compatible with a number of other polymers whose T_g is lower than that of Kapton. When a silicon thin film is grown on plastic foil, the contribution of the semiconductor to the mechanical behavior of the film/substrate couple becomes comparable to that of the plastic foil. This is very noticeable during processing, when the structure is observed to bend inward or outward with varying radii of curvature after each process step. We are entering a regime in which the film controls the mechanical properties of the structure as much as the substrate.

Before proceeding with a-Si:H TFT fabrication, the deposition of all layers was re-optimized for 150°C to obtain TFT electrical properties comparable to those obtained at higher temperature. The gate SiN_x was deposited from a mixture of $Si:H_4$, NH_3 and H_2, the undoped a-Si:H from a mixture of $Si:H_4$ and H_2, and the (n^+) a-Si:H from a mixture of $Si:H_4$, PH_3 and H_2. The TFTs have a bottom gate, back-channel etch structure. We began by passivating the polyimide substrate on both sides with a 0.5 μm thick layer of SiN_x. These layers serve as a barrier against the solvents, bases and acids used during photolitography. An ≈ 100 nm thick Cr layer was thermally evaporated and wet etched to create the gate electrode. Then we deposited a sequence of: ≈ 400 nm of SiN_x, ≈ 200 nm of undoped a-Si:H, and ≈ 50 nm of (n^+) a-Si:H. An ≈ 100 nm thick Cr layer was thermally evaporated. The Cr source-drain pattern was wet etched and the (n^+) a-Si:H was dry etched in CF_4 gas. Then the undoped a-Si:H was dry etched to define the transistor island. In the last photolithographic step, dry etched windows into the SiN_x were opened to access the gate contact pad. Arrays of six TFTs were fabricated on 1.5×1.5 in.^{2} substrates. Four substrates were processed simultaneously. Figure 5.20 shows the cross section of one TFT [5.28].

The dependence of the source-drain current I_{ds} on the gate voltage V_{gs} for $V_{ds} = 0.1$ V and 10 V is shown in Fig. 5.21 [5.28]. The off-current is $\approx 1 \times 10^{-12}$ A ($< 1 \times 10^{-14}$ A/µm gate width at $V_{DS} = 10$ V) and the on-off current ratio is 10^7. At $V_{ds} = 0.1$ V we obtain $V_T \approx 3.5$ V. The dielectric constant of our 150°C nitride measured at 1 MHz is 7.46. The electron mobility in the linear regime is ≈ 0.5 cm^2 V^{-1} s^{-1}. These characteristics are comparable to those of a-Si:H TFTs fabricated on glass substrates at temperatures between 250 and 350°C. They suggest that a low-temperature a-Si:H technology can be developed that is fit for a large variety of plastic substrates.

5.7 3-D Integration on a Foil Substrate: OLED/TFT Pixel Elements on Steel

An active-matrix thin-film emissive display requires the integration of a light-emitting device with a switch. Organic light-emitting diodes can be integrated with amorphous silicon thin film transistors on steel foil, in a circuit that demonstrates the monolithic 3-D integration of very different thin-film devices on a foil substrate [5.24, 25, 29, 30]. While the devices were fabricated by conventional, non-printing techniques, the nature of the thin-film structure and the foil substrate suggests that large-area circuits based on such devices can be made by printing. A schematic cross-section of the integrated TFT/OLED structure on the steel foil is shown in Fig. 5.22a, and its equivalent circuit is shown in Fig. 5.22b [5.29]. The as-rolled stainless steel foil is planarized with 0.5 µm thick spin-on glass to remove the short-wavelength roughness of 0.3 µm rms. This planarization functions as primary insulation. Further insulation is provided by a 0.5 µm thick plasma-enhanced CVD SiN_x layer. The TFTs are made in the inverted-staggered, back-channel etch configuration with 120 nm thick Cr gates, 400 nm PECVD gate SiN_x dielectric, 150 nm a-Si:H channel layer, and 50 nm (n^+) a-Si contacts, followed by 120 nm Cr source/drain contacts. The channel length and width are 42 µm and 776 µm, respectively. The TFTs and contact pads are made large, to ease probing and diagnosis on bent or rolled substrates.

Following the fabrication of TFTs, OLEDs were made on the surface of the 2×2 mm^2 Cr source/drain contact pads. Conventional OLEDs are built

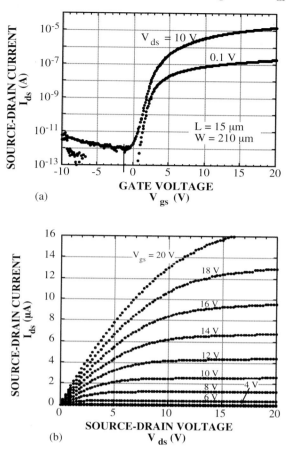

Fig. 5.21. (a) Transfer characteristics and (b) current-voltage characteristics of a TFT made at 150°C on a polyimide substrate (from [5.28])

on transparent substrates coated with a transparent hole-injecting anode contact such as indium tin oxide (ITO), so that light can be emitted through the substrate, because the top contact is an opaque electron-injecting metal cathode. Because of the opacity of the steel substrate, we developed the top-emitting structure in which the high work function metal Pt functions as the reflective bottom anode and a semi-transparent cathode is applied on top. OLEDs were fabricated by sequential electron-beam deposition and patterning of 40 nm Pt anode contacts, spin-coating of a continuous layer of 170 nm active luminescent molecularly doped polymer (MDP), followed by the electron beam evaporation of a 14 nm semi-transparent Ag top cathode. The overlap of the anode and cathode contact areas determines the active OLED device area, a 250 μm diameter dot, without the need to separately

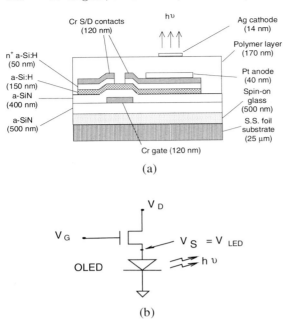

Fig. 5.22. (a) Cross section of the integrated OLED/TFT structure made on steel foil. (b) Equivalent circuit of the OLED/TFT (from [5.29])

isolate the organic layers. All OLED fabrication steps were performed at room temperature and are compatible with finished TFTs.

The active organic material used is a single-layer thin film. The hole-transport matrix polymer poly(N-vinylcarbazole) (PVK) contains dispersed electron-transport molecules of 2(-4-biphenyl)-5-(4-tert-butyl-phenyl)-1,3,4-oxadiazole and a small amount of the green fluorescent dye, coumarin 6 (C6), which provides efficient emission centers. The OLED luminescence depends linearly on device current as already shown for the spin-coated OLED of Fig. 5.12 [5.19]. The factor limiting the luminance in this device is not the polymer material itself, but rather the poor transparency of the top contact and its work function mismatch with the organic material. Here, we obtain $50\,\mathrm{cd/m^2}$ at $40\,\mathrm{mA/cm^2}$ ($20\,\mathrm{\mu A/device}$) with $V_\mathrm{G} = V_\mathrm{D} = 40\,\mathrm{V}$. Figure 5.23a shows the I–V characteristics of the integrated TFT/OLED device on a linear current scale [5.25]. Figures 5.23b,c illustrate how the voltage is applied to obtain the OLED and TFT/OLED curves, respectively. The shift in turn-on voltage results from the additional 3.5 V required to turn on the TFT. This functional OLED/TFT pixel circuit illustrates the ease of integrating very different materials into a 3-D thin-film configuration that could be fabricated by printing.

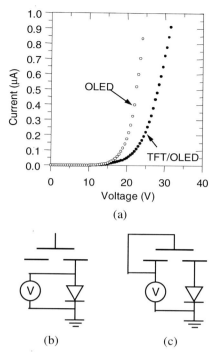

Fig. 5.23. (a) Current-voltage characteristics of the OLED and of the OLED/TFT; (b) and (c) show the circuits used for these measurements (from [5.25])

5.8 Outlook

We saw in this chapter that laser and jet printing techniques can be used to pattern thin-film transistors and organic light-emitting diodes. The resolution and registration achievable by conventional printing techniques will provide a packing density of more than 10 000 transistors per square centimeter. These transistors can be made on glass, steel, and plastic foils. The OLED/TFT structure suggests that they can be integrated into three-dimensional thin-film circuits. Direct printing will enable the manufacture of such circuits with very large-area. We also saw that these circuits are very rugged by virtue of the mechanics of thin structures.

The direct printing of macroelectronic circuits is receiving growing and worldwide attention. This newly found focus will facilitate the marriage of electronics and printing that is needed to develop the manufacturing technology for macroelectronics. An important ingredient is the development of printable materials for active circuits, which will draw on chemistry, materials science, and surface physics.

Early applications can be expected to make us of existing printing technology to make large-area circuits for displays, sensor arrays, and large-area micro-electromechanical circuits. The pattern of increased resolution and enhanced performance that is well known from the development of silicon integrated circuits will be repeated in large-area electronics as more and more macroelectronic products are introduced.

Acknowledgement. We thank the DARPA HDS program for supporting our research on macroelectronics.

References

5.1 E. Kaneko, Displays 14 (1993) 125–130. E. Kaneko, in Thin Film Transistor Technologies III, Electrochemical Society PV96-23, p. 8 (1997).

5.2 D.M. de Leeuw, P.W.M. Blom, C.M. Hart, C.M.J. Mutsaers, C.J. Drury, M. Matters and H. Termeer, Tech. Digest IEDM 1997, p 331. IEEE, New York 1997. C.J. Drury, C.M.J. Mutsaers, C.M. Hart, M. Matters, D.M. de Leeuw, Appl. Phys. Lett. **73**, 108 (1998).

5.3 J. Yang, A. Banerjee, T. Glatfelter, S. Sugiyama, and S. Guha, Conf. Record 26th IEEE PVSC, IEEE, New York (1997), p 563.

5.4 D.E. Carlson, R.R. Arya, M. Bennett, L.-F. Chen, K. Jansen, Y.-M. Li, J. Newton, K. Rajan, R. Romero, D. Talenti, E. Tweseme, F. Willing and L. Yang, Conf. Record 25th IEEE PVSC, IEEE, New York (1996), p 1023.

5.5 H. Gleskova, S. Wagner, and D.S. Shen, J. Non-Cryst. Solids **227–230**, 1217 (1998).

5.6 H. Gleskova, S. Wagner, and D.S. Shen, IEEE Electron Devices Letters **16**, 418 (1995).

5.7 Pocket Pal, 16th ed., International Paper Company, Memphis, Tennessee (1995).

5.8 The values for λ and δ are a consensus on the capability of high-quality printing equipment. Agreement on the feasibility of $\lambda = 10\,\mu m$ is better than on $\delta = \pm 5\,\mu m$. The value for δ is considered less certain, because the two alternatives for obtaining registration, mechanical or by optical alignment, are so different. The consensus values were reached in discussions with experts in three laboratories of the printing industry: Mr. Shinichi Hikosaka and colleagues of the Central Research Institute of Dainippon Printing Co., Ltd., Dr. Kaneki Yoshida and colleagues of the Technical Research Institute of Toppan Printing Co., Ltd., and of Mr. Russell Fling and colleagues of the Technical Center of R.R. Donnelley Printing Co.

5.9 We thank the researchers of Dai-Nippon and Toppan for discussions of alignment techniques.

5.10 A point made by Mr. Russell Fling of R.R. Donnelley.

5.11 S.Y. Chou, P.R. Krauss, and P.J. Renstrom, Science **272**, 85 (1996).

5.12 H. Gleskova, R. Könenkamp, S. Wagner, and D.S. Shen, IEEE Electron Devices Lett. **17**, 264 (1996).

5.13 H. Gleskova, S. Wagner, and D.S. Shen, MRS Symp. Proc. **467**, 869 (1997).

5.14 B. Green, "A New Way to Make PC Boards," Electronics Now (November 1997), p. 52.

5.15 For introductions to xerography and laser printing, see L.B. Schein, "Electrophotography and Development Physics'", Springer, New York (1992), and R.M. Schaffert, "Electrophotography," Halstead Press, New York (1975).

5.16 H. Gleskova and S. Wagner, unpublished results.

5.17 H. Gleskova, S. Wagner, and D.S. Shen, Proc. AMLCDs '95, Lehigh University, 25–26 (Sep. 1995), p 16.

5.18 E.Y. Ma and S. Wagner, MRS Symp. Proc. **508**, 18 (1998).

5.19 T. R. Hebner, C. C. Wu, D. Marcy, M. H. Lu, and J. C. Sturm, Applied Physics Letters **72**, 519 (1998).

5.20 T.R. Hebner and J.C. Sturm, Applied Physics Letters **73**, 1775 (1998).

5.21 D.M. Moffat, MRS Symp. Proc. **377**, 871 (1995).

5.22 S.D. Theiss and S. Wagner, MRS Symp. Proc. **424**, 65 (1996).

5.23 E.Y. Ma, Ph.D. thesis, Princeton University (1998).

5.24 S.D. Theiss, C.C. Wu, M. Lu, J.C. Sturm and S. Wagner, MRS Symp. Proc. **471**, 26 (1997).

5.25 E.Y. Ma, S.D. Theiss, M.H. Lu, C.C. Wu, J.C. Sturm and S. Wagner, IEEE (1997) Internat. Electron Devices Meeting Tech Digest p 535.

5.26 T. Dragone, S. Wagner and T.D. Moustakas, Tech Digest PVSEC-1, Kobe, Japan (Nov 13–16, 1984); p 711.

5.27 S.P. Timoshenko and J.N. Goodier, Theory of Elasticity, McGraw-Hill, New York (1970).

5.28 H. Gleskova, S. Wagner, and Z. Suo, MRS Symp Proc. **508**, 73 (1998).

5.29 C.C. Wu, S.D. Theiss, G. Gu, M.H. Lu, J.C. Sturm, S. Wagner and S.R. Forrest, Society for Information Display, Intern. Symp. Digest, Vol. XXVIII, SID, Santa Ana, CA (1997), 67.

5.30 C.C. Wu, S.D. Theiss, G. Gu, M.H. Lu, J.C. Sturm, S. Wagner and S.R. Forrest, IEEE Electron Devices Lett. **18**, 609 (1997).

5.31 Z. Suo, E.Y. Ma, H. Gleskova and S. Wagner, Appl. Phys. Lett. **74**, 1177 (1999).

6 Multijunction Solar Cells and Modules

Subhendu Guha

United Solar Systems Corp., 1100 West Maple Road, Troy, MI 48084, USA
E-mail: sguha@ic.net

Abstract. Significant progress has been made in improving the stable efficiency of amorphous silicon alloy solar cells and modules. This has been achieved through a better understanding of the effect of plasma chemistry and growth kinetics on material characteristics and incorporation of the material in optimum device configurations. The R&D results have now been translated to production, and amorphous silicon alloy photovoltaic products are now being used for a variety of applications ranging from consumer products to grid-connected large power systems. In this chapter, we discuss the science and technology of amorphous silicon alloy solar cells and modules.

6.1 Introduction

The conversion of sunlight into electricity using single crystal silicon was first demonstrated at the Bell Laboratories in 1954 [6.1]. This was essentially of academic curiosity until the space program started in the sixties when there was a need for recharging the batteries used for running the equipment on board the satellites. Solar cells that can convert sunlight into electricity were ideally suited for this purpose, and for more than thirty years solar cells are being used extensively for space applications. With the advent of the oil crisis during the seventies, attention was focused on using this technology for large-scale terrestrial applications. Low cost is an essential requirement for solar cells to compete with more conventional sources of electricity. While efforts were made to reduce cost by devising new ways of growing and processing single crystal silicon for the fabrication of solar cells, new materials that could lead to reduction of cost also received a great deal of attention.

It was in this decade that the possibility of using thin films of amorphous silicon (a-Si) as a viable candidate for the manufacture of solar cells was first explored. In the 1960s, Chittick and his coworkers at the Standard Telecommunication Laboratories in the UK had shown [6.2] that thin films of a-Si can be deposited by glow discharge decomposition of silane. Spear and his coworkers [6.3] showed that, when deposited under optimum conditions, these films have very low defect density. This led to the successful demonstration of substitutional doping of this material by the addition of small amounts of phosphine or diborane [6.4] and the fabrication of the first p-n junction [6.5]. The first solar cells using a-Si that were made [6.6] showed light-to-electricity conversion efficiencies in the range of 2–3%. Extensive research in the last two decades has resulted in the achievement of efficiencies close to 15% [6.7]. This, together with the development of a novel, low-cost manufacturing technology [6.8], has made a-Si solar cells a viable candidate for the generation of electricity for terrestrial applications [6.9].

The use of a-Si for solar cells attracted attention mainly for two reasons. First, since long-range order is absent in amorphous materials, selection rules for conservation of momentum during photon absorption is not relevant for a-Si and it absorbs light very efficiently. A thin film (less than 1 μm thick) is, therefore, adequate to absorb the sunlight as opposed to hundreds of micrometers needed for the crystalline counterpart. The material cost for a-Si cells is thus very low. Second, a-Si films are very easy to deposit. As mentioned earlier, the films are deposited from glow discharge decomposition of suitable gas mixtures. Typical deposition temperature is less than 300°C and the pressure is about 1 torr. The manufacturing method can, therefore, be simple.

Even though the low material cost and the ease of deposition make a-Si a very attractive candidate as a low-cost solar cell material, there are several challenges that needed to be overcome before large-scale use of a-Si could be feasible. A-Si is a very intriguing material. The absence of long-range order, which facilitates efficient photon absorption, unfortunately adversely affects the material quality since it causes tailing of the conduction and the valence band edges. The material also contains defects such as dangling, strained and weak bonds that act as recombination centers for the electrons and the holes. The efficient operation of a solar cell depends on a two-step process - photon absorption for the generation of the carriers and subsequent carrier transport and collection. The presence of the recombination centers arising from the defects and the bandtails hurts the transport and lowers the efficiency even though the photon absorption is efficient. A major challenge facing the researchers was to develop high quality material with low defect density. It was quickly realized that incorporation of hydrogen plays a key role in reducing the defect density. The best quality material is, therefore, a hydrogenated alloy of a-Si [6.10]. Understanding of both plasma chemistry and growth kinetics helped us in defining optimum growth conditions to obtain materials with low defect density. New device designs were developed using the multijunction approach in which several cells are connected in tandem, each cell incorporating an alloy of a different bandgap to facilitate light absorption of a broader spectrum. All these developments led to increased confidence in the potential of this material as a candidate for low-cost manufacture of solar cells resulting in the establishment of an annual production capacity of 25 MW by the end of 1997. a-Si alloy technology has thus taken a dominant role in the world photovoltaic (PV) market.

In this chapter, we shall discuss the advances made in the development of a-Si alloy photovoltaic technology both in terms of improving the cell efficiency and devising low- cost manufacturing processes. The discussions will center on solar cells and modules only; for detailed description of physics of the materials, the readers are referred to other treatises [6.11, 12]. In Sect. 6.2, we describe the deposition methods that are used to obtain a-Si alloy materials and solar cells. In Sect. 6.3, we discuss the simplest a-Si alloy solar cell structure and its theory of operation. The phenomenon of light-induced degradation in a-Si alloys and its effect on cell performance are also discussed. In Sect. 6.4, we outline the multijunction approach and the advantages of using alloys with different bandgaps to capture photons of a wider spectrum. The device design and the materials used to achieve the highest efficiency triple-junction cell are discussed. Manufacturing methods are outlined in Sect. 6.5, and some recent developments to obtain high efficiency cells by using microcrystalline materials are discussed in Sect. 6.6. This section also includes a discussion on future directions.

6.2 Deposition Methods

6.2.1 Glow-Discharge Deposition Technique

Although a variety of techniques has been used [6.13] to deposit a-Si alloy materials and solar cells, the most common method is still the glow-discharge deposition (also referred to as plasma-enhanced or plasma-assisted chemical vapor deposition). The glow discharge is initiated by applying a direct current (DC) or an alternating current power input to a pair of electrodes in a vacuum chamber to which silane is introduced. A typical reactor (Fig. 6.1) consists of a gas inlet arrangement, a deposition chamber that holds the pair of electrodes and a substrate heating assembly, a pumping system and a source of power for the discharge. The growth of the film on the substrate takes place through four different stages [6.14].

In the first stage, electrons collide with silane to dissociate the silane molecules into a mixture of reactive species of ions and free radicals. The second stage is the drifting or diffusion of these species to the surface of the substrate during which time there is a multiplicity of secondary reactions. In the third stage, the different species are adsorbed onto or react with the growing surface, and finally, these species or their reaction products are incorporated into the growing film or are reemitted from the surface into the gas phase.

6.2.2 Plasma Chemistry and the Growth Process

Detailed analysis of the deposition process has been provided by Gallagher [6.15] and Matsuda and Tanaka [6.16]. Electron–molecule collisions dissociate the molecules into ions and neutral radicals. These radicals undergo various secondary reactions during their transport to the substrate. Some of the

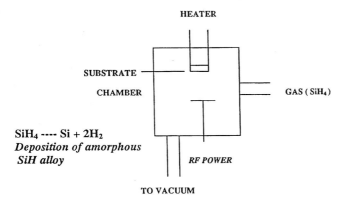

Fig. 6.1. Schematic diagram of a glow-discharge deposition reactor

Table 6.1. Some of the species generated by primary and secondary collisions

$e + SiH_4 \rightarrow SiH_4 + e$	$SiH_4 + H \rightarrow SiH_3 + H_2$
$SiH_2 + H_2 + e$	$SiH_4 + SiH_2 \rightarrow Si_2H_6$
$SiH_3 + H + e$	$SiH_3 + SiH_4 \rightarrow SiH_4 + SiH_3$
$SiH + H_2 + H + e$	$SiH_4 + Si_2H_6 \rightarrow Si_nH_m$
$SiH_2^+ + H_2 + e + e$	
$SiH_3^+ + H + e + e$	
$SiH_3^- + H$	
$SiH_2^- + H_2$	
Primary	**Secondary**

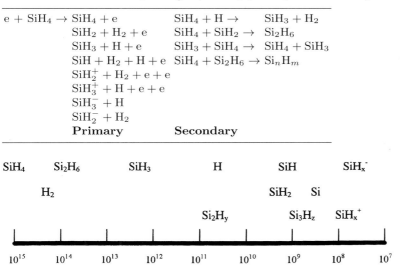

Fig. 6.2. Number density cm^{-3} of species in a typical rf glow-discharge plasma

species generated in the plasma as a result of primary and secondary reactions are shown in Table 6.1.

Amongst the various species generated by the primary impact process, SiH_3 has the longest lifetime since SiH_2 reacts with the parent SiH_4 molecule to form SiH_4 and SiH_3 again. On the other hand, successive collisions of SiH_4 with SiH_2 can create higher silane-related species that could continue to grow causing powder formation. The number density of the various species in the plasma that have been detected in a typical glow-discharge reactor is shown in Fig. 6.2 [6.17].

All the species shown in Fig. 6.2, together with some higher silane-related species, arrive at the substrate. Most of the growing surface is terminated with hydrogen and will not take up SiH_3 radicals, which is the most abundant species. Bonding of SiH_3 onto the surface needs dangling bonds, and removal of hydrogen from the surface is, therefore, a necessary step in the deposition of films from SiH_3. Hydrogen can be released from the surface by thermal excitation, or it can be stripped by SiH_3 reacting with SiH to form a dangling bond and silane. Another SiH_3 molecule migrating along the surface or arriving directly can then be incorporated in the film. The other radicals like SiH or SiH_2 can be incorporated directly on the hydrogen-terminated surface and thus have high sticking coefficient. Under normal deposition conditions, these radicals do not contribute much to the film growth since their density is small. They, however, have an adverse effect on cell or film quality.

6.2.3 Factors that Influence Film and Cell Quality

The deposition parameters /conditions that control the property of the material are as follows.

Pressure and Electrode Separation. The voltage necessary to sustain a plasma is defined by the Paschen curve which determines the sustaining voltage as a function of the pressure and the electrode separation. In order to have a low sustaining voltage, the pressure is maintained between 0.1 to 1 torr and the electrode separation between 4 to 6 cm. Higher pressure or larger electrode separation result in many secondary reactions which may cause formation of powder and higher silanes affecting material property.

Temperature. The substrate temperature plays a key role in determining the adatom mobility of the impinging species on the substrate, and a higher substrate temperature results in higher adatom mobility allowing more surface diffusion. This allows the species to find an energetically favorable site resulting in a denser material. Higher temperature, however, causes loss of hydrogen from the surface. The dangling bonds generated thereby increase the sticking coefficient and lower surface mobility. The optimum substrate temperature is thus between 200 to 300°C for typical rf glow-discharge deposition.

Power Density. With increasing power density, more polymeric radicals are formed in the plasma that may lead to powder formation. The heavier radicals also have low adatom mobility as they impinge on the surface of the growing film. This results in films grown with higher density of microvoids [6.18] and leads to poorer cell performance. The optimum power density is just above the value at which the plasma can be sustained and is typically 10 to 100 mW/cm^2.

Gas Flow Rate. As the gas flow rate decreases, the residence time increases, which may result in depletion of the active species. Matsuda [6.17] has shown that under these conditions, the contribution of the short-lifetime radicals (including the heavy radicals) to the growth rate increases. This gives rise to poorer quality material. An adequate flow rate that results in a linear increase in the deposition rate as the power density increase is, therefore, recommended.

Hydrogen Dilution. Guha et al. [6.19] showed in 1981 that films grown with a gas mixture of silane diluted with hydrogen have improved quality. Hydrogen dilution is now used extensively for obtaining high quality material with many different active gases. The most important role of hydrogen dilution is to improve surface coverage that results in increased surface diffusion. We have discussed earlier that an increase in the surface diffusion results in the impinging species finding more energetically favorable sites. In fact, very high hydrogen dilution results in the growth of microcrystallites, and the best material is grown at the threshold between amorphous to microcrystalline transition [6.20]. This material is also characterized by a more ordered structure.

Impurities. Impurities can give rise to states in the gap that will act as recombination centers. The most common impurities in a vacuum system are oxygen, nitrogen and carbon. It has been shown [6.21] that the quality of the material is unaffected if the impurity content is below $10^{19}\,\mathrm{cm}^{-3}$. This can be routinely achieved when conventional precautions applicable to vacuum systems are taken while designing the deposition reactor.

The simple reactor shown in Fig. 6.1 suffers from two limitations. First, the system has to be vented to atmosphere after every deposition run, and the atmospheric water vapor and oxygen that cling to the wall of the chamber take a very long time to get rid of. Second, when dopant gases are used to make the doped layers in the solar cells, a trace amount of those impurities gets incorporated in the intrinsic layer as well. This affects the quality of the intrinsic layer where phosphorus or boron concentration exceeding $10^{16}\,\mathrm{cm}^{-3}$ is not desirable. A multichamber load-locked system (Fig. 6.3) is, therefore, usually used for deposition of high quality solar cells. By using dedicated chambers separated by gate valves for growing the doped and undoped layers, cross contamination can be reduced to acceptable levels. After the deposition is completed, only the load chamber needs to be vented to atmosphere, and the growth chambers are not exposed to atmosphere in between deposition runs.

Frequency. Although most of the deposition studies reported in the literature for growing high quality devices use a frequency of 13.56 MHz, high

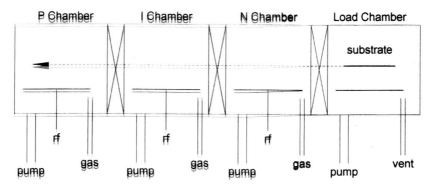

Fig. 6.3. Schematic diagram of a multichamber glow discharge deposition system

quality cells have been made using both DC and higher frequencies. Use of higher frequencies leads to higher deposition rates without formation of excessive polymeric radicals, and both very high frequency, (vhf), [6.22, 23] and microwave frequency [6.24, 25] have been used to obtain high efficiency cells at high deposition rates.

We have so far focused our discussion on only the most commonly used method of deposition, namely, plasma-enhanced chemical vapor deposition. There are many other deposition methods that have been tried to obtain high quality materials and cells. Some are minor variations of the conventional method; others explore totally different concepts. Results on solar cells obtained using some of these methods will be presented in Sect. 6.6 when we discuss the future trends. A detailed description of these techniques may be found in [6.13].

6.3 Single-Junction Cells

6.3.1 Cell Structure

A-Si alloy solar cells can be fabricated using a variety of structures such as metal-semiconductor (Schottky diode), metal-insulator-semiconductor diode, p-n junction and p-i-n junction. As in the case of single crystals, the first two structures do not produce high efficiency devices since the voltage generated is determined by the interface properties and is usually low [6.26]. Since the incorporation of the dopant atoms in the intrinsic material increases the defect density and hurts carrier transport, it is desirable that the photon absorption and carrier transport take place in the intrinsic (i) layer only, and the highest efficiencies are obtained only with the p-i-n structures. The two structures that are commonly used are shown in Fig. 6.4 In one configuration (case a, also known as substrate-type structure), the a-Si alloy layers are deposited on a metal substrate (e.g., stainless steel) or on an insulating substrate coated with a metal. The metal forms one electrode of the cell and

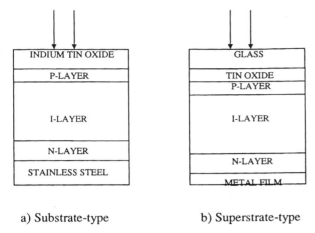

Fig. 6.4a,b. Schematic diagram of (**a**) substrate-type and (**b**) superstrate-type a-Si alloy p-i-n solar cell structures

it forms an ohmic contact to the heavily doped n-type layer. The top contact is the transparent conducting oxide layer through which the light enters. This oxide layer, which is typically indium tin oxide (ITO), forms an ohmic contact to the heavily doped p-type layer and also serves as an anti-reflection coating. In the other configuration (case b, also known as the superstrate-type structure), the transparent conducting oxide layer is deposited on glass through which the light enters. The a-Si alloy layers are deposited consecutively on top of this oxide which is typically zinc oxide or tin oxide. A metal film is finally deposited to complete the cell structure. For reasons to be discussed later, the highest cell efficiencies have been achieved using the substrate-type structure.

6.3.2 Cell Characteristics

Light entering into the i-layer is absorbed to create the electron–hole pairs. These carriers move toward the electrodes by a process of drift and diffusion. A typical current-voltage characteristic of a solar cell under illumination is shown in Fig. 6.5. Three parameters define the performance: the open-circuit voltage (V_{oc}), the short-circuit current (I_{sc}), and the fill factor (FF). The fill factor is the ratio of the product of the voltage and the current obtained at the maximum power point to that of V_{oc} and I_{sc}, where V_{oc} and I_{sc} are the values of the voltage and the current obtained under illumination when $I_{sc} = 0$ and $V_{oc} = 0$, respectively.

The efficiency of the cell that equals the ratio of the power output to the power input is given by,

$$\eta = V_{oc} J_{sc} FF / P_{input}, \tag{6.1}$$

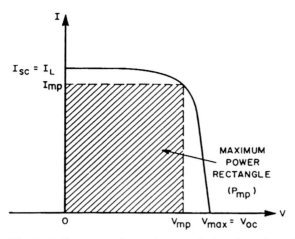

Fig. 6.5. Current-voltage characteristics of a solar cell

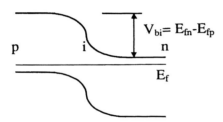

Fig. 6.6. Band diagram of a simple p-i-n structure

where J_{sc}, the short-circuit current density, is I_{sc} per unit area. Before discussing a detailed theory of the operation of a-Si alloy solar cells, we shall briefly evaluate the material parameters that determine the cell performance. Considering the simple band diagram of the p-i-n structure as shown in Fig. 6.6, the built-in potential, V_{bi}, is defined as the difference between the Fermi levels of the two doped layers. In this example, V_{bi} is close to the bandgap of the intrinsic material. The open circuit voltage is determined by V_{bi} and, hence, the larger the bandgap, the higher will be the value of V_{oc}. However, as discussed earlier, the gap states present in the material act as recombination centers, and V_{oc} is governed by the separation of the electron and hole quasi-Fermi levels under illumination. Since this separation is determined by the gap state density, V_{oc} is determined both by the bandgap and the gap state density.

The short-circuit current density is determined by the number of the absorbed photons which, in turn, depends on the bandgap of the material and the thickness of the i-layer. As the bandgap is lowered, J_{sc} increases. The thickness of the i-layer can not be increased indefinitely for increased photon absorption to obtain higher J_{sc}. The gap states limit the effective length that the carriers can move before recombining, and increasing the thickness beyond a certain "collection length" does not result in any further improvement of J_{sc}. The fill factor determines how efficiently the carriers are collected. This also is determined by the i-layer thickness and the gap state density. In order to design a cell to obtain the highest efficiency, all the above factors need to be considered carefully. Computer simulation that takes into account the material parameters and the cell dimensions can play a very valuable role in cell optimization.

6.3.3 Numerical Modeling

In order to have a proper understanding of the various material parameters on the cell performance, numerical models have been developed [6.27–35] which solve the Poisson and current continuity equations across the p-i-n structure. Knowledge about the gap state distribution and other material parameters are necessary for the simulation studies. As mentioned earlier, the absence of long-range order causes tailing of the band edges. The presence of dangling, weak and strained bonds, impurities and defect-impurity complexes can also give rise to gap states. The net effect is a continuous distribution of states in the mobility gap. A typical gap state distribution is shown in Fig. 6.7. The valence band tail has a characteristic energy of about 50 meV and is wider than the conduction band tail that has a characteristic energy of about 25 meV. The mid-gap states, defined as D_0 are caused by the neutral dangling bonds. Depending on the position of the Fermi level, these states can get charged occupying new positions D^- and D^+. In addition, impurities and other defects can give rise to a continuous distribution of gap states as shown in Fig. 6.7.

Any accurate prediction of the performance of the solar cell depends on the precise knowledge of the gap state distribution. In addition to the density of these defect states, the capture cross-sections of these states, both in the charged and the neutral states, need to be known. Also needed are the band mobility of the electrons and the holes and density of the states at the band edges. A precise knowledge of the defect densities in the doped layers and

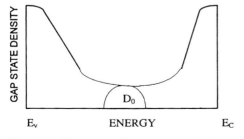

Fig. 6.7. Gap state distribution of a-Si alloy

the transition regions between the doped and the intrinsic layers (interface states) are also necessary. In spite of many years of efforts in determining these material parameters using many different experimental technique [6.13], there is still not a clear consensus as to what these numbers are. All the numerical studies make reasonable assumptions about these parameters, and based on the fit to the experimental data for a variety of film and cell performances, try to arrive at a self-consistent set of parameters. Even then, rather than looking for a quantitative prediction of device performance, the numerical studies should be looked at more for predicting general trends and to give an insight into device physics. Most of the studies have been remarkably successful in this regard.

The first comprehensive model of device operation was presented by Swartz [6.27] who solved the transport equations by assuming a single-level recombination process. Schwartz et al. [6.28] also used a density-of-states model where only dangling bonds at the center of the gap were assumed to be responsible for recombination. In the studies of Hack and Shur [6.29], an asymmetric density-of-state distribution with two exponentials was used. The minimum gap state density was kept at $10^{16}\,\mathrm{cm}^{-3}\,\mathrm{eV}^{-1}$. The electron band mobility was assumed to be five times higher than the hole mobility, and the ratio of the charged to neutral cross-section was assumed to be 100. The model was first applied to explain photoconductivity of a-Si alloy films [6.36, 37] and could accurately predict the sensitization of photoconductivity with respect to the position of dark Fermi level and its temperature and intensity dependence. Based on calculations of the free and trapped carrier densities, as well as space charge and electrical field distribution, Hack and Shur concluded that hole transport limits the performance of the devices and bulk recombination plays the dominant role in determining the fill factor. The computed values [6.29] of V_{oc}, J_{sc}, FF and η as a function of i-layer thickness are shown in Fig. 6.8.

Cases when the light enters through the n- or the p-layers were considered. It is apparent that light entering through the p-layer is more desirable. Since the hole transport is poor, improved characteristic is obtained if the carriers are generated close to the p-i interface which is the case for nonuniform illumination when the light enters through the p-layer. In accordance with the intuitive prediction that was presented earlier, J_{sc} increases as the thickness increases up to a certain thickness beyond which it does not increase any more. The fill factor, however, decreases as the thickness increases while the open circuit voltage is fairly independent of thickness. The efficiency of the device, therefore, peaks at a certain thickness, the value of which depends on the quality of the material. These results are in good agreement with experimental observations that are shown in Fig 6.9.

Several other studies [6.30–32] have considered more elaborate gap state distribution. Pawlikiewicz and Guha [6.30] have extended the analysis of Hack and Shur to consider a more realistic gap state distribution. The effect

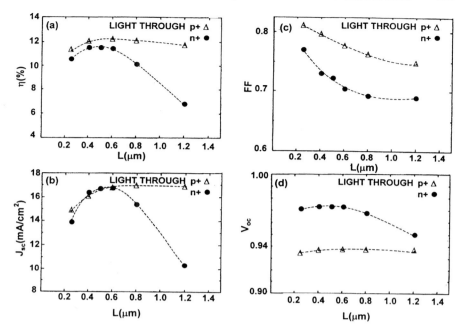

Fig. 6.8. Computed values of (**a**) efficiency, (**b**) short-circuit current density, (**c**) fill factor and (**d**) open-circuit voltage as a function of intrinsic layer thickness [6.29]

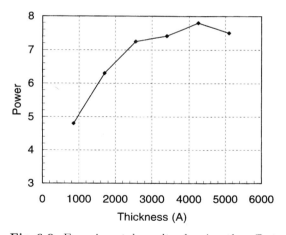

Fig. 6.9. Experimental results showing the effect of i-layer thickness on solar cell performance. Power is measured in mW/cm^2

of the doped layer on the performance of the solar cell was carefully investigated and it was concluded [6.31] that p-i junction is the dominant junction which determines V_{oc}. Recently, a comprehensive numerical model has been developed [6.32] at the Pennsylvania State University that allows use of many different parameters in each of the regions of the solar cell to arrive at a self-consistent set of parameters to explain many different properties. The simulation program (AMPS model) is very user-friendly and has been used by several organizations to explain cell performance under different conditions [6.33–35].

The simple cell structure discussed above suffers from the limitation that the cell efficiency is rather low (typically 8%). Use of light trapping within the cell by the use of a suitable back reflector can increase the efficiency to about 12%. The phenomenon of light-induced degradation to be discussed later, however, lowers this efficiency by about 50% when these high efficiency single-junction cells are exposed to light. The highest efficiency devices, therefore, use the multijunction approach in conjunction with a well-designed back reflector. Before we discuss the design considerations for these devices, we present below a discussion on light-induced degradation and its effect on cell performance.

6.3.4 Light-Induced Degradation

It was first reported by Staebler and Wronski [6.38] in 1977 that there are significant changes in the dark and photoconductivity of a-Si alloy films when subjected to light exposure. The phenomenon, named the Staebler-Wronski effect after the authors, is reversible. Annealing at temperatures around 150°C for an hour restores the original values. The metastable changes in the material properties that take place under light exposure also have a pronounced effect on solar cell performance.

In the last twenty years, extensive work has been carried out both to obtain a fundamental understanding of light-induced degradation (LID) and also to develop materials and devices that will be immune to this degradation. Although we have a much better understanding of this effect today, LID is always observed in high quality materials and remains one of the most intensely investigated subjects [6.39, 40]. Early work revealed several interesting features of this phenomenon:

1. The metastability is caused by changes in bulk properties. For example, the density of neutral dangling bonds increases after light exposure [6.41].
2. Metastable defects can be created by charge injection, e.g., by forward biasing a p-i-n diode in the dark [6.42].
3. Recombination of electron–hole pairs is necessary to trigger LID [6.43, 44]. A Schottky diode when forward-biased does not show LID nor does a p-i-n junction when it is reverse-biased under illumination.

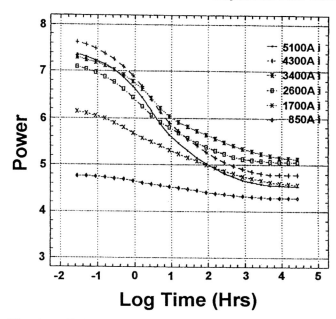

Fig. 6.10. Light-induced degradation in single-junction cells deposited on stainless steel. Power is measured in mW/cm^2

4. Reciprocity in light intensity and time is not observed [6.45]. High intensity exposure for a short time causes more degradation than low-intensity exposure for a longer time even though the product of intensity and time is kept constant.
5. Exposure to light reduces cell efficiency. The maximum amount of degradation is in the fill factor. Degradation is more when the cell thickness is large. We should note that to absorb light efficiently, thick cells are needed, and these cells degrade the most.
6. Degradation does not go on indefinitely. As new defects are created, they also get annealed out so that equilibrium is reached after a few hundred hours. Degradation is less at high temperature.

A typical plot of the degradation characteristic of a single-junction cell when exposed to light of one-sun intensity at 50°C is shown in Fig. 6.10. The cells are deposited on stainless steel with different i-layer thickness. It is clear that degradation is higher as the cell thickness increases. The highest stable efficiency is obtained for i-layer thickness of about 300 nm.

The defect-creation kinetics has been studied by Stutzmann et al. [6.46] who showed that the defect density (N_d) depends on the illumination intensity G and time t as,

$$N_d(t) = \text{const.}\, G^{2/3} t^{1/3}, \tag{6.2}$$

when $N_d(t)$ is much higher than the equilibrium defect density. This relationship obviously explains the lack of reciprocity between intensity and time. Stutzmann and his coworkers explain this nonlinear dependence in the following way. Defects are created by nonradiative recombination of electron–hole pairs at spatially correlated tail states that correspond to the bonding and antibonding states of weak Si–Si bonds. The recombination releases about 1.5 eV energy that breaks a weak bond and generates a defect. To prevent the two neighboring bonds from recombining, a neighboring hydrogen atom moves in to separate the two dangling bonds. The defect pair may separate further by additional bond switching. In the annealing process, the hydrogen atoms revert back to their original positions. Hydrogen is thus directly involved both in the defect creation and annealing.

The above model can explain the self-limiting nature of defect generation. As the defect density increases, the additional recombination through the new states reduces the band edge carriers and suppresses the creation mechanism. There are still many open questions regarding the universal applicability of the above model. No spatial correlation between dangling bond defects and between those and hydrogen have been observed [6.47]. The dangling bond defects are separated from each other by about 10 nm and from hydrogen atoms by about 0.4 nm. Moreover, the observation [6.48] that annealing of defects created at low temperature (4 K) starts at 150 K where hydrogen diffusion is negligible suggests that the bond-breaking model together with hydrogen diffusion is not adequate to explain all the different features of this phenomenon.

A new model for LID has recently been proposed [6.49] that postulates the formation of a metastable complex containing two Si–H bonds. Light exposure generates the electron–hole pairs that recombine through multiphonon excitation of Si–H vibrations. The recombination breaks the Si–H bonds and promotes H to a transport state. Molecular dynamics calculations show [6.50] that a diffusing H-atom successively breaks Si–Si bonds creating Si–H bonds and a neighboring dangling bond. When this mobile hydrogen in the form of Si–H and dangling bond pair meets another pair of the same kind, the two dangling bonds annihilate each other leaving behind a metastable complex of two Si–H bonds. Biswas and Pan showed [6.51] that formation of such a defect complex is indeed feasible, and they are the lowest energy configurations other than the initial state. Branz [6.49] has been successful in explaining the kinetics of many of the defect creation and annealing behaviors using this model. To explain the kinetics of defect creation at low temperature, enhanced hydrogen diffusion due to carrier capture effects was postulated; further studies will be necessary to explain the annealing behavior in the dark at low temperatures when, ordinarily, H diffusion will be negligible.

There have been suggestions that impurities such as oxygen, carbon or nitrogen may be associated with LID. While this may be the case when the concentration is high so as to change the microstructure, LID has been observed [6.52] in a-Si alloy with impurity content as low as $2 \times 10^{15}\,\mathrm{cm}^{-3}$ for oxygen, $7-10 \times 10^{15}\,\mathrm{cm}^{-3}$ for carbon and $5 \times 10^{15}\,\mathrm{cm}^{-3}$ for nitrogen. Since the light-induced defect density was found to be $10^{17}\,\mathrm{cm}^{-3}$, this clearly demonstrates the intrinsic nature of light-induced degradation.

It is increasingly believed that light exposure causes structural changes in the material. There is much experimental evidence to indicate [6.53] that just an increase in the defect density can not explain several features of LID. The most direct proof of structural changes taking place in the material came from the bending experiments of Gotoh et al. [6.54]. Light exposure was found to change the compressive stress in the film indicating metastable expansion of the film. The time constants for changes in the stress and the annealing behavior are similar to light-induced photoconductivity changes, indicating that the generation and removal of the defects correspond to structural changes.

How does one reduce LID? A correlation between microstructure of the material and LID was demonstrated by Guha et al. [6.18]. For cells of similar thickness, both the initial efficiency and the stability were found to be poorer when the i-layer of the cell had poorer microstructure (Table 6.2). In this experiment, the microstructure was made poorer by depositing the i-layer at a high deposition rate by increasing the rf power density.

How does one improve the structure? Hydrogen dilution [6.19] seems to be the most effective method of doing that. As discussed earlier, excess hydrogen will result in increased passivation of the growing surface, allowing the impinging species to have larger surface diffusion to find an energetically favorable site. Hydrogen can also be helpful in etching away the weak bonds. Studies on cells prepared with low and high hydrogen dilution clearly show the beneficial role of hydrogen dilution on cell stability [6.55]. As shown in Table 6.3, both the initial efficiency and the stability improves when the intrinsic layer is prepared with high hydrogen dilution.

There has been recent evidence that films prepared with high hydrogen dilution have a more ordered structure. Based on high resolution transmission electron microscope studies, Tsu et al. [6.20] reported that films grown with high hydrogen dilution are characterized by a heterogeneous mixture of

Table 6.2. Degradation of a-Si alloy single-junction cell prepared at different deposition rates

	0.14 nm/s	1.35 nm/s
Deposition rate		
Void fraction	1%	4%
Predominant void diameter	-	0.9 nm
Hydrogen content	8%	12%
Microstructure fraction (R)	8.4%	18.4%
Monohydride contents	6.4%	6.3%
Initial efficiency	7.85%	6.31%
Degraded efficiency	6.53%	3.5%

Table 6.3. Effect of hydrogen dilution on a-Si alloy single-junction cell performance

Description	State	J_{sc} (mA/cm^2)	V_{oc} (V)	FF	P_{max} (mW/cm^2)
300°C, low dilution	Initial	12.3	0.94	0.65	7.5
	Degraded	11.6	0.91	0.55	5.8
300°C, high dilution	Initial	11.6	0.96	0.68	7.6
	Degraded	11.2	0.94	0.61	6.4
175°C, low dilution	Initial	11.4	0.96	0.64	7.0
	Degraded	9.5	0.91	0.46	4.0
175°C, high dilution	Initial	10.9	1.00	0.69	7.5
	Degraded	10.5	0.97	0.60	6.1

amorphous network and linear-like objects that show evidence of order along their length. These "linear-like" objects, typically few tens of nm long and 2–3 nm wide, meander randomly through the material. Raman studies show that as the hydrogen dilution increases, the volume fraction of the ordered regions increases. When the hydrogen dilution is extremely high, the material becomes microcrystalline. In fact, the best material that demonstrates the highest solar cell performance is grown right at the threshold when microcrystalline grains are just beginning to get formed.

In spite of significant improvement in material quality, the highest stable efficiency that could be attained from a single-junction cell using the structures shown in Fig. 6.4 is only between 5 to 7%, and new developments were needed to improve the stable efficiency further.

6.4 High Efficiency Cells

6.4.1 Introduction

An increase of the cell thickness results in more photon absorption and increases the efficiency. Thick cells, however, degrade more on light-exposure. In a thinner cell, LID is small and the cells are more stable. The initial efficiency, however, is low, resulting from inadequate photon absorption. There does appear an inherent efficiency barrier for a-Si alloy single-junction solar cells. Two important developments resulted in significant improvement of stable cell efficiency. One is the use of the multijunction structure where several cells are stacked one on top another. The cells use intrinsic layers of different bandgaps so as to capture a wider spectrum of sunlight. Individual cells are kept thin so that LID is small. The second important development is the use of back reflector. In a substrate-type structure (Fig. 6.4), stainless steel has poor reflectivity and only a small fraction of the photons that are not absorbed in the cell get reflected. Incorporation of a reflecting layer

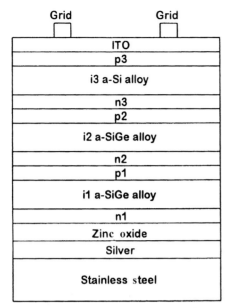

Fig. 6.11. Schematic diagram of a triple-junction cell

between the stainless steel and the cell can lead to multiple optical passes, thereby increasing efficiency.

6.4.2 Multijunction Cell

Although any number of cells can be stacked together to form the multijunction structure, from a practical point of view, the losses in the junction between the cells limit the maximum number that can be used without affecting the efficiency. A triple-cell structure has been found to be optimal for obtaining the highest efficiency. A schematic diagram of a triple cell structure in the substrate-type configuration is shown in Fig. 6.11. Light enters from the top, and the thin layer (60 nm) of indium tin oxide (ITO) serves as the antireflection coating and also forms a low-resistance contact to the top doped layer. Typical sheet resistivity of ITO is 50–100 Ω/cm^2, and silver grid lines are evaporated on this layer to reduce series resistance losses. The three component p-i-n cells use i-layers of different bandgaps. The top cell has an optical gap of about 1.8 eV and captures the blue light. The intrinsic layer of the middle cell incorporates about 10% Ge to lower the gap to about 1.6 eV. This cell captures the green photons. The i-layer of the bottom cell has about 40% Ge and has a gap of about 1.4 eV. This captures the red light.

The longer wavelength photons that are not absorbed get reflected from the back reflector that consists of a bilayer of silver and zinc oxide. The back reflector usually is textured so that the light is reflected at an angle to facilitate multiple reflections.

The above cell design has been used to obtain the highest initial cell efficiency of 14.6% [6.7] using a multijunction structure. Much higher efficiencies are expected to be reached by using even lower bandgap materials. Kuwano et al. [6.56] have computed cell efficiency as a function of bandgap of the i-layers of the component cells both for double- and triple-junction structures. A cell efficiency of 21% can be achieved for a double-junction structure using bandgaps of 1.75 and 1.15 eV whereas using bandgaps of 2.0, 1.7 and 1.45 eV for the three component cells, an efficiency of 24% can be achieved. We should point out that the thermodynamic limit of efficiency for a multijunction cell when only radiative recombination is present is 71% [6.57].

6.4.3 Key Requirements for Obtaining High Efficiency

In order to obtain high efficiency in a multijunction cell structure, the following are the requirements: (1) high quality back reflector for efficient light trapping, (2) doped layers with high conductivity and low optical loss, (3) high quality intrinsic layers of different bandgaps, (4) high quality component cells with proper current matching, (5) low-resistance tunnel junctions and (6) antireflection coating or window layer of high conductivity and transparency. We give below a discussion on how these different requirements can be fulfilled.

6.4.4 Back Reflector

Multiple passes within the cell by the use of suitable back reflector increase the optical path and improve photon absorption especially for the long-wavelength photons. Yablonovitch and Cody [6.58] have shown that for random scattering, the path length can be increased by a factor of $4n^2$ where n is the refractive index of the material. For a-Si alloy, this amounts to about 50 passes or 25 reflections. Figure 6.12 shows the calculated value of the fraction of absorbed light as a function of wavelength for different number of internal reflections. The calculation assumes an a-Si alloy solar cell with an optical gap of 1.68 eV and i-layer thickness of 500 nm. The series of curves show the significant enhancement in photon absorption that can be obtained in the red/infrared part of the spectrum with an increasing number of reflections within the cell.

Referring to Fig. 6.13, which shows the schematic of a simple one-layer back reflector, the incident light is reflected back into the cell for a second and subsequent passes. This phenomenon results in enhanced absorption in the cell. Thus, the back reflector should possess high reflectance in the solar part of the spectrum, and Al and Ag are good candidates. The random or

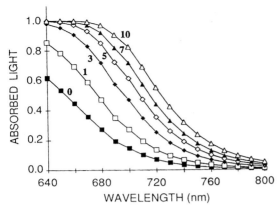

Fig. 6.12. Light absorbed in a-Si alloy solar cell as a function of wavelength for different number of reflections

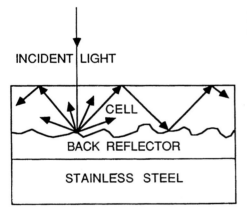

Fig. 6.13. Schematic diagram of a one-layer textured back reflector

textured surface of the back reflector leads to scattering of light on reflection at the cell/back reflector interface. Consequently, light is reflected in various directions inside the cell, and the optical path length is further enhanced. Also, the obliquely reflected light can undergo total internal reflections inside the cell at the top surface at the cell/TCO interface and/or the TCO/air interface. The condition for total internal reflection is given by $(\theta_2)_{\text{crit}} = \sin^{-1}(n_1/n_2)$, where $(\theta_2)_{\text{crit}}$ is the critical angle of incidence inside the higher refractive index medium, and n_2 and n_1 are the refractive indices of the higher and lower refractive indices media. Also, the wavelength λ of light changes when it goes from medium 1 to 2 and vice versa according to the equation $\lambda_1 n_1 = \lambda_2 n_2$. Thus, weakly absorbing light will be totally internally reflected at the top surface of the cell (irrespective of the actual interface) as long as the angle of incidence inside the a-Si alloy at the a-Si alloy/TCO interface is greater than $\approx 16.6°$ as deduced from the above equations using a refractive index of 3.5 for a-Si alloy. In other words, the critical angle for total internal reflection is $\approx 16.6°$ inside the a-Si alloy material.

The biggest problem of using a single-layer back reflector as shown in Fig 6.13 is that the a-Si alloy material and the underlying metal surface may react to form a thin interfacial layer which would reduce the reflectance of the back reflector. This deleterious effect can be overcome by sandwiching a thin buffer dielectric layer such as a transparent conducting oxide in between the cell and the metal reflector. ZnO can be used as the buffer layer in a double-layer back reflector. In addition, ZnO possesses its own surface morphological texture and internal texture, both of which result in increased scattering of the reflected light. Thus, the role of the ZnO film is extremely important. A back reflector using the ZnO/textured Ag configuration yields a value of J_{sc} which is typically $\approx 15\%$ higher than that obtained without the ZnO. The gain in J_{sc} is predominantly due to an increased red response of the cell.

To assess the role of back reflector on cell performance, Banerjee and Guha [6.59] carried out systematic studies of reflections from many different back reflectors deposited on stainless steel and correlated those data with cell performance. Ten different back reflectors were prepared for evaluation and are listed as follows: (1) specular Ag, (2) textured Ag, (3) specular Al, (4) specular Al–Si, (5) textured Al–Si, (6) 2 μm ZnO/textured Ag, (7) 1000 Å ZnO/4000 Å Ag/textured Al–Si, (8) 1000 Å ZnO/textured Al–Si, (9) 2 μm ZnO/textured Al–Si, and (10) Cr/textured Al–Si.

The specular back reflectors were obtained by sputtering at substrate temperature less than 100°C. The textured back reflectors and ZnO were deposited at 250–350°C. A careful comparison was made between the reflection characteristics from these back reflectors and their performance in a cell configuration. Table 6.4 shows the solar cell performance for some selected back reflectors. The highest J_{sc} was obtained from the back reflector that has textured silver and 2 μm ZnO. This back reflector also shows the best reflection characteristic in the long wavelength range. The dispersion characteristic of this back reflector is shown in Fig 6.14. Note that in spite of the fact that there is a considerable amount of light that gets scattered at large angles, there is a reasonable amount of specular reflection, too.

Table 6.4. Summary of performance of solar cells deposited on different back reflectors

No Back Structure	Reflector	Thick (Å)	V_{oc} (V)	FF	J_{sc} from Q (mA/cm^2)	Q_{700}
6	2 μm ZnO/textured Ag	5100	0.936	0.628	17.50	0.52
7	1000 Å ZnO/4000 Å Ag/textured Al–Si	4900	0.929	0.620	16.54	0.43
8	1000 Å ZnO/textured Al–Si	4900	0.912	0.631	15.50	0.35
9	2 μm ZnO/textured Al–Si	4900	0.943	0.558	15.48	0.32

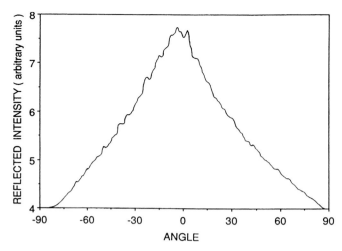

Fig. 6.14. Dispersion characteristic of 2 μm ZnO/textured Ag deposited on stainless steel

The above results show that substantial gain in J_{sc} can be obtained by using a suitable back reflector. Back reflectors using specular silver and thick ZnO or textured silver and thin ZnO both have shown good results. The increase in J_{sc} is about $5\,\text{mA cm}^{-2}$ when a suitable back reflector is incorporated in the a-Si alloy cell structure. The gain is even more impressive for cells with lower bandgap. Figure 6.15 hows the quantum efficiency plot of an amorphous silicon germanium (a-SiGe) alloy cell deposited on stainless steel with and without the back reflector. There is an increase in J_{sc} from 18.5 to $24.9\,\text{mA cm}^{-2}$ when the back reflector is incorporated. The optimum back reflector shows pyramidal structure with submicron features. About 20% of the light is reflected specularly, and the rest of the reflection is diffused. Special back reflector textures were designed [6.60] where the specular component was further reduced and the light scattering was Lambertian. In some other cases [6.61], a grating-like texture was used so that most of the light was scattered at an angle larger than the critical angle. In none of these cases was the long wavelength response improved beyond what could be obtained with the scattering surface discussed earlier. It appears that beyond a certain number of reflections, there is a point of diminishing return and there is no further photon absorption. This could be caused by some loss associated with each reflection. Theoretical calculation [6.62] indicates that there is a loss associated with reflection from any textured surface. This could be the reason for reaching the limit in photon absorption since beyond a certain texture, the loss associated with each reflection neutralizes the gain from increasing number of reflections. Even though the gain in J_{sc} that has been achieved to date by the use of suitable back reflector is substantial, the magnitude of J_{sc} is still much lower than what is expected if we have total internal reflection

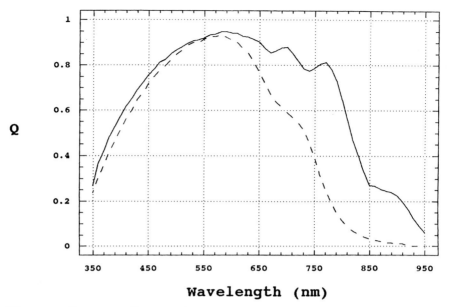

Fig. 6.15. Quantum efficiency plot of a-SiGe alloy cell on stainless steel (*broken line*) and back reflector (*solid line*)

without any loss. Research activities on the design of high performance back reflector will play a key role in further improvement of cell efficiency.

The above discussion has been mostly on the optimization of the back reflector for a substrate-type structure. The design considerations for superstrate-type structures are also very similar. Morris et al. [6.63] have shown that use of the rear contact consisting of ITO/silver improves J_{sc} by about $3\,\mathrm{mA\,cm^{-2}}$ over that of a simple molybdenum back reflector for a-Si alloy cells. Contacts involving ZnO/silver have also been used. We should mention that for this configuration, the texture is obtained at the front contact itself, which is the conducting oxide deposited on glass.

A variety of deposition techniques has been used for the deposition of the back reflector. For the substrate-type back reflector, both the metal and the oxide layers are usually sputtered; magnetron sputtering has been extensively used. Other methods such as atmospheric chemical deposition from chlorides [6.64], deposition from organometallic feedstock [6.65] and use of spray pyrolysis [6.66] have been reported. Gordon [6.67] has used a variety of techniques to obtain different oxide layers for back reflector applications. Under optimized conditions, all these methods give comparable results in terms of device performance.

6.4.5 Doped Layer

The doped layers in the cell serve the purpose of providing low resistance contacts to the conducting oxide layers and also to each other in between the junctions. High electrical conductivity is a requirement to perform these functions. Moreover, in order to obtain high V_{oc}, it is necessary that the doped layers should have low electrical activation energy that also requires that the doped layers should be highly conducting. Another requirement for the doped layer is transparency to light. Since the doped layers are inactive in the photovoltaic process, light absorbed in these layers do not contribute to photocurrent. In order to obtain high efficiency with single- or multijunction devices, high bandgap, high electrical conductivity materials with low optical loss are necessary.

Use of phosphorus doping can produce an n-type layer with high conductivity. Films with electrical conductivity of $10^{-2}\,(\Omega\,\text{cm})^{-1}$ with an activation energy of 100 meV can be easily obtained by mixing phosphine with silane in the gas stream. The bandgap is close to that of a-Si alloy and the optical loss is, therefore, minimal. Phosphorus doping, however, increases the defect concentration in the material. For high levels of phosphorus doping, the defect density increases as the square root of the density of phosphorus atoms [6.68, 69]. This results in higher sub-bandgap absorption in the films [6.70], limiting the long wavelength response.

Boron doping in a-Si alloy is not as successful in moving the Fermi level close to the band edge because of the high defect density associated with the valence band tail. The best conductivity is only around $10^{-4}(\Omega\,\text{cm})^{-1}$ with an activation energy of 300–400 meV. Moreover, incorporation of B results in a lowering of the gap of a-Si alloy, resulting in a higher optical absorption in the material. Solar cells made with p-type a-Si alloy as the doped layer are thus characterized by low V_{oc} and J_{sc}. There have been two main approaches to alleviate the above problem. One is to incorporate carbon to widen the bandgap. In the other approach, microcrystalline alloys are used. Both the approaches are used extensively to make high efficiency devices.

Incorporation of C in the film increases the bandgap of the material. A carbon content of about 15% is enough to increase the optical bandgap to about 2.1 eV [6.71]. Boron doping lowers the bandgap, and to obtain p-type amorphous silicon carbon (a-SiC) alloy of about 2 eV bandgap, a carbon concentration of 30–40% is needed [6.72]. Incorporation of such a large quantity of C increases the defect density in the material and makes the doping efficiency poorer. The conductivity of the material is lower than $10^{-6}(\Omega\,\text{cm})^{-1}$, and the activation energy is about 0.5 eV. This poses a problem in obtaining low resistance contact to the conducting oxide layers and also to the adjacent n-layers in the multijunction. Use of trimethyl boron rather than diborane as the dopant source has been shown [6.73, 74] to improve conductivity and reduce optical absorption. Attempts have been made to make p-type a-SiC alloy using a large number of feedstock gases containing carbon. The device performance has been found to be similar [6.75] once the deposition conditions are optimized for the appropriate gas mixture.

Boron doped a-SiC alloy films as the window layer through which light first enters has been used in many laboratories [6.76] to obtain single-junction devices with high V_{oc} and J_{sc}. Quantum efficiency at 400 nm has been demonstrated to exceed 60% [6.77], indicating low absorption in the p-layer. High efficiency multijunction cells have also been fabricated, but the highest efficiency cells have been achieved using microcrystalline p-layers.

Microcrystalline doped a-Si alloys are characterized by low optical absorption and high conductivity. Guha et al. [6.78] demonstrated that thin layers of microcrystalline p-type materials can be made showing high conductivity and low optical loss. The alloy was grown using a very dilute mixture of silane in hydrogen at a higher power density than is normally used for growing a-Si alloy. We have discussed before the use of a heavy dilution of hydrogen to facilitate grain growth to obtain microcrystalline material. Both diborane and boron trifluoride were used as the dopant gas and were used to make high efficiency single- and multijunction devices. Transmission electron microscopy and Raman studies showed grain sizes between 8 to 12 nm with the microcrystalline phase greater than 80%. The electrical and optical characteristics of the microcrystalline film are compared with its amorphous counterpart in Table 6.5. The amorphous-to-micro-crystalline transition results in an increase in conductivity by three to four orders of magnitude. The optical gap is also increased by about 400 meV, lowering absorption losses.

Incorporation of microcrystalline p-layers in single-junction cells has resulted in V_{oc} greater than 1 V and quantum efficiency at 400 nm exceeding 75%. This has also resulted in very low resistance at the tunnel junctions between the component cells. These results will be discussed later.

A natural extension of this work is to grow microcrystalline SiC layers. Such layers have been made in several laboratories [6.79–81]. Although high V_{oc} has been demonstrated using this alloy [6.79, 80], studies involving incorporation of these material are rather limited. We should mention that in order to reduce absorption losses, the doped layer needs to very thin. The requirement for growing a thin microcrystalline layer is much more stringent than that for a thick layer, and the substrate also plays a critical role. In general, it is easier to grow microcrystalline layers on a-Si alloy rather than

Table 6.5. Dark conductivity (σ_d), activation energy (ΔE), optical gap (E_0), and absorption coefficient (α) at 5500 Å

	Amorphous	Microcrystalline
σ_d $(\Omega\,cm)^{-1}$	10^{-4}–10^{-3}	1–20
ΔE (eV)	0.3–0.4	0.02–0.05
E_0 (eV)	1.5–1.6	1.9–2
α (cm^{-1}) at 5500 Å	1×10^5	3×10^4

on an oxide, and this explains why microcrystalline p-layers have not been used successfully in a superstrate-type cell for the window layer.

In contrast to the p-type layer, microcrystalline n-layers are easier to grow, and both fluorinated and nonfluorinated n-type layers have been made [6.82, 83]. As we have discussed before, V_{oc} of the cell is primarily determined by the major junction [6.31], and use of the microcrystalline n-layer does not increase V_{oc}. The optical loss, however, is lower, and, recently, microcrystalline n-layer has been used successfully to obtain high efficiency multijunction devices [6.7].

6.4.6 Intrinsic Layers

High Bandgap Alloys for the Top Cell. In order to capture the blue light of the solar spectrum, it is necessary to use i-layers with optical gap greater than 1.8 eV. There are two main approaches to obtain the higher bandgap. The first approach uses a-SiC alloy [6.84–87]. In the second approach, a-Si alloy deposited at low temperature and high hydrogen dilution is used [6.52, 88–90].

We have mentioned before that addition of carbon-containing feedstock gases to silane can increase the bandgap of the alloy. Introduction of C, however, introduces new states in the gap and makes the transport properties poorer. This is caused by the poor microstructure of a-SiC alloys [6.91] which, in turn, may be attributed to the presence of high density of C–H$_x$ bonds. A wide range of feedstock gases has been explored [6.84] to improve the material properties. Materials with the best minority carrier transport property, which leads to superior solar cell performance, are made from a hydrogen mixture of methane and silane diluted in hydrogen and deposited at $\approx 200°C$. As the ratio of the flow of methane to silane increases, V_{oc} appears to saturate at 1.03–1.04 V. The fill factor, however, goes down with the increasing flow of methane, indicating poorer material quality with increasing carbon incorporation. For cells with an intrinsic layer thickness of ≈ 100 nm, the following initial cell characteristics were obtained: $V_{oc} = 1.01$ V, $J_{sc} = 7$ mA cm^{-2} and FF $= 0.74$. The poorer microstructure of a-SiC alloy results in larger LID. For the very thin cells needed for the top of the triple-cell structure, stable cell performance of the following characteristics [6.84] has been achieved: $V_{oc} = 0.98$ V, $J_{sc} \approx 7$ mA cm^{-2} and FF $= 0.68$. This performance is reasonable for the design of high efficiency triple-junction cells. For double-junction cells, however, one needs $J_{sc} > 11$ mA cm^{-2} for the top cell. The large amount of light-induced degradation associated with thicker cells made from a-SiC alloy makes this material impractical for use in double-junction cells unless dramatic improvement in material properties are made.

The cells made with a-SiC alloy as the intrinsic layer use p-layers of a-SiC also. A buffer layer between the doped p and the intrinsic layer has been found to be useful in improving device efficiency. The buffer layer is typically 10 to 15 nm thick and has a graded bandgap [6.92] that can be fabricated by changing the C-content. The buffer layer reduces interface recombination that may otherwise occur at the band discontinuity at the p-i interface; the graded bandgap also increases the electric field at the interface, resulting in better device performance. We should mention that a similar buffer layer is also used to improve efficiency of cells where the p-type layer is a-SiC alloy and the i-layer is a-Si alloy.

The second approach for making the top cell is to use hydrogen-diluted a-Si alloy deposited at a low temperature for the i-layer and microcrystalline silicon alloy for the p-layer. The highest stable fill factor for a cell producing more than $7\,\text{mA}\,\text{cm}^{-2}$ is 0.72 [6.88] which is much higher than that reported for top cells producing comparable current density using a-SiC alloy in the intrinsic layer. With the rapid progress that is being made in increasing the efficiency of multijunction solar cells, there is a need for designing top cells that will produce in excess of $8\,\text{mA}\,\text{cm}^{-2}$. This has been reported [6.93] recently, and the following characteristic in the stable light-degraded state has been obtained: $V_{oc} = 0.953\,\text{V}$, $J_{sc} = 8.78\,\text{mA}\,\text{cm}^{-2}$ and $FF = 0.684$. This amounts to an efficiency of 5.93% which is the highest reported in the literature for the top cell. Increasing the hydrogen content of the i-layer can give even higher V_{oc}, and the highest V_{oc} reported is $1.054\,\text{V}$ with a fill factor of 0.76 [6.88].

High bandgap alloys have also been made by incorporating oxygen or nitrogen in the alloy; the minority transport properties are, however, poor and solar cell results are not available [6.94].

Low Bandgap Alloys for Middle and Bottom Cells. The middle and the bottom cells need to absorb the green and the red photons, and the bandgap of the i-layer needs to be lower than what is typically obtained for a-Si alloy (1.7–1.8 eV). Addition of germane to the gas mixture can continuously change the bandgap from 1.7–1.1 eV. a-SiGe alloys are, therefore, used extensively for the middle and the bottom cells.

A wide range of gas mixture has been used to deposit a-SiGe alloys [6.13]. The most commonly used active gases are silane and germane. These two gases have very different dissociation rates in the plasma. This causes a deterioration in the material quality and also compositional nonuniformity along the direction of the gas flow. While high hydrogen dilution can improve the film quality, the problem of nonuniformity in deposition remains. Guha et al. [6.95] have used a mixture of disilane and germane to deposit a-SiGe alloy. Since disilane and germane have similar dissociation rates in the plasma, the problem of compositional inhomogeneity is significantly reduced. Both disilane and germane are easily dissociated in the plasma and in order to eliminate powder formation and to reduce deposition rate under normal plasma conditions, hydrogen is used as a diluent. Earlier work has also used SiF_4 in the gas mixture. Even though the film properties improve remarkably when deposited from a gas mixture of disilane and germane diluted in hydrogen, they are still poorer than those of a-Si alloy only. There are many reasons for this. The compositional disorder in a-SiGe alloy makes the conduction band tail wider [6.96, 97]. The microstructure is also poorer [6.91], probably due to lower adatom mobility of GeH_3 molecules. The presence of Ge–Ge bonds also makes the material quality poorer. Although significant advances have been made [6.96] in improving the material properties of a-SiGe alloys, there is almost a continuous degradation of transport properties as Ge content in the alloy is increased. Cells incorporating a-SiGe alloy in the i-layer thus have poorer fill factor than the corresponding a-Si alloy cell.

A novel way of improving the efficiency of a-SiGe alloy cells was reported by Guha et al. [6.98] by profiling the composition of a-SiGe alloy throughout the bulk of the intrinsic material so as to have a built-in electrical field in a substantial portion of the intrinsic material. Different types of profiling configurations are shown in Fig. 6.15. Two simple cases were considered first: (1) the bandgap is minimum at the p-i interface through which light enters and increases linearly away from the interface (normal profiling), and (2) the bandgap is maximum at the p-i interface and decreases away from it (reverse profiling). From band structure considerations, assuming that both the conduction and the valence band edge positions shift equally as Ge in incorporated in the material, intuitively one would expect that normal profiling will help hole transport and make electron transport more difficult. The opposite will be the case for reverse profiling. Since hole transport is the limiting factor determining a-Si alloy solar cell performance, one would therefore expect an improvement in the fill factor for the case of normal profiling. Note that in this case, the maximum number of carriers is generated right at the p-i interface and the holes have a very short distance to move. However, since the open-circuit voltage of the solar cell is determined by the major junction that is at the p-i interface, one expects a higher open-circuit voltage for the reverse profiling where the bandgap for the i-layer at the p-i interface is larger.

Using a computer simulation model developed by Pawlikiewicz and Guha [6.30] for amorphous silicon alloy p-i-n solar cells, Guha et al. [6.98] studied the performance of the devices using structures with and without profiling. Results for red-light incident on the cells are shown in Table 6.6. In agreement with the previously mentioned intuitive arguments, the computer simulation demonstrates that, FF is enhanced for normal profiling and V_{oc} is increased for reverse profiling. One can then take advantage of the two situations and use a double-profiled structure as shown in Fig. 6.16d. Computer simulation results show that this gives the best performance. Both V_{oc} and FF improve over the case when the composition is uniform. The improvement in V_{oc} depends on the thickness of the region of reverse profiling. As the thickness increases, V_{oc} increases; however, FF starts decreasing at larger thickness since the holes now have to move against a potential barrier over a larger distance.

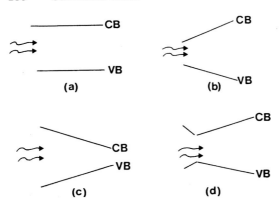

Fig. 6.16. Different types of profiling configurations; (**a**) no profiling; (**b**) normal profiling; (**c**) reverse profiling; (**d**) double profiling

Table 6.6. Simulated results for a-SiGe alloy solar cells with different types of profiling. The results are for red-light illumination

	Constant bandgap 1.52 eV	Normal profiling 1.52–1.72 eV	Reverse profiling 1.72–1.52 eV	Double profiling
J_{sc} (mA/cm^2)	1.64	1.51	1.57	1.55
V_{oc} (V)	0.67	0.70	0.76	0.72
FF	0.55	0.65	0.44	0.64
Power (mA/cm^2)	0.62	0.68	0.52	0.71

Experimental results for V_{oc}, FF and J_{sc} for three structures with no profiling, normal profiling and reverse profiling are shown in Table 6.7. The results are for red-light illumination. As theoretically predicted, normal profiling is seen to give better fill factor, and reverse profiling gives higher open-circuit voltage. Further evidence that the improvement in fill factor is caused by better hole transport came from dynamic internal collection efficiency (DICE) measurements. Significant enhancement in the collection efficiency for carriers generated near the n-i interface was found when normal profiling was done. The holes generated there would have to traverse the full length of the sample in order to be collected. Computer simulation results also showed that the internal electric field for holes is higher for normal profiling, and it is therefore understandable that the probability for hole collection in this case would be better.

The highest efficiency devices for the middle and the bottom cells use a multiple-profiled a-SiGe alloy in the i-layer. Several groups have [6.99–6.101] used this method to design high efficiency cells. The middle cells typically use 1.6 eV bandgap material at the minimum position. For the bottom cell, the corresponding bandgap is ≈ 1.4 eV.

Table 6.7. Experimental results for a-SiGe alloy cells with different types of profiling. The results are for red-light illumination

	Constant bandgap 1.55 eV	Normal profiling 1.45–1.72 eV	Reverse profiling 1.72–1.45 eV
J_{sc} (mA/cm^2)	3.43	3.68	3.48
V_{oc} (V)	0.73	0.68	0.74
FF	0.57	0.68	0.37
Power (mW/cm^2)	1.43	1.72	0.93

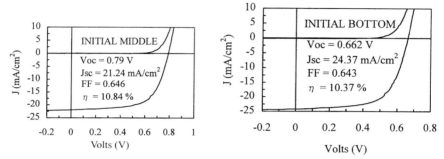

Fig. 6.17a,b. Characteristics of a-SiGe alloy cells with Ge content suitable for (**a**) the middle and (**b**) the bottom cells

The highest initial efficiency of a-SiGe alloy cells where the i-layer uses an alloy with Ge content suitable for the middle cell is shown in Fig. 6.17a. The efficiency under AM1.5 illumination is 10.84%. When an alloy of Ge content suitable for the bottom cell is used, the highest initial efficiency is 10.37% (Fig. 6.17). The high level of Ge in the i-layer results in a quantum efficiency of 45% at 850 nm, indicating excellent red response characteristic [6.90].

6.4.7 Optimization of the Component Cells and Current Matching

In order to optimize the component cells, they have to produce the desired current density from the light that will be absorbed in that cell. The top and the middle cell will not see much reflected light, and hence, these component cells are deposited on stainless steel without any back reflector to simulate their performance in the actual triple-cell structure. Moreover, the middle cell will not see any blue light, and hence, its performance is measured with light filtered with a 530 nm cut-on filter. Degradation studies are also carried out using the same filtered light. The bottom cells are deposited on the back reflector, but their performance is measured, and degradation studies are carried out using light filtered with a 630 nm cut-on filter.

Degradation studies carried out [6.102] on the component cells of a triple-junction structure are shown in Fig. 6.18. The top cell is found to degrade by about 10–11%, the middle and the bottom cells by about 17–18%. Since the top cell has the largest fill factor both in the initial and the degraded state, it is desirable to design a multijunction structure in which the current is limited by the top cell. We should mention that all the three component cells show saturation after prolonged light exposure [6.102].

6.4.8 Tunnel Junction

A multijunction cell incorporates one or more internal "tunnel" junctions at the interface of the n- and p-layers of the adjacent component cells. Any parasitic junction loss at the "tunnel" junction, electrical or optical, leads to a deleterious effect on the overall characteristics of the device [6.103]. The property of the doped layers can have significant effect on the "tunnel" junction and thereby may affect the multijunction device performance.

A new technique was developed by Banerjee et al. [6.104] to characterize the internal "tunnel" junction of a multijunction cell. The device structure used for this purpose, referred to as the NIPN structure, consists of the bottom $n_1i_1p_1$ cell followed by the n_2 layer. The NIPN device is coated with ITO dots for measurement. The I–V characteristics of the NIPN structure have been found to be sensitively dependent on the quality of the p_1/n_2 "tunnel" junction. Evaluation of the NIPN device, therefore, provides a powerful way to investigate the quality of the internal junctions in a multijunction cell.

The NIPN studies were used to investigate the effect of thickness of the microcrystalline p-layer on the quality of the "tunnel" junction. The n-layer was amorphous. For this purpose, NIPN devices incorporating a-Si:H i-layer were fabricated, employing different thicknesses of the p-layer. A corresponding set of a-Si alloy n i p cells were also fabricated for comparison. Figure 6.19 shows a plot of V_{oc} and FF of both the NIPN and n i p devices as a function of p-layer deposition time. The values of both V_{oc} and FF initially increase with deposition time and finally saturate in both cases. For the n i p structure, the values of V_{oc} and FF are higher than those for the NIPN structure at low deposition time. A deposition time for the p-layer of as low as 20 s is found to give V_{oc} of 0.92 V; only a 5 s deposition time is adequate to give a FF of 0.72. In contrast, much thicker p-layers with deposition time greater than 40 s are required to give comparable values of V_{oc} and FF for the NIPN structure. The highest values of V_{oc} and FF, corresponding to the longest p-layer deposition time of 80 s, are similar for both cases and are 0.95 V and 0.75, respectively. The study shows that relatively thick p-layers are required to form good quality "tunnel" junctions. Thus, a multijunction cell requires thicker p-layers for efficient operation of the internal junctions than that required on the top for the p/ITO junction.

Banerjee et al. also used this tool to evaluate the quality of the p-layer necessary to obtain optimum performance from the tunnel junction. They

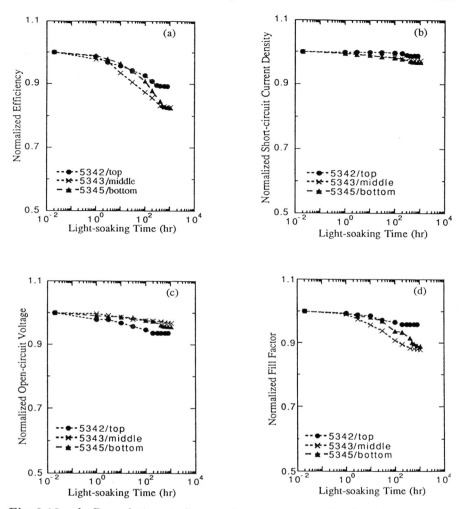

Fig. 6.18a–d. Degradation studies on the component cells of a triple-junction structure

concluded that even though microcrystalline p-layers grown by low frequency glow discharge are adequate to get high V_{oc} in a single-junction device, good tunnel junction characteristic is obtained only by using p-layers grown using high (rf) frequency glow discharge. They suggest that in addition to having an adequate recombination, avoidance of intermixing of the p- and n-layers is critical to obtain good tunnel junction characteristics.

Since the microcrystalline n-layer is more transparent and conducting than its amorphous counterpart, attempts were made by Yang et al. [6.17] to use this material to obtain improved tunnel junction characteristics. A pair

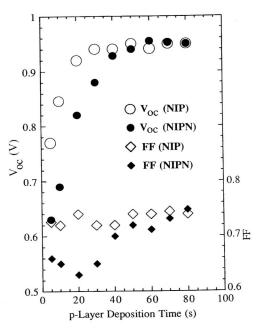

Fig. 6.19. Open circuit voltage (V_{oc}) and fill factor (FF) of n i p and n i p n devices as a function of p-layer thickness

of a-Si/a-Si double-junction cells was made under substantially the same deposition conditions. The only difference between the two devices is that one uses an amorphous n-layer, while the other uses a microcrystalline n-layer in the tunnel junction. Even though the use of microcrystalline n-layer allows more light to go to the bottom cell, it was found that the V_{oc} value for the tandem cell having a microcrystalline n-layer is much lower than the corresponding cell with an amorphous n layer in the tunnel junction. The lowering of V_{oc} may be due to band edge discontinuity between the amorphous i and microcrystalline n or the intermixing of dopants in thin microcrystalline layers, or both.

Yang et al. have developed a thin buffer layer that can be inserted between the microcrystalline p- and n-layers and also between the microcrystalline n- and the adjacent amorphous i-layers to alleviate the problem. The results are shown in Table 6.8. Incorporation of the new tunnel-junction structure gives rise to improved cell characteristics, as evidenced by the higher total current and lower series resistance. Incorporation of such tunnel junction in the triple-cell structure has resulted in obtaining high efficiency.

Table 6.8. Characteristics of a-Si/a-Si tandem cells with different tunnel junction structures

Tunnel Junction structure	V_{oc} (V)	J_{sc} (mA/cm^2)	FF (%)	η	Q_{top} (mA/cm^2)	Q_{btm}	Q_{total} (mA/cm^2)	R_s (Ω cm^2)
Micro-crystalline p/ Amorphous n	1.901	7.80	0.752	11.15	7.97	7.8	15.77	15.0
Micro-crystalline p/ Multi-layered n	1.919	8.06	0.766	11.85	8.06	8.28	16.34	14.3

6.4.9 Top Conducting Oxide

The top conducting oxide in the substrate-type configuration is usually ITO. Both reactive evaporation and sputtering have been used successfully. A quarter wavelength thickness is used to serve the purpose of antireflection coating. Typical sheet resistivity is 50–100 Ω cm^{-2}, and evaporated grid lines are deposited onto the ITO to reduce series resistance losses. In order to reduce optical losses, the ITO layer needs to be very transparent. Typical loss associated with state-of-the-art ITO lowers J_{sc} by about 0.5 mA cm^{-2} [6.105].

For the superstrate-type structure, the light enters through the conducting oxide that is directly deposited on the glass. Both tin oxide and zinc oxide have been used. The conducting oxide is usually thick, 500 to 1000 nm, to obtain the necessary texture, and this leads to lowering of J_{sc} by 3–4 mA cm^{-2}.

6.4.10 Cell and Module Performance

There are many groups in the world working in the industries, universities and other research institutions that are involved in a-Si alloy solar cell research. Performance of single-, double- and triple-junction cells has been reported by many groups. Some of the highest stable cell efficiencies achieved are shown in Table 6.9. We report only the stable efficiencies since the initial values do not reflect the performance in the real world. It is interesting to note that as one moves from single-junction to the same bandgap double-junction structure, the stabilized efficiency increases from 9.2% to 10.2%. Incorporation of Ge in the bottom cell increases the efficiency further to 11.2%. Use of a triple-junction, triple-bandgap structure results in the highest stable efficiency of 13%. We should point out that unless there is a fundamental solution to light-induced degradation, the triple-junction structure is the only option by which further improvement in efficiency can be made. In order to improve the stable efficiency of the double-junction cell, the top cell has to be made thicker to increase current density, and that will lead to further degradation.

The highest stable cell efficiencies have been achieved using the substrate-type structure. As discussed earlier, it is easy to incorporate the microcrys-

Table 6.9. Highest stable cell efficiencies reported by different groups

Cell structure	Efficiency	Group	Reference
Single-junction	9.2%	United Solar, USA	[6.106]
	9.1%	TIT, Japan	[6.107]
	8.9%	Sanyo, Japan	[6.108]
	8.5%	APS, USA	[6.109]
	7.3%	Utrecht, The Netherlands	[6.110]
Double-junction, same bandgap	10.1%	United Solar, USA	[6.106]
	9.2%	Jülich, Germany	[6.111]
	9.2%	Neuchatel, Switzerland	[6.114]
	9.0%	Rome, Italy	[6.112]
	8.7%	Kyung Hee Univ., Korea	[6.113]
Double-junction, dual bandgap	11.2%	United Solar, USA	[6.115]
	10.6%	Sanyo, Japan	[6.108]
	9.6%	Solarex, USA	[6.116]
	9%	Kyung Hee Univ., Korea	[6.113]
	8.7%	Mitsui Toatsu, Japan	[6.117]
Triple junction, multibandgap	13%	United Solar, USA	[6.7]
	10.2%	Sharp, Japan	[6.118]
	9.15%	Solarex, USA	[6.119]

talline p window layer in this structure since it is deposited on an a-Si alloy layer rather than on an oxide layer. Moreover, the first cell through which the light enters is usually deposited at a lower temperature to increase the bandgap. In the superstrate structure, this is the cell that is deposited first, and it is exposed to higher temperature during subsequent operation that is not desirable. In the substrate-type structure, however, this is the top cell and is deposited last.

The highest cell efficiencies reported by United Solar take advantage of all the innovations reported in Sect. 6.4. The design of the triple-junction cells depended on the component cell performance. This is shown in Table 6.10. Also shown in the table is the performance of the component cells needed to obtain 16% stable cell efficiency [6.120]. It is apparent that improvement in the performance of all the component cells is necessary to meet this goal. There has been steady progress in improving the efficiency of the triple-junction cell, and this has been achieved through improvements of the component cells and the tunnel junctions. Figure 6.20 shows the progress in stable cell efficiency, and Table 6.10 shows the corresponding progress in the component cell performance. There is a great deal of activity going on to improve materials and devices, and it is expected that progress in efficiency will continue.

The cell results have been translated into large-area modules also. For improving module efficiency, one has to pay attention to uniformity over

Table 6.10. Efficiency of component cells and the triple-junction structure

Cell	Status	J_{sc} (mA/cm^2)	V_{oc} (V)	FF	P_{max} (mW/cm^2)
top[d]	initial[a]	8.97	0.980	0.761	6.69
middle[e]	initial[b]	9.30	0.753	0.687	4.81
bottom[c]	initial[c]	12.2	0.631	0.671	5.17
triple[d]	initial[c]	8.57	2.357	0.723	14.6
top[d]	1994 stable[b]	7.3	0.97	0.72	5.1
	1997 stable	8.78	0.953	0.709	5.93
	Goal	8.3	1.10	0.75	6.8
middle[e]	1994 stable[b]	7.6	0.76	0.60	3.5
	1997 stable[b]	8.92	0.717	0.587	3.75
	Goal	8.4	0.89	0.70	5.2
bottom[f]	1994 stable[c]	9.84	0.67	0.55	3.60
	1996 stable[c]	11.4	0.60	0.56	3.80
	1997 stable[c]	11.12	0.609	0.622	4.21
	Goal	> 8.6	0.68	0.68	> 4.0
triple[d]	1995 stable[c]	6.87	2.38	0.68	11.1
	1996 stable[c]	7.49	2.283	0.692	11.8
	1997 stable[c]	8.27	2.294	0.684	13.0
	Goal	8.3	2.67	0.72	16.0

[a] deposited on a bare stainless steel substrate, [b] deposited on a textured substrate coated with Cr, [c] deposited on textured Ag/ZnO back reflector, [d] measured under AM1.5 illumination, [e] measured under AM1.5 with a $\lambda > 530$ nm filter, [f] measured under AM1.5 with a $\lambda > 630$ nm filter

the large-area and grid and encapsulation losses. The best results for stable module efficiencies [9] are shown in Table 6.11. At this time, there is a large gap between the highest cell and module efficiencies (13% vs. 10.2%). This is mostly contributed by nonuniformity issues, and with a focused effort on large-area research, the gap can be narrowed down substantially.

6.5 Manufacturing Technology

6.5.1 Manufacturing Process

Early introduction for a-Si alloy solar cells was for calculators and other consumer applications. It is only recently that large-volume production plants have been built to address the terrestrial power market [6.121]. Two different approaches have been adopted for production for the substrate- and superstrate-type solar cell structures. Detailed discussions are given below.

A roll-to-roll manufacturing process has been developed [6.8] for the production of solar panels using the substrate-type structure. The manufacturing

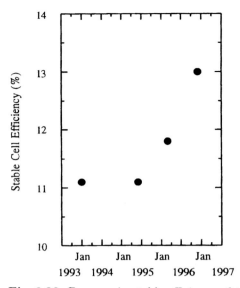

Fig. 6.20. Progress in stable efficiency of triple-junction cells

Table 6.11. Stable efficiency of a-Si alloy PV modules

		Efficiency (%)	Area (cm^2)
United Solar	Si/SiGe	9.5	902
	Si/SiGe/SiGe	10.2[a]	903
	Si/SiGe/SiGe[b]	7.9[a]	9276
ECD	Si/Si/SiGe[b]	7.8[a]	3906
Solarex	Si/Si/Ge	9.1[a]	842
	Si/Si/SiGe	8.8	863
Fuji	Si/Si	8.9[a]	1200
Sanyo	Si/SiGe	9.5	1200
APS	Si[b]	4.4[a]	11634
	Si/Si	7.1	848
	Si/Si	4.6	11522

[a] Verified by measurements at NREL; [b] made using production machine

process for a recently constructed facility of 5 MW annual production capacity [6.9] consists of the following steps. A roll of stainless steel, half-a-mile long, 14 inches wide and 5 mil thick, moves in a continuous manner at a speed of 2 foot a minute in four machines that serve the purpose of (*i*) washing, (*ii*) depositing the back reflector, (*iii*) depositing the a-Si and a-SiGe alloy layers, and (*iv*) depositing indium tin oxide (ITO) which serves as an antireflection coating. Both the transport of the web and the process parameters are computer-controlled ensuring reliable and low-cost operation. The coated web is next processed to make a variety of lightweight, flexible and rugged products. The processing steps involve (1) cutting of the web into slabs, (2) short and shunt passivation and etching of ITO to define strip-cell area, (3) attaching electrodes and grids, and (4) final assembly involving strip cutting, interconnection of the strips and lamination.

A brief description of the various operations is given below.

Wash Machines. The roll of stainless is washed in a roll-to-roll processing system which transports the web through a detergent cleaning station, multiple deionized water rinsing baths, and an infrared drying oven. The clean, dry and dust-free web is next loaded into the back reflector machine.

Back Reflector Machine. The back reflector machine sequentially deposits a reflective metal layer and a metal oxide "buffer" layer onto the cleaned stainless steel web by magnetron sputtering. The back reflector layers provide the ohmic contact between the stainless steel and the a-Si alloy. The layers are deposited at a high temperature to obtain a textured surface so as to facilitate multiple reflections.

Amorphous Silicon Alloy Deposition Machine. The web coated with the back reflector is next loaded into the a-Si alloy processor (Fig. 6.21) which is 140 feet long and has, in addition to the pay-off and take-up chambers, nine chambers to deposit the nine layers of the triple-cell structure. The adjacent chambers are separated by proprietary gas gates to eliminate contamination of the dopant gases in the intrinsic layers. The individual layers are grown by plasma-enhanced chemical vapor deposition process at a pressure of about 1 torr, and all the layers are deposited simultaneously and consecutively on the moving web (Fig. 6.22) to complete the triple-junction cell structure. Special cathode designs ensure improved gas utilization and uniformity of the deposited layers. The gas manifolds for introducing disilane and germane for the a-SiGe alloy component cells are specially designed to facilitate bandgap profiling [6.98]. The process conditions used for the deposition of the p-type layer facilitate microcrystalline growth [6.78].

Indium Tin Oxide Deposition Machine. The final step in the deposition process is reactive magnetron sputtering of ITO which serves as the antireflection coating and also provides the top conducting contact. Typical deposition temperature is 200°C.

Module Assembly Operation. The module assembly operation is semiautomated to facilitate flexibility in the choice of the product line while ensuring low cost and reliability. The finished roll of the coated web is first cut into $9.4'' \times 14''$ slabs using a semiautomated press; small coupons ($3.8'' \times 14''$) are also cut during the same operation at preset intervals along the length of the web. These coupons are processed off-line for QA/QC evaluation which will be discussed later. The slabs are then processed to define cell size, passivated to remove shunts and shorts [6.122] and tested to ascertain quality.

Fig. 6.21. A-Si alloy triple-junction cell deposition machine

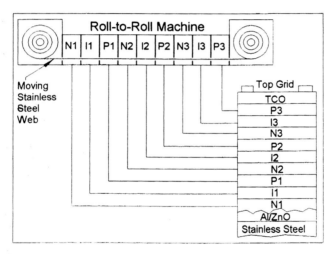

Fig. 6.22. The roll-to-roll operation for the deposition of the triple-junction cell

Grid wires and contact pads are next applied, and the slabs are cut into predetermined cell sizes for the various product requirements. The cells are next interconnected and the cell block laminated to provide protection against outside atmosphere. Depending on the application, frames and junction boxes are added and the finished modules undergo a highpot test and performance measurement under global AM1.5 illumination before they are shipped out.

In order to reduce product cost, several important departures from the R&D design were made: (*i*) The deposition rates for the intrinsic alloys were

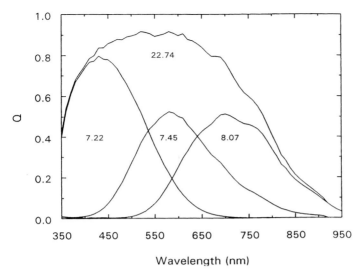

Fig. 6.23. Quantum efficiency of a triple-junction cell deposited on an Al/ZnO back from the production line

increased and the component cell thicknesses reduced to improve machine throughput, (*ii*) the back reflector uses Al rather than Ag, and (*iii*) deposition regimes were chosen in which the flow rates of the active gases such as germane and disilane were kept low. Imposition of the above constraints lowered the conversion efficiency of the cells and modules, but helped in improving the dollar per watt cost criterion.

The back reflector layers are deposited by sputtering from Al and ZnO targets; the optimization process involved changing the texture of the Al/ZnO by varying the process parameters and the layer thicknesses. Quantum efficiency measurements on both the bottom a-SiGe alloy cell and the triple-junction structure were used for evaluating the quality of the back reflector. The quantum efficiency plot for a triple-junction cell on an optimized back reflector made on a production run is shown in Fig. 6.23. The total current density of $22.74\,\text{mA/cm}^2$ and the quantum efficiency of 0.21 at 850 nm indicate the superior quality of the back reflector.

The optimization process for the a-Si alloy processor involved extensive experimentation on component cells, tunnel junctions and triple-cell structures to improve efficiency without sacrificing the machine throughput and gas utilization. Measurements on the QA coupons provide information about the quality and yield. An array of 2×10 cells of $10\,\text{cm}^2$ area are defined on each coupon, and the average subcell efficiency and the yield for a typical run are shown in Fig. 6.24. Many production runs have been made to date, and the efficiency and the yield numbers are extremely consistent.

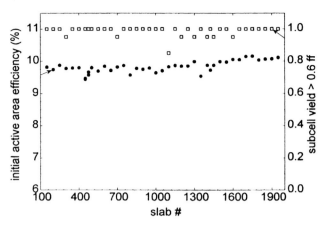

Fig. 6.24. Initial active-area efficiency (●) and sub-cell yield (□) versus slab number for a 2000 ft production run

For the superstrate-type solar panels, the manufacturing steps are as follows. Window glass plates are first washed; a thin (≈ 50 nm) layer of silicon dioxide is deposited onto the glass followed by about 600–1000 nm of tin oxide (Fig. 6.25). Usually these layers are deposited by atmospheric chemical vapor deposition. A silver frit is next applied to the tin oxide and cured in a furnace to form the conducting bus bars. The tin oxide is then laser scribed to form the front contacts of the individual cell segments (Fig. 6.26). The glass plate is next loaded into the deposition systems where the different layers of the a-Si alloys are deposited followed by the deposition of the back reflector layer (Al) or layers (Al and ZnO). Two more laser scribing steps complete the module integration. The first step is to scribe the contact layer(s), and the second step fuses the metal back reflector to the tin oxide to make the series interconnect. Final encapsulation is done by bonding another glass plate to the cell structure.

Instead of using glass or stainless steel, polymeric substrates (Kapton) have also been used [6.123, 124]. Since the substrate is insulating, monolithic modules can be made, and the flexible nature of the substrates allows the roll-to-roll manufacturing process. Since the top conducting oxide in the substrate-type modules is thin, grid lines or wires are necessary to carry the current, and this could give rise to shadow losses of about 5%. Two new approaches have recently been suggested [6.124, 125] for the fabrication of substrate-type solar modules which reduce the shadow losses due to the grid lines significantly. In the first approach [6.125], an insulating oxide is first deposited on the stainless steel to isolate the substrate from the back reflector. Using a two-step laser-drilling approach, the top conducting oxide can be connected to the substrate without touching the back reflector. The stainless steel and the back reflector then form the two electrodes. The laser-drilled

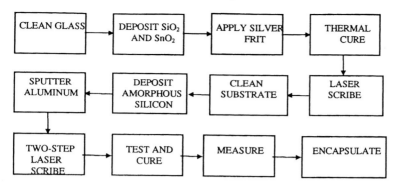

Fig. 6.25. The flowchart of a manufacturing line for making superstrate-type solar modules

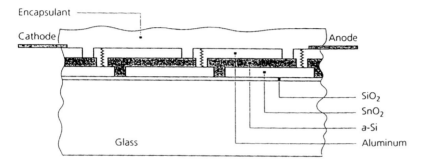

Fig. 6.26. Schematic diagram of a series-interconnected superstrate-type solar module

holes can be made very small, and since there are no other grid lines, the shadow losses can be less than 1%. The second approach [6.124] is also somewhat similar except that holes are drilled onto the Kapton substrate, and the conducting oxide makes contact with a metal layer deposited onto the back of the Kapton.

6.5.2 Production Status and Product Advantage

Amorphous silicon alloy solar cells and panels have been in use for consumer applications for more than a decade. The spectral response of a-Si alloy to fluorescent light is very good, and hence a-Si alloy solar cells have been used in calculators. These are usually single-junction solar cells, although same-bandgap double-junction cells have also been used. A number of manufacturers have also been producing a-Si alloy solar panels for battery-charging applications. A significant increase in a-Si alloy production capacity took place in 1997 when three manufacturers completed construction of production plants of total capacity of 25 MW per year. The total sales of a-Si alloy PV products for 1997 is 15 MW [6.121]. The products range from small cells for calculators to large modules ($20' \times 1'$) for PV roofing applications.

For any new product, it has to compete against the more mature products that have been around for a long time, and a-Si alloy products are no exception. During the last few years, however, several new, attractive features of a-Si alloy products have been established resulting in a wider acceptance of these products. Since the bandgap of a-Si alloy is much higher than that of crystalline or polycrystalline silicon, the power output from an a-Si alloy panel is remarkably independent of temperature. For example, for a typical ambient temperature of $30°$C, the module temperature can be $60°$C. The power output from an a-Si alloy panel barely changes in this temperature range [6.126], whereas that for a crystalline silicon panel can change by about 20%. For the same power rating, a-Si alloy PV products, therefore, perform much better than the crystalline counterparts under normal outdoor conditions. This is a significant advantage both for grid-connected and battery charging applications. Independent studies carried out in Phoenix, Arizona by Photocomm, Inc. [6.127] on United Solar a-Si alloy panels show that the battery charging capacity is 30% higher than that of a polycrystalline panel of the same power rating. As can be seen from Fig. 6.27 the daily output from the polycrystalline panel is 45 Ah, whereas that for the United Solar triple-junction module is 60 Ah. These advantages of a-Si alloy technology are not very widely known. Thin-film products can, of course, be made lightweight and flexible, and there is a tremendous opportunity for these products in the building-integrated PV market.

The stability of a-Si alloy modules under normal outdoor condition is also not an issue any longer. NREL has been monitoring the performance of an array consisting of 102 United Solar modules since April 1993. Another array consisting of 64 modules is being evaluated since January 1994. The long-term data show degradation less than 1% per year [6.126] which is "equal to or better than crystalline silicon systems." Similar stability data have also been reported on a-Si alloy modules made by other manufacturers [6.128–130].

Although we have discussed until now use of a-Si alloy for terrestrial applications only, a-Si alloy technology is receiving attention for space applications as well. Historically, cost has not been an issue while deciding on a PV technology for space application, and the main emphasis has been on efficiency. The trend, however, is changing rapidly. The communication revolution demands greater bandwidth, and a constellation of satellites and spaceships is to be launched to meet this demand. Low cost and light weight are the two key requirements for these space applications, and a-Si alloy solar panels on thin stainless steel or Kapton are ideally suited for this purpose. a-Si alloy is also extremely radiation hard, and it has been shown recently [6.131] that the radiation in a typical space environment can be annealed out at the actual temperatures prevailing there. Conventional crystalline silicon solar cells, on

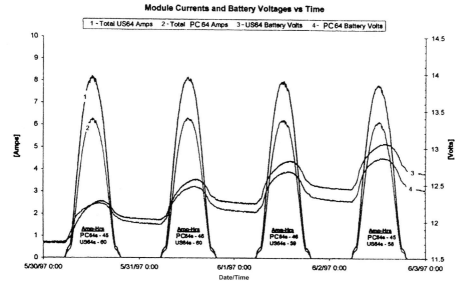

Fig. 6.27. Comparison of battery charging performance of United Solar a-Si alloy solar panel with that of a commercial polycrystalline solar panel with the same standard power rating (64 W)

the other hand, degrade by as much as 20% for the same doses of radiation. The superior performance of a-Si alloy cells at high temperature (the actual ambient temperature in space) is also an additional advantage. There is a great deal of international activity going on now to qualify a-Si alloy solar cells for space applications.

6.6 Alternative Technologies and Future Trends

Several other approaches have been used to fabricate cells using alternative materials and deposition techniques. Microcrystalline silicon is a promising candidate to replace a-SiGe alloy in a multijunction cell since it absorbs the infrared photons efficiently. The conventional microcrystalline material contains many grain boundaries that hurt the transport of carriers. The Neuchatel group has pioneered the use of very high frequency (vhf) for making high efficiency solar cells using microcrystalline silicon [6.132]. High quality material could be obtained by using hydrogen dilution and by lowering the level of oxygen contamination. Addition of small quantities of B also improve the minority carrier transport by shifting the Fermi-level near the mid-gap. A single-junction cell with a stabilized efficiency of 7.7% has been reported. Using a double-junction structure where the top and the bottom cells use a-Si and microcrystalline alloys respectively, a stable cell efficiency of 10.7% has been reported. Yamamoto et al. [6.133] have reported an efficiency of

9.8% using microcrystalline silicon grown by rf glow discharge at a temperature less than 550°C. Use of a high quality back reflector has resulted in the achievement of J_{sc} of 27 mA cm^{-2} using an i-layer of 4.7 μm thickness. Both double-junction and triple-junction cells were made with a-Si alloy in the top cell, and preliminary results indicated a stable efficiency of 11.5%.

The above results are very encouraging and open up new research areas. The major issue that needs to be addressed before this technology is commercialized is the improvement of the deposition rate. In contrast to the case of a-SiGe alloy for the bottom cell where an i-layer thickness of 100 nm is adequate, use of a microcrystalline i-layer needs a thickness that is 30 times more. Unless the deposition rate is increased 30 times, the throughput will be low. As discussed earlier, a-SiGe alloy is typically deposited at 0.3 nm/s in production machines. A deposition rate of about 9 nm/s is, therefore, required for microcrystalline silicon to achieve the same throughput. Recent results indicate dilution of Ar in the gas mixture improves the deposition rate when used in conjunction with vhf [6.134]. This is an area that merits further attention.

As mentioned earlier, the highest quality a-Si alloy solar cells using rf glow discharge are made at around 0.1 nm/s. Increasing the deposition rate of a-Si alloy is also extremely important to throughput and to reduce capital cost. The deposition rate can be easily increased by using higher power density, but this results in the formation of higher silanes in the plasma with subsequent poor cell performance. Higher deposition rates also give less time for the impinging species on the growing surface to find energetically favorable sites, thereby giving rise to poorer microstructure. vhf has been used extensively to increase deposition rate [6.22]. The increased growth rate has been attributed to improved dissociation of the process gas, to the decreased sheath thickness and also to an enhancement of surface reactivity by positive ions. Growth rates of 3 nm/s have been achieved without significant deterioration of material quality, and it has been suggested that the film quality does not deteriorate at these high deposition rates because of high doses of low energy bombardment [6.135] which may improve structure. Although preliminary cell results have been presented, a systematic study of cell stability as a function of deposition rate is yet to be done. Further increase in deposition rates can be achieved by going to even higher frequencies as, for example, microwaves [6.24, 25]. Both a-Si and a-SiGe alloys have been grown at deposition rates exceeding 5 nm/s using microwaves. Even though an initial efficiency of 11.2% has been obtained [6.24] using a double-junction structure, stability is poorer perhaps because of the increased void density in the materials grown at these high rates. Positive ion bombardment lowers the void density and improves the cell performance [6.136]. The stability is still poorer than that of cells grown at low deposition rates. An increase of deposition rate by hot-wire chemical vapor deposition has also been reported by several workers [6.137, 138]. This method was first used to grow high quality materials and devices at low deposition rates [6.138] and showed promising results in terms of improved stability. Another interesting technique is the use of the gas jet concept where a source gas is directed at high speeds through a jet nozzle pointed at the heated substrate [6.139]. The gas is activated during the transit and allows deposition at high rates. Preliminary cell results presented appear to be promising. We should mention that studies on materials and cells deposited by different methods as a function of different deposition rates are important because they can help us distinguish between the effects of plasma chemistry and growth kinetics on cell performance.

Where does one go from here? The progress in our understanding of a-Si alloy materials and devices has been spectacular in the last ten years, and this has resulted in significant improvement in cell and module efficiencies and, perhaps more importantly, large-scale expansion of production capacity. Many challenges still remain. The progress in cell and module efficiency needs to be maintained. We have mentioned before that the structure of the material plays the most important role in improving cell performance. The best material shows a more ordered structure and has been grown with high hydrogen dilution. Recent results indicate improved structure and cell stability with deuterium dilution as well [6.140]. Materials grown with hot wire also show more ordered structure [6.141]. While the conventional rf glow-discharge method still produces the highest efficiency cell and has not exhausted its potential, there are many new growth techniques that appear to be promising [6.142]. These include deposition using chemical annealing [6.143, 144], fluorinated or chlorinated precursors [6.145, 146], electron cyclotron resonance [6.147], and also involve a triode configuration [6.148]. Based on the studies of material properties using photoconductivity and sub-bandgap absorption, many of these materials exhibit quality comparable to that of the best quality rf-glow discharge alloy. Efforts to incorporate these materials in actual cell structures have not yet resulted in the achievement of a higher efficiency. This is primarily due to two reasons. Information about the minority carrier transport is most important in predicting solar cell performance. Photoconductivity is a measure of the majority carrier transport; sub-bandgap measurement provides information about the defect density in a certain region of the energy of the mobility gap, but does not give an unambiguous determination of the minority carrier transport property. High photoconductivity and low sub-bandgap absorption are, therefore, necessary but not sufficient conditions to obtain high efficiency. Moreover, both these measurements are carried out on films in a coplanar configuration, whereas solar cells need optimization of transport properties in the transverse direction. Capacitance measurements indeed give information about the material in the solar cell structure; hence, this technique has been used extensively [6.149] to investigate the properties of both a-Si and a-SiGe alloys. One limitation of this method is the requirement of rather thick films ($\approx 1\,\mu m$).

The evaluation of the material in the actual solar cell structure, therefore, seems to be the most desirable method. This also brings out unique problems that may exist in incorporating a material grown using a novel technique in a solar cell. A good example is the attempt of hot-wire experimentalists to make solar cells. The required deposition at a high temperature needed substantial changes in the deposition methods to make them compatible with cell fabrication [6.150].

New tools are also being developed or old tools modified to improve our understanding of materials and devices [6.151]. Use of these measurement techniques give valuable information about material and device parameters which, in turn, lead to improved device modeling [6.152]. The device models are based on one-dimensional transport, whereas since the devices are grown on the textured back reflectors, the current transport is actually two-dimensional. New models are being developed [6.153, 154] to take this into account, and this will help us in designing improved devices.

The physics and technology of a-Si alloy have come a long way since the first solar cells were made. The recent scale-up of production has already demonstrated the viability of the product for the highly competitive terrestrial PV market. There is a tremendous economy-of-scale in this technology, and projections by industry analysts [6.155] show a fully-loaded cost of less than 50 c/W when plants of annual capacity of 100 MW will be built. This will result in PV plants generating electricity at a cost that is competitive with that obtained from fossil fuels. Large-scale use of the clean PV energy will, of course, reduce pollution and mitigate global warming, both of which will have enormous societal benefits.

Acknowledgements

The author is grateful to J. Yang, A. Banerjee, T. Glatfelter, H. Fritzsche and S. R. Ovshinsky for discussions and to V. Trudeau for preparation of the manuscript. This work was supported in part by the National Renewable Energy Laboratory under Subcontract No. ZAK-8-17619-09.

References

6.1 D. M. Chapin, C. S. Fuller, and G. L. Pearson, J. Appl. Phys. **25**, 676 (1954).
6.2 R. C. Chittick, J. H. Alexander, and H. F. Sterling, J. Electrochem. Soc. **116**, 77 (1969).
6.3 W. E. Spear and P. G. LeComber, J. Non-Cryst. Solids **8–10**, 727 (1972).
6.4 W. E. Spear and P. G. LeComber, Phil. Mag **33**, 935 (1976).
6.5 W. E. Spear, P. G. LeComber, S. Kinmond, and M. H. Brodsky, Appl. Phys. Lett. **28**, 105 (1976).
6.6 D. E. Carlson and C. R. Wronski, Appl. Phys. Lett. **28**, 671 (1976).
6.7 J. Yang, A. Banerjee, and S. Guha, Appl. Phys. Lett. **70**, 2975 (1997).
6.8 S. R. Ovshinsky, in *Intl. Tech. Digest PVSEC-1* (Tokyo, Japan, 1984), p. 577; P. Nath, K. Hoffman, J. Call, C. Vogeli, M. Izu, and S. R. Ovshinsky, in *Intl. Tech. Digest PVSEC-3* (Tokyo, Japan, 1987), p. 395; M. Izu, X. Deng, A. Krisko, K. Whelan, R. Young, H. C. Ovshinsky, K. L. Narasimhan, and S. R. Ovshinsky, in *Conference Record of the Twenty Third IEEE Photovoltaic Specialist Conference 1993* (IEEE, New York, 1993), p. 919; S. Guha, Mater. Res. Soc. Symp. Proc. **336**, 645 (1994).
6.9 S. Guha, J. Yang, A. Banerjee, K. Hoffman, S. Sugiyama, J. Call, S. J. Jones, X. Deng, J. Doehler, M. Izu and H. C. Ovshinsky, in *Conference Record of the Twenty Sixth IEEE Photovoltaic Specialist Conference 1997* (IEEE, New York, 1997), p. 607.
6.10 A. Triska, D. Dennison, and H. Fritzsche, Bull. Am. Phys. Soc. **20**, 392 (1975).
6.11 R. A. Street, Hydrogenated Amorphous Silicon (Cambridge University Press, Cambridge and New York, 1991).
6.12 Amorphous Silicon and Related Materials, edited by H. Fritzsche (World Scientific, Singapore, 1988).
6.13 W. Luft and Y. S. Tsuo, Hydrogenated Amorphous Silicon Alloy Deposition Processes (Marcel Dekker, New York, 1993).
6.14 J. C. Knights, Mater. Res. Soc. Symp. Proc. **38**, 372 (1985).
6.15 A. Gallagher, Annual Subcontract Report, 15 April 1986 June 1987, STR-211-3288, Solar Energy Research Inst., 1987.
6.16 A. Matsuda and K. Tanaka, Thin Solid Films **92**, 171 (1982).
6.17 A. Matsuda, in *Conference Record of the Twenty Fifth IEEE Photovoltaic Specialist Conference 1996* (IEEE, New York, 1996), p. 1029
6.18 S. Guha, J. Yang, S. J. Jones, Y. Chen, and D. L. Williamson, Appl. Phys. Lett. **61**, 1444 (1992).
6.19 S. Guha, K. L. Narasimhan, and S. M. Pietruszko, J. Appl. Phys. **52**, 859 (1981).
6.20 D. V. Tsu, B. S. Chao, S. R. Ovshinsky, S. Guha, and J. Yang, Appl. Phys. Lett. **71**, 1317 (1997).
6.21 C. C. Tsai, M. Stutzmann, and W. B. Jackson, Amer. Inst. of Physics Conf. Proc. **120**, 242 (1984).
6.22 H. Curtins, N. Wyrsch, M. Favre, and A. V. Shah, Plasma Chem. Plasma Process **7**, 267 (1987).
6.23 A. Shah, U. Kroll, H. Keppner, J. Meier, P. Torres, and D. Fischer, in *Intl. Tech. Digest PVSEC-9* (Tokyo, Japan, 1996), p. 267.
6.24 S. Guha, X. Xu, J. Yang, and A. Banerjee, Appl. Phys. Lett. **66**, 595 (1995).

6.25 K. Saito, M. Sano, J. Matsuyama, M. Higasikawa, K. Ogawa, and I. Kajita, in *Intl. Tech. Digest PVSEC-9* (Tokyo, Japan, 1996), p. 579.
6.26 S. J. Fonash, Solar Cell Device Physics (Academic Press, New York, 1981).
6.27 G. A. Swartz, J. Appl. Phys. **53**, 712 (1982).
6.28 R. J. Schwartz, J. L. Gray, G. B. Turner, D. Kanani, and H. Ullal, in *Conference Record of the Seventeenth IEEE Photovoltaic Specialist Conference 1984* (IEEE, New York, 1984), p. 369.
6.29 M. Hack and M. Shur, J. Appl. Phys. **59**, 998 (1985).
6.30 A. H. Pawlikiewicz and S. Guha, IEEE Trans. Electron Devices **ED-37**, 403 (1990).
6.31 P. Sichanugrist, M. Konagai, and K. Takahashi, J. Appl. Phys. 55, 1155 (1984).
6.32 J. K. Arch, F. A. Rubinelli, J. Y. Hou, and S. J. Fonash, J. Appl. Phays. **69**, 7057 (1991).
6.33 Q. Wang, R. S. Crandall, and E. A. Schiff, in *Conference Record of the Twenty Fifth IEEE Photovoltaic Specialist Conference 1996* (IEEE, New York, 1996), p. 1113.
6.34 L. Jiao, H. Liu, S. Semoushika, Y. Lee, and C. R. Wronski, ibid, (1996), p. 1073.
6.35 K. Vasanth, M. Nakata, S. Wagner, and M. Bennett, Mater. Res. Soc. Symp. Proc. **297**, 827 (1993).
6.36 M. Hack, S. Guha, and M. Shur, Phys. Rev. B **30**, 6991 (1984).
6.37 M. Hack, S. Guha, and M. Shur, in AIP Conf. Proc. **120**, 40 (1984)
6.38 D. L. Staebler, C. R. Wronski, Appl. Phys. Lett. **31**, 292 (1977).
6.39 *Amorphous Silicon materials and Solar Cells*, edited by B. L. Stafford (American Institute of Physics, New York, 1991). Conf. Proc. 234.
6.40 See, for example, Mater. Res. Symp. Proc. (1997).
6.41 H. Dersch, J. Stuke, and J. Beichler, Appl. Phys. Lett. **38**, 456 (1980).
6.42 W. den Boer, M. J. Geerts, M. Ondris, and H. M. Wentinick, J. Non-Cryst. Solids **60**, 268 (1984).
6.43 S. Guha, J. Yang, W. Czubatyj, S. J. Hudgens, and M. Hack, Appl. Phys. Lett. **42**, 589 (1983).
6.44 D. L. Staebler, R. S. Crandall, and R. Williams, Appl. Phys. Lett. **39**, 733 (1981).
6.45 S. Guha, Appl. Phys. Lett. **45**, 569 (1984).
6.46 M. Stutzmann, W. B. Jackson, and C. C. Tsai, Phys. Rev. B **32**, 23 (1985).
6.47 S. Yamasaki and J. Isoya, J. Non-Cryst. Solids **164–166**, 169 (1993).
6.48 P. Stradins and H. Fritzsche, Phil. Mag. B **69**, 121 (1994).
6.49 H. M. Branz, Solid State Comm. **105/6**, 387 (1998).
6.50 R. Biswas, Q. Li, B. C. Pan, and Y. Yoon, Mater. Res. Symp. Proc. **467**, 135 (1997).
6.51 R. Biswas and B. C. Pan, Appl. Phys. Lett. **72**, 371 (1998).
6.52 T. Kamei, N. Hata, A. Matsuda, T. Uchiyama, S. Amano, K. Tsukamoto, Y. Yoshioka, and T. Hirao, Appl. Phys. Lett. **68**, 2380 (1996).
6.53 H. Fritzsche, Mater. Res. Soc. Symp. Proc. **467**, 19 (1998).
6.54 T. Gotoh, S. Nonomura, M. Nishio, N..Masui, S. Nitta, M. Kondo, and A. Matsuda, in *Proceedings of the 17th Conf. Amer. Microcryst. Semiconductors*, Budapest, 1997 (to appear).
6.55 X. Xu, J. Yang, and S. Guha, J. Non-Cryst. Solids 198-200, 60 (1996).

6.56 Y. Kuwano, M. Ohnishi, H. Nishiwaki, S. Tsuda, T. Fukatsu, K. Enomoto, Y. Nakashima, and H. Tarui, in *Conference Record of the Sixteenth IEEE Photovoltaic Specialist Conference* 1982 (IEEE, New York, 1982), p. 1338.

6.57 G. D. Cody and T. Tiedje, in *Energy and the Environment*, edited by B. Abeles, J. Jacobson, and P. Sheng (World Scientific, Singapore, 1991), p. 147.

6.58 E. Yablonovitch and G. D. Cody, IEEE Trans. Electron Devices ED-29, 303 (1982).

6.59 A. Banerjee and S. Guha, J. Appl. Phys. **69**, 1030 (1991).

6.60 A. Banerjee, J. Yang, K. Hoffman, and S. Guha, Appl. Phys. Lett. **65**, 472 (1994).

6.61 S. Guha, Optoelectronics 5, 201 (1990).

6.62 B. Sopori (private communication).

6.63 J. Morris, R. R. Arya, J. G. O'Dowd, and S. Wiedeman, J. Appl. Phys. **67**, 1079 (1990).

6.64 J. Kane, Photodetector with Enhanced Light Absorption, U.S. Patent No. 4,532,537 (30 July 1985).

6.65 R. G. Gordon, J. Proscia, F. B. Ellis, and A. E. Delahoy, Sol. Energy Mater. 18, 263 (1989).

6.66 K. L. Chopra, R. C. Kainthla, D. K. Pandya, and A. P. Thakoor, Physics of Thin Films, Vol. 12 (Academic Press, New York, 1982).

6.67 R. G. Gordon, AIP Conf. Proc. **394**, 39 (1997).

6.68 R. A. Street, Phys. Rev. Lett. **49**, 1187 (1982).

6.69 R. A. Street, D. K. Biegelsen, W. B. Jackson, N. M. Jackson, and M. Stutzmann, Phil. Mag. B 52, 235 (1985).

6.70 N. M. Amer and W. B. Jackson in *Semiconductors and Semimetals*, Vol. 21, Part B, edited by J. Pankove (Academic Press, New York 1984), p. 93.

6.71 Y. Tawada, K. Tsuge, M. Kondo, H. Okamoto, and Y. Hamakawa, J. Appl. Phys. **53**, 5273 (1982).

6.72 A. Catalano, R. V. D'Aiello, J. Dresner, B. Faughman, A. Firester, J. Kane, H. Schade, Z. E. Smith, G. Schwartz, and A. Triano, in *Conference Record of the Sixteenth IEEE Photovoltaic Specialist Conference* 1982 (IEEE, New York, 1982), p. 1421.

6.73 H. Tarui, T. Matsuyama, S. Okamoto, Y. Hishikawa, H. Dohjo, N. Nakamura, T. Tsuda, S. Nakanao, M. Ohnishi, and Y. Kuwano, in *Intl. Tech. Digest PVSEC-3* (Tokyo, Japan, 1987), p. 41.

6.74 B. Fieselmann and B. Goldstein, Mater. Res. Soc. Symp. Proc. **149**, 441 (1989).

6.75 Y. M. Li, F. Jackson, L. Yang, B. F. Fieselman, and L. Russel, Mater. Res. Soc. Symp. Proc. **336**, 663 (1994).

6.76 See, for example, A. Catalano in Amorphous & Microcrystalline Semiconductor Devices: Optoelectronic Devices, edited by J. Kanicki (Artech House, Norwood, MA 1991), p. 9.

6.77 A. Catalano and G. Wood, J. Appl. Phys. **63**, 1220 (1988).

6.78 S. Guha, J. Yang, P. Nath, and M. Hack, Appl. Phys. Lett. **49**, 218 (1986).

6.79 S. Guha and J. Kulman, U.S. Patent No. 4,600,801 (15 July 1986).

6.80 Y. D. Kruangam, K. Katoh, Y. Nitta, H. Okamoto, and Y. Hamakawa, in *Conference Record of the Nineteenth IEEE Photovoltaic Specialist Conference 1987* (IEEE, New York, 1987), p. 689.

6.81 B. Goldstein, C. R. Dickson, I. H. Campbell, and P. M. Fauchet, Appl. Phys. Lett. **53**, 2672 (1988).

6.82 R. Tsu, S. S. Chao, M. Izu, S. R. Ovshinsky, G. J. Jan, and F. H. Pollack, J. Phys. (France), C **4**, 269 (1981).

6.83 K. Nakatani, M. Yano, K. Suzuki, and H. Okaniwa, J. Non-Cryst. Solids, 59-60, 827 (1983).

6.84 Y. M. Li, Mater. Res. Soc. Symp. Proc. **297**, 803 (1993).

6.85 T. Yoshida, K. Maruyama, O. Nabeta, Y. Ichikawa, H. Sakai, and Y. Uchida, Mater. Res. Soc. Symp. Proc. **149**, 477 (1989).

6.86 K. Nomoto, Y. Takeda, S. Moriuchi, H. Sannomiya, T. Okuno, A. Yokota, M. Kaneiwa, M. Itoh, Y. Yamamoto, Y. Nakata, and T. Inoguchi, in *Intl. Tech. Digest PVSEC-4* (Australia, 1989), p. 85.

6.87 K. Hanaki, Y. Hattori, H. Nakabayashi, S. Yamaguchi, and Y. Hamakawa, Mater. Res. Soc. Symp. Proc. **148**, 385 (1989).

6.88 J. Yang, X. Xu, A. Banerjee, and S. Guha, in *Conference Record of the Twenty Fifth IEEE Photovoltaic Specialist Conference 1996* (IEEE, New York, 1996), p. 1041.

6.89 S. Guha, J. Yang, A. Banerjee, T. Glatfelter, and S. Sugiyama, in *Intl. Tech. Digest PVSEC-9* (Tokyo, Japan, 1996), p. 283.

6.90 J. Yang, A. Banerjee, T. Glatfelter, K. Hoffman, X. Xu, and S. Guha, in *Conference Record of the Twenty Fourth IEEE Photovoltaic Specialist Conference 1994* (IEEE, New York, 1994), p. 380.

6.91 D. L. Williamson, Mater. Res. Soc. Symp. Proc. **377**, 251 (1995).

6.92 R. R. Arya, A. Catalano, and R. S. Oswald, Appl. Phys. Lett. **49** (17), 1089 (1986).

6.93 J. Yang, A. Banerjee, S. Sugiyama, and S. Guha, Proceedings of the 2nd World Conference and Exhibition on Photovoltaic Solar Energy Conversion, Vienna, Austria, 1998 (to be published).

6.94 M. J. Williams, S. M. Cho, and G. Lucovsky, Mater. Res. Soc. Symp. Proc. **297**, 759 (1993).

6.95 S. Guha, J. S. Payson, S. C. Agarwal, and S. R. Ovshinsky, J. Non-Cryst. Solids **97–98**, 1455 (1987).

6.96 S. Aljishi, Z. E. Smith, and S. Wagner in Amorphous Silicon and Related materials, edited by H. Fritzsche (World Scientific, Singapore, 1989), p.887.

6.97 Q. Wang, H. Antoniadis, E. A. Schiff, and S. Guha, Phys. Rev. B 47, 9435 (1993).

6.98 S. Guha, J. Yang, A. Pawlikiewicz, T. Glatfelter, R. Ross, and S. R. Ovshinsky, Appl. Phys. Lett. **54**, 2330 (1989).

6.99 Y. Nakata, H. Sannomiya, S. Moriuchi, A. Yokota, Y. Inoue, M. Itoh, and H. Itoh, Mater. Res. Soc. Symp. Proc. **192**, 15 (1990).

6.100 J. Zimmer, H. Stiebig, J. Folsch, F. Finger, T. Eickhoff, and H. Wagner, Mater. Res. Soc. Symp. Proc. **467**, 735 (1997).

6.101 V. L. Dalal and G. Baldwin, Mater. Res. Soc. Symp. Proc. 297, 833 (1993).

6.102 X. Xu, J. Yang, and S. Guha, in *Conference Record of the Twenty Third IEEE Photovoltaic Specialist Conference 1993* (IEEE, New York, 1993), p. 971.

6.103 J. H. Hou, J. K. Arch, S. J. Fonash, S. Wiedeman, and M. Bennett, in *Conference Record of the Twenty Second IEEE Photovoltaic Specialist Conference 1991* (IEEE, New York, 1991), p. 1260.

6.104 A. Banerjee, J. Yang, T. Glatfelter, K. Hoffman, and S. Guha, Appl. Phys. Lett. **64**, 1517 (1994).
6.105 K. Hoffman and T. Glatfelter, in *Conference Record of the Twenty Third IEEE Photovoltaic Specialist Conference 1993* (IEEE, New York, 1993), p. 986.
6.106 S. Guha, J. Yang, A. Banerjee, T. Glatfelter, and S. Sugiyama, in *Proceedings of the Fourteenth European Photovoltaic Solar Energy Conference and Exhibition* (Barcelona, Spain, 1997), p. 2679.
6.107 M. Konagai (private communication).
6.108 Y. Hishikawa, K. Ninomiya, E. Maruyama, S. Kuroda, A. Terakawa, K. Sayama, H. Tarui, M. Sasaki, S. Tsuda, and S. Nakano, in *Conference Record of the Twenty Fourth IEEE Photovoltaic Specialist Conference 1994* (IEEE, New York, 1994), p. 386.
6.109 J. Xi, T. Liu, V. Iafelice, K. Si, and F. Kampas, Mater. Res. Soc. Symp. Proc. **336**, 681 (1994).
6.110 R. E. I. Schropp (private communication).
6.111 B. Rech, S. Weider, F. Siebke, C. Beneking, and H.Wagner, Mater. Res. Soc. Symp. Proc. **420**, 33 (1996).
6.112 E. Tarzini, A. Rubino, R. de Rosa, in *Proceedings of the Fourteenth European Photovoltaic Solar Energy Conference and Exhibition* (Barcelona, Spain, 1997), p. 570.
6.113 K. C. Park, T. G. Kim, S. K. Kim, S. C. Kim, M. H. Hwang, J. M. Jun, and J. Jang, Mater. Res. Soc. Symp. Proc. **297**, 767 (1993).
6.114 R. Platz, D. Fischer, C. Hof, S. Dubaii, J. Meier, U. Kroll, and A. Shah, Mater. Res. Soc. Symp. Proc. **420**, 51 (1996); B. Rech (private communication).
6.115 J. Yang and S. Guha, Appl. Phys. Lett. **61**, 2917 (1992).
6.116 Y. M. Li, L. Yang, M. S. Bennett, L. Chen, F. Jackson, K. Rajan, and R. R. Arya, Mater. Res. Soc. Symp. Proc. **336**, 723 (1994).
6.117 M. Sadamoto, H. Tanaka, N. Ishiguro, N. Yanagawa, and S. Fukuda, in *Intl. Tech. Digest PVSEC-9* (Tokyo, Japan, 1996), p 645.
6.118 K. Nomoto, H. Saitoh, A. Chida, H. Sannomiya, M. Itoh, and Y. Yamamoto, in *Intl. Tech. Digest PVSEC-7* (Nagoya, Japan, 1993), p. 275.
6.119 R. R. Arya, L. Yang, M. Bennett, J. Newton, Y. M. Li, B. Fieselmann, L. Chen, K. Rajan, G. Wood, C. Poplawski, and A. Wilczynski, in *Conference Record of the Twenty Third IEEE Photovoltaic Specialist Conference 1993* (IEEE, New York, 1993), p. 790.
6.120 S.Guha, in *Intl. Tech. Digest PVSEC-7* (Nagoya, Japan, 1993), p. 395.
6.121 P. Maycock, PV News, February 1998.
6.122 P. Nath, K. Hoffman, C. Vogeli, and S. R. Ovshinsky, Appl. Phys. Lett. **53**, 986 (1988).
6.123 J. J. Hanak, E. Chen, C. Fulton, A. Myatt, and J. Woodyard, Proc. 8th. PV Res. Tech. Conf., NASA Conf. Pub. **2475**, 99 (1986); F. R. Jeffrey, D. P. Grimmer, S. Brayman, B. Scandrett, and M. Noak, AIP Conf. Proc. **394**, 451 (1996).
6.124 K. Tabuchi, S. Fujikake, H. Sato, S. Saito, A. Takano, T. Wada, T. Yoshida, Y. Ichikawa, H. Sakai, and F. Natsume, in *Conference Record of the Twenty Sixth IEEE Photovoltaic Specialist Conference 1997* (IEEE, New York, 1997), p. 611.

6.125 T. Glatfelter, M. Lycette, E. Akkashian, and K. Hoffman, in *Conference Record of the Twenty Fifth IEEE Photovoltaic Specialist Conference 1996* (IEEE, New York, 1996), p. 1173.
6.126 B. Kroposki and R. Hansen, in *Conference Record of the Twenty Sixth IEEE Photovoltaic Specialist Conference 1997* (IEEE, New York, 1997), p. 1359.
6.127 Data obtained from tests done by Photocomm. Inc. at its Scottsdale, Arizona facility.
6.128 M. Kondo, H. Nishio, S. Kurata, K. Hayashi, and Y. Tawada, in *Intl. Tech. Digest PVSEC-9* (Tokyo, Japan, 1996), p 557.
6.129 M. Kameda, S. Sakai, Y. Hishikawa, S. Tsuda, and S. Nakano, in *Conference Record of the Twenty Fifth IEEE Photovoltaic Specialist Conference 1996* (IEEE, New York, 1996), p. 1049.
6.130 K. Fukae, C. Lim, M. Tamechika, N. Takehara, K. Saito, I. Kajita, and E. Kondo, in *Conference Record of the Twenty Fifth IEEE Photovoltaic Specialist Conference 1996* (IEEE, New York, 1996), p. 1227.
6.131 S. Guha, J. Yang, and A. Banerjee, Proceedings of the 2nd World Conference and Exhibition on Photovoltaic Solar Energy Conversion, Vienna, Austria, 1998 (to be published).
6.132 A. Shah, J. Meier, H. Keppner, P. Torres, and D. Fischer, in *Conference Record of the Twenty Sixth IEEE Photovoltaic Specialist Conference 1997* (IEEE, New York, 1997), p. 569.
6.133 K. Yamamoto, M. Yoshimi, T. Suzuki, and A. Nakajima, in *Conference Record of the Twenty Sixth IEEE Photovoltaic Specialist Conference 1997* (IEEE, New York, 1997), p. 575.
6.134 H. Meiling, J. Bezemar, R. E. I. Schropp, and W. F.van der Weg, Mater. Res. Soc. Symp. Proc. **467**, 459 (1997).
6.135 M. Heize, ibid **467**, 471 (1997).
6.136 S. Sugiyama, X. Xu, J. Yang, and S.Guha, Mater. Res. Soc. Symp. Proc. **420**, 197 (1996).
6.137 K. F. Feenstra, C. H. M. van der Werf, E. C. Molenbroek, and R. E. I.Schropp, Mater. Res. Soc. Symp. Proc. **467**, 645 (1997).
6.138 A. H. Mahan, J. Carapella, B. P. Nelson, R. S. Crandall and I. Balberg, J. Appl. Phys. **69**, 6728 (1991); A. H. Mahan and M. Vanecek, AIP Conf. Proc. **234**, 195 (1991).
6.139 S. J. Jones, A. Myatt, H. Ovshinsky, J. Doehler, M. Izu, A. Banerjee, J. Yang, and S. Guha, in *Conference Record of the Twenty Sixth IEEE Photovoltaic Specialist Conference 1997* (IEEE, New York, 1997), p. 659.
6.140 S. Sugiyama, J. Yang and S. Guha, Appl. Phys. Lett. **70**, 378 (1997).
6.141 A. H. Mahan, D. L. Williamson, and T. E. Furtak, Mater. Res. Soc. Symp. Proc. **467**, 657 (1997).
6.142 For a recent review, please see, S. Guha, Curr. Opinion. in Sol. State & Mat. Sc. **2**, 425 (1997).
6.143 K. Yoshino, W. Futako, Y. Wasai, and I. Shimizu, Mater. Res. Soc. Symp. Proc. **420**, 335 (1996).
6.144 W. Futako and I. Shimizu, Mater. Res. Soc. Symp. Proc. **420**, 431 (1996).
6.145 S. R. Ovshinsky, Solar Energy materials and Solar Cells **32**, 443 (1994).
6.146 M. Payne and S. Wagner, Mater. Res. Soc. Symp. Proc. **420**, 883 (1996).
6.147 V. Dalal, S. Kaushal, R. Girvan, S. Hariasra, and L. Sipahi, in *Conference Record of the Twenty Fifth IEEE Photovoltaic Specialist Conference 1996* (IEEE, New York, 1996), p. 1069.

6.148 G. Ganguly, T. Ikeda, I. Sakata, A. Matsuda, K. Kato, S. Iizuka, and N. Sato, Mater. Res. Soc. Symp. Proc. **420**, 347 (1996).

6.149 C. C. Chen, F. Zhong, and J. D. Cohen, Mater. Res. Soc. Symp. Proc. **420**, 581 (1996).

6.150 A. H. Mahan, E. Iwaniczko, B. P. Nelson, R. C. Reedy, Jr., R. S. Crandall, S. Guha, and J. Yang, in *Conference Record of the Twenty Fifth IEEE Photovoltaic Specialist Conference 1996* (IEEE, New York, 1996), p. 1065.

6.151 R. W. Collins, J. Koh, Y. Lu, S. Kim, J. S. Burnham, and C. R. Wronski, in *Conference Record of the Twenty Fifth IEEE Photovoltaic Specialist Conference 1996* (IEEE, New York, 1996), p. 1035; C. N. Yeh, D. Han, Q. Wang, and Y. Q. Xu, Mater. Res. Soc. Symp. Proc. **420**, 63 (1997); Y. Tang, R. Braunstein, B. von Roedern, and F. R. Shapiro, Mater. Res. Soc. Symp. Proc. **297**, 407 (1993).

6.152 S. Bae and S. J. Fonash, in *Conference Record of the Twenty Fourth IEEE Photovoltaic Specialist Conference 1994* (IEEE, New York, 1994), p. 484.

6.153 Y. Hishikawa, E. Maruyuma, S. Yata, M. Tanaka, S. Kiyama, and S. Tsuda, in *Intl. Tech. Digest PVSEC-9* (Tokyo, Japan, 1996), p. 639.

6.154 M. Zeman, J. H. van den Berg, L. L. A. Vostun, J. A. Willemen, J. W. Metselaar, and R. E. I. Schropp, Mater. Res. Soc. Symp. Proc. **467**, 671 (1997).

6.155 P. Maycock, in *Proceedings of the Fourteenth European Photovoltaic Solar Energy Conference and Exhibition* (Barcelona, Spain, 1997), p. 869.

7 Multilayer Color Detectors

Fabrizio Palma

Università di Roma "La Sapienza", Dipartimento di Ingegneria Elettronica,
Via Eudossiana 18, I-00184 Roma, Italy
E-mail: palma@die.ing.uniroma1.it

Abstract. Amorphous silicon technology can be favorably employed to achieve a new generation of integrated color sensitive detectors. The high absorption coefficient, together with the ease of fabricating multi-layered structures, offer many opportunities for new devices. The structures take further advantage of two particular properties of a-Si:H, that it can be deposited on large area by glow discharge and at relatively low temperature. This chapter describes the amorphous silicon color detectors, along with their device structure and addressing architecture. A brief introduction to related photo-detectors for infrared and ultraviolet detection is also presented.

7.1 Introduction

Amorphous silicon technology is attractive for large area image detection because the materials can be deposited on large area substrates by glow discharge or hot wire deposition. The low deposition temperature allows the use of low cost substrates such as glass or plastic. Hydrogenated amorphous silicon (a-Si:H) has a high absorption coefficient so that very thin films absorb visible radiation efficiently, which offsets the relatively poor electronic quality of the material. Large area arrays of detectors have been developed to achieve page size document imagers without mechanical scanning [7.1] and X-ray image detectors [7.2].

The low deposition temperature also makes this technology compatible with integrated circuits, which can thus be used as substrates, coupling the properties of a-Si:H detectors with microelectronics circuit capable of handling information. This technique has already been used to implement a-Si:H color detectors [7.3]. Artificial retina arrays, able to be implanted inside the human eye, are currently under development [7.4].

This chapter discusses the design and fabrication of a-Si:H detectors with the capability to distinguish color, and the possibility to extend the range of detectivity into the UV and IR.

Color sensitivity in crystalline silicon video cameras is obtained by using different detection channels with three CCD arrays, or by using a mosaic of filters deposited directly onto one single solid state sensor (the latter is especially used in the consumer product market) [7.5]. This method can also be used for large area amorphous silicon sensor arrays and requires three sensors for each pixel. This choice is not simple in large area application due to the complexity of addressing a very large number of detectors, and the decreasing device yield as the number of pixels increases.

Several attempts to achieve structures capable of modifying their sensitivity spectrum by simply changing the applied bias have been reported in the literature and are describe in this chapter. This approach dramatically simplifies the interconnections since only two terminals are necessary, although greater complexity is demanded in the driving sequence.

These detectors are based on two fundamental properties of amorphous silicon structures. First, glow discharge deposition allows the design of complex thin film structures with a high degree of freedom in the choice of layers and the junction sequence. Second, to a certain extent, the absorption coefficient of the deposited material can be chosen by changing the deposited material. The optical gap of amorphous semiconductors based on a-Si:H technology, ranges from approximately 3.0 eV to 1.2 eV by using amorphous carbon, amorphous silicon, amorphous germanium, and/or alloys of these materials. The choice of the absorber material naturally must take into account its electronic properties. Among amorphous semiconductors, intrinsic amorphous silicon has the lowest defect density. The defect density rapidly increases as the concentration of other species increases in the alloy composition. The degradation of the electronic properties of the device is balanced only by using a very thin junction absorber layer, thus limiting the freedom to choose the structure dimensions.

The possibility to choose an arbitrary sequence of junctions, to design the absorption profile of the device and attention to effects of these choices on the electronic properties are the basic ingredients of the color detectors described here.

7.1.1 Applications of a-Si:H Color Sensors

There are several application opportunities for a-Si:H color sensors. The main characteristics of this technology are the large area substrates and low temperature deposition. It is thus possible to have a large area scanner of colored pictures without mechanical motion. This concept can be extended to robotics applications obtaining simple sensors capable of detecting the position and color of an object, thus providing more information for its recognition.

A number of industrial applications require spectral analysis; as an example we may recall flame sensors [7.6]. A-SiH technology provides a relatively inexpensive solution to this problem. Spectral analysis obtained by numerical processing of measurements performed by a three color detector, with 30 nm resolution, has been demonstrated [7.7]. Multi-point detection could be obtained by directly depositing the a-Si:H detector on the end portion of an optical fiber.

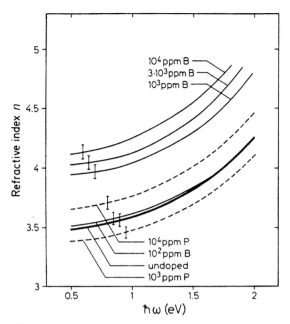

Fig. 7.1. Refractive index $n(\omega)$ of a-Si:H at various doping levels [7.11]

Alloys with germanium are employed in multijunction solar cells, where they form the bottom device, which is sensitive to long-wavelength light. The red-shifted spectral sensitivity provided by alloying with Ge is also used for making electro-photographic layers sensitive to light from GaAs light-emitting diodes [7.8], and in photodetectors sensitive to GaAs laser light [7.9].

7.2 Optical Properties of Amorphous Silicon

The optical design of planar layered structures requires knowledge of the complex refractive index of the materials. In this section we summarize the optical properties of a-Si:H, with particular attention to the influence of deposition condition, and a more complete review can be found in reference [7.10]. We also describe some analytical tools for the optical design of the structures.

Values of the refractive index, refractive index, n of a-Si:H range from 2.9 to 3.7, with a pronounced frequency dispersion. Generally, samples with high hydrogen contents or those prepared from SiH_4 diluted in Ar or He show a lower value of n. The refractive index of high-quality glow discharge films made from pure silane is close to the crystalline value of 3.4 [7.11]. As shown in Fig. 7.1 there is a pronounced influence of doping on n, in particular with boron. This behavior is related to the fact that the mass density of the

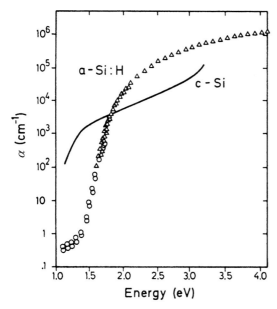

Fig. 7.2. The absorption coefficient of a-Si:H compared to the absorption coefficient of c-Si [7.12]

films increases with boron doping [7.11]. Inside a multilayer structure these differences may play a relevant role, since the junction structure is made up of a sequence of doped and intrinsic layers.

A second relevant quantity in optical characterization is the absorption coefficient, $\alpha(E)$. Figure 7.2 shows the absorption coefficient of a-Si:H deposited at 240°C, as measured by optical transition and photoconductivity, compared to crystalline silicon. A-Si:H has an absorption edge at higher energy than crystalline silicon, and above this energy, its absorption coefficient is about an order of magnitude higher than the crystal. This property is due to the loss of momentum conservation in the disordered structure and ensures that visible light is absorbed within 1 µm of film thickness. This high absorption is the basis of the application of a-Si:H in photovoltaics (Chap. 6).

Although there is no specific structure in the absorption coefficient around the absorption edge, it is nevertheless common to define an optical gap in order to characterize the film properties. The optical gap is defined in different ways, for example E_{03} or E_{04} are the photon energies where $\alpha = 10^{-3}$ or 10^{-4} cm^{-1}, respectively. Alternatively, the Tauc definition is used [7.13], for which it is assumed that optical transitions take place between the extended states of the valence band and of the conduction band, whose density of states have a square root energy dependence and that the momentum matrix element is constant. This last assumption has been demonstrated by Jackson to be valid up to 3.4 eV. The Tauc model gives the following description of α,

$$(\alpha \hbar \omega)^{1/2} = B(\hbar \omega - E_g), \tag{7.1}$$

where $\hbar$ is Planck's constant divided by 2π, ω is the frequency, and B a proportionality constant. This expression defines the energy of the optical gap, such that in a plot of $(\alpha \hbar \omega)^{1/2}$ versus photon energy, the linear extrapolation to zero yields the value of E_g.

The value of E_g depends on the hydrogen concentration, C_H, in the film [7.14]. E_g amounts to 1.2–1.5 eV in evaporated and sputtered material and increases linearly with C_H in hydrogenated films. This result relates the position of the absorption edge to the removal of states from the top of the valence band [7.15], due to bonding with hydrogen atoms which form a more stable complex, generating states with energy deeper in the extended states band [7.16].

In the range $\alpha < 10^3$ cm^{-1} the absorption coefficient depends exponentially on the energy, forming the so called Urbach tail, described by,

$$\alpha = \alpha_0 \exp\left(E/E_0\right), \tag{7.2}$$

where E is the photon energy and E_0 the parameter of the Urbach tail. E_0 depends on both the temperature and the disorder of the material, as result of combined effect of thermal and structural disorder.

Both E_g and C_H depend on the doping concentration and on the defect density, so that the general structure of the material rather than simply the hydrogen concentration is responsible for the position of the absorption edge. Hydrogen itself influences the material structure, and in particular it reduces the mechanical stress, while doping increases it. Nevertheless, a general description of E_g in a-Si:H is not available, and optical design must rely on measurements on single films. Such data must be treated cautiously, since the material inside a structure may be different from material deposited with the same recipe in a single film.

The energy dependence of the absorption coefficient decreases at low energy, and α depends also sensitively on the details of the deposition of the films. Specifically, the absorption is proportional to the density of defects, and this shoulder is related to absorption due to transitions between extended and localized states.

7.2.1 Optical Properties of Amorphous Silicon Alloys

Using the same gas phase deposition techniques, amorphous silicon can be alloyed with other materials. In particular, carbon (a-SiC:H), nitrogen (a-SiN:H), germanium (a-SiGe:H) alloys are used in device applications. Carbon and nitrogen increase the semiconductor band gap while germanium reduces it. Low-concentration carbon alloys have a low defect density and are largely

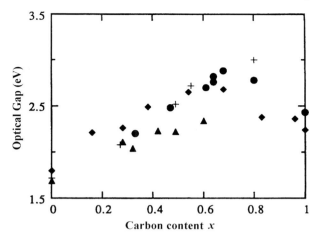

Fig. 7.3. Optical gap as a function of carbon content for a-Si$_{1-x}$C$_x$:H [7.17]

employed as semiconductors. Nitrogen alloys of composition near Si$_3$N$_4$ are used as insulators in thin film transistors, but are of minor importance in detector applications.

Significant variations of the shape of the optical absorption edge are reported for amorphous silicon alloys, so that often E_{04} is used to characterize the material instead of E_g. The assumptions of the Tauc model appear to be less applicable in the alloy materials.

The composition dependence of the optical gap of a-Si$_{1-x}$C$_x$:H is shown in Fig. 7.3. The maximum gap occurs around $x = 0.6$ to 0.7 for glow discharge material which coincides with the maximum in the number of Si–C bonds [7.17].

The increase of the band gap unfortunately also corresponds to an increase of disorder in the material and to an increase of electronic defects. Hydrogenation during the deposition can reduce this effect which ultimately limits the range of variation of the gap in practical application. Figure 7.4 shows the relationship between E_{04} and the Urbach energy (E_0) and defect density [7.18].

Alloys of a-Si:H with germanium give a reduction of the band gap and a shift to long wavelength of the spectral response. The presence of hydrogen offsets this reduction due to its tendency to increase the band gap. An empirical formulation of the dependence of E_g on hydrogen (C_H) and germanium (C) content is given by [7.19],

$$E_g = 1.6 + C_H - 0.7C . \tag{7.3}$$

The dependence of E_g on germanium concentration (denoted in the figure as E_{Tauc}) and E_{04} [7.16] are shown in Fig. 7.5.

The density of dangling bonds increases exponentially with decreasing E_g, as shown in Fig. 7.6, in which the integrated excess sub gap absorption,

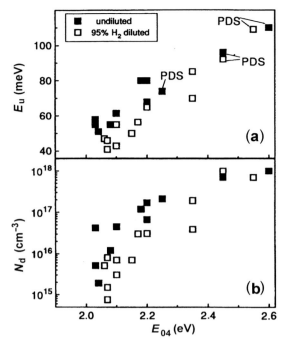

Fig. 7.4. (a) Urbach energy and (b) defect density of a-SiC:H films grown in undiluted and 95% diluted gas mixture as a function of the optical gap E_{04} [7.18]

measured by CPM, is plotted as a function of E_g for both dc and rf a-$Si_{1-x}Ge_x$:H alloy films. The exponential increase indicates that the possibility to modify E_g while maintaining good electronic properties is also limited in this material and must be carefully considered in designing the detector structure.

7.2.2 Optical Design of Layered a-Si:H Structures

In order to obtain an optical model of a multijunction device, it is necessary to know the dielectric constant of each material of the structure. The dielectric function is complex, with frequency dependent real and imaginary parts. The frequency dependence is a specific characteristic of the material and the real and imaginary parts are related each other by the Kramers–Kronig relationship [7.21]. In a semiconductor the dielectric function is equivalent to knowing the frequency dependence of the refractive index, n, and extinction factor, k which are also connected by Kramers–Kronig relations,

$$\varepsilon'(\omega) = 1 + \frac{2}{\pi}\int_0^\infty \frac{\omega'\varepsilon''(\omega')}{\omega'^2 - \omega^2}d\omega',$$

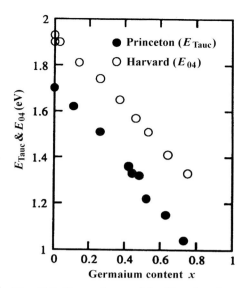

Fig. 7.5. Dependence of the Tauc band gap (E_{Tauc}) and E_{04} of a-Si$_{1-x}$Ge$_x$:H alloys on the germanium content [7.17]

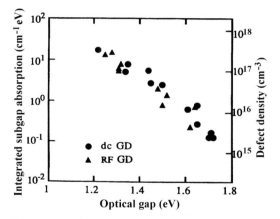

Fig. 7.6. Defect density of a-Si$_{1-x}$Ge$_x$:H as a function of the optical gap

$$\varepsilon''(\omega) = \frac{2}{\pi}\omega \int_0^\infty \frac{\varepsilon'(\omega')}{\omega'^2 - \omega^2} d\omega'. \tag{7.4}$$

A practical way to obtain these relationship was presented by Adachi who proposed a model in which $\varepsilon''(\omega)$ is obtained from the absorption coefficient measurements in the form of (7.1), but limited to the energy range $E_{\text{g}} < \hbar\omega < E_{\text{c}}$, where E_{c} is a high-energy cutoff of the parabolic bands [7.22].

Applying the Kramers–Kronig relationship, the real part of the dielectric constant is obtained from $\varepsilon''(\omega)$. A damping factor Γ is then introduced to take care of a lifetime broadening effect in a phenomenological manner, by replacing $\hbar\omega$ by $E = \hbar\omega + j\Gamma$ obtaining:

$$\varepsilon'(\omega) = 1 + \frac{2B^2}{\pi}\left[-\left(\frac{E_\mathrm{g}}{E}\right)^2 \ln\left(\frac{E_\mathrm{C}}{E_\mathrm{g}}\right) + \frac{1}{2}\left(1 + \frac{E_\mathrm{g}}{E}\right)^2\right.$$
$$\left. \times \ln\left(\frac{E + E_\mathrm{C}}{E + E_\mathrm{g}}\right) + \frac{1}{2}\left(1 - \frac{E_\mathrm{g}}{E}\right)^2 \ln\left(\frac{E - E_\mathrm{C}}{E - E_\mathrm{g}}\right)\right]. \quad (7.5)$$

The refractive index and the extinction factor can be computed from the real and imaginary parts of $\varepsilon(\omega)$ by,

$$n(\omega) = \sqrt{\left[\sqrt{\varepsilon'^2(\omega) + \varepsilon''^2(\omega)} + \varepsilon'^2(\omega)\right]/2}$$
$$k(\omega) = \sqrt{\left[\sqrt{\varepsilon'^2(\omega) + \varepsilon''^2(\omega)} - \varepsilon'^2(\omega)\right]/2} \quad (7.6)$$

E_g and B are evaluated by using the simple Tauc model fit to (7.1), while Γ and E_c are obtained by fitting data for the frequency dependence of $\varepsilon(\omega)$. The values of 0.6 eV and 4.6 eV, respectively, for Γ and E_c, are obtained for amorphous silicon [7.22].

As will be presented in the following, a-Si:H photodetectors are mainly constituted of plane thin film layered structures. Knowledge of the refractive index and the extinction coefficient allow a solution of the one-dimensional electromagnetic problem. Plane wave light radiation propagates in each layer with the complex propagation constant:

$$\beta = (2\pi/\lambda)(n - jk), \quad (7.7)$$

We assume that the propagation direction, z, is normal to the layered structure, and thus with no loss of generality, we may assume a TEM mode. The transverse electric field in the i-th layer can be described as:

$$E_{xi}(z) = A_i e^{-j\beta_i z} + B_i e^{+j\beta_i z}, \quad (7.8)$$

where A_i and B_i are respectively the amplitude of the forward and backward propagating waves.

The continuity conditions on the tangential component of electric and magnetic fields must be satisfied at every interface, leading to a relationship between amplitudes of the forward and backward propagating modes in the two adjacent layers. Each of these boundary conditions and the relationship between amplitudes at the two edges of a layer can be expressed in a matrix form. This formulation can be easily applied to any sequence of layers as given by,

$$\begin{bmatrix} A_i \\ B_i \end{bmatrix} = \prod_{j=1}^{i} T_j M_j^{-1} M_{j-1} \begin{bmatrix} A_0 \\ B_0 \end{bmatrix}, \quad (7.9)$$

where the two matrix are defined as follows, with d_i, ε_i, μ_i, respectively the thickness, dielectric constant and magnetic permeability of the i-th layer:

$$M_i = \begin{bmatrix} 1 & 1 \\ \sqrt{\varepsilon_i/\mu_i} & -\sqrt{\varepsilon_i/\mu_i} \end{bmatrix}; \quad T_i = \begin{bmatrix} e^{-j\beta d_i} & 0 \\ 0 & e^{+j\beta d_i} \end{bmatrix}. \quad (7.10)$$

Expression (7.9) relates the vector $[A_i, B_i]$ of electric field components at the inner interface of the i-th layer, with the vector $[A_0, B_0]$ of the electric field components outside the structure, practically with amplitudes of the incident and reflected ray at the structure surface. Once all the layers are included in (7.9) one final boundary condition must be applied. Usually the last, N-th, layer is assumed to have infinite thickness, which is equivalent to assuming a null reflected ray. Assuming the amplitude of the incident ray, A_0, is also known, (7.9) calculated on the whole structure, gives two linear conditions in two unknowns, which can be easily solved to obtain B_0, the reflected ray, and amplitude of the electric field in the last layer, A_N, as given by

$$\begin{bmatrix} A_N \\ 0 \end{bmatrix} = M_N^{-1} M_{N-1} \prod_{j=1}^{N-1} T_j M_j^{-1} M_{j-1} \begin{bmatrix} A_0 \\ B_0 \end{bmatrix}. \quad (7.11)$$

By applying (7.9) again, limited to the i-th layer, the component of the electric field in that layer can be obtained. The sum of the two electric field components at each point, squared, is proportional to the light intensity and gives the rate of generation of charge when multiplied by the absorption coefficient.

7.3 Two-Color Sensors

Red, green and blue (RGB) are the fundamental perceptual components of the visible spectrum and can be easily separated by using appropriate detectors and/or bandpass filters. Light filtering, employing distinct wavelength-optimized structures, in an array containing many integrated detectors is rather complex and is an expensive solution. Therefore, the possibility to obtain the RGB signals directly from one detector is attractive. Optical filters may be eliminated by using a-Si:H multilayer stacked devices, in which the detector structure itself behaves as the filter. The complexity of the electrical interconnection is reduced, while the sequence of test voltages applied to the device provides the color information.

In the stacked structure, information about the spectrum corresponds to information about where the radiation is absorbed. Unfortunately, the photocurrent at the two terminal loses this information. On the other hand, time resolved measurements which retain the information, such as time-of-flight are difficult to perform on large arrays and under continuous illumination. The solution consists of sampling the absorption region with different bias voltages, and extracting separately the integrated information about the radiation absorbed in each region.

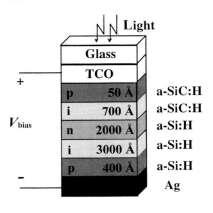

Fig. 7.7. p-i-n-i-p Si/SiC structures of two color detector

It is important to note that only a portion of the structure may contribute to the detection. In structures with typical layer thicknesses of 300–5000Å, photogenerated carrier pairs can only be separated in depletion regions where there is an electric field, because the difffusion lengths in a-Si:H are very small. In doped and highly defective layers, the electric field is usually low and the carrier lifetime is extremely short, so that generated carrier pairs recombine rapidly without the possibility to be detected.

The two color detector is based on this concept. Two junctions, with two depletion regions, give information about the light absorbed at two different depths in the structure.

The first tunable photodetector structure was a two color n-i-p-i-n detector fabricated from amorphous Si/SiC, capable of obtaining a separation of the spectrum into two bands centered in the green and in the red [7.23, 24] The structure was deposited on a glass substrate covered with transparent conductive oxide (TCO).

The two color device is further improved by using a p-i-n-i-p Si/SiC structure [7.25] as illustrated in Fig. 7.7.

In order to maximize the blue light transmittance, a thin (50 Å) a-SiC:H p^+ window layer, with a large band-gap is deposited on the TCO. Then, an a-SiC:H intrinsic (i) layer follows with a band-gap of approximately 2 eV. The thickness of this i-layer (600 Å) is a trade-off between sensitivity and selectivity. The need for high efficiency, blue light, photo-generation of carriers calls for thicker layers, while high transparency to red photons requires thinner films. The n layer is rather thick (2000 Å) in order to absorb green, and thus separate the blue and red components. The conversion efficiency in the red is maximized in the rear diode by depositing a 3500Å-thick a-Si:H intrinsic layer ($E_g = 1.7$ eV). The rear p layer is about 400 Å thick. Finally, a 1 µm thick silver electrode is evaporated through a 1 cm^2 mask.

By using a combination of amorphous and crystalline junctions, the two color detection can be extended from the visible to the near-infrared [7.26].

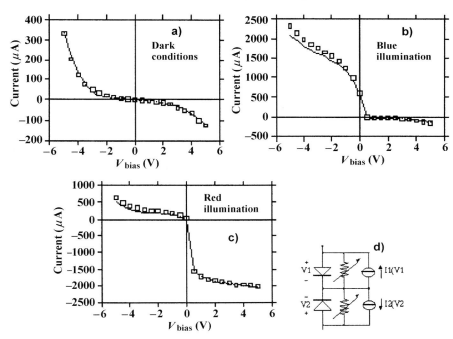

Fig. 7.8a–d. Experimental (*markers*) and simulated (*continuous line*) I–V curves of p-i-n-i-p two color detector: (**a**) dark condition, (**b**) blue illumination, (**c**) red illumination. In (**d**) the equivalent circuit used in SPICE simulation is sketched [7.25]

7.3.1 Steady State and Transient Operation

Figure 7.8 shows the I–V characteristics of the p-i-n-i-p photodetector [7.25] illuminated with an AM1.5 source filtered below 500 nm and above 600 nm. The light intensity is adjusted to 32 mW/cm^2 in both cases. Good independent responses under either blue or red illumination are observed, although the red-curve exhibits photocurrents of a few hundreds of μA when the rear diode is forward biased. This is due to the poor isolation of the front diode, since no mesa etch was performed on the device.

The device behavior can be explained as follows. For negative $V_{bias} = -3$ V (referring to the polarity indicated in Fig. 7.7), the front p-i-n diode, in reverse bias, and the rear n-i-p diodes, in forward bias, operate as detector and load respectively, as shown in Fig. 7.9. High energy photons (blue light) are absorbed in the front diode (F.D.). As a result, photo-generated carriers created by blue illumination are collected (working point P in Fig. 7.9). If the back diode (B.D.) is illuminated by red light, the current flowing in the structure does not change much since it is determined by the reverse characteristic of the front diode (P′). With the front diode in a dark condition

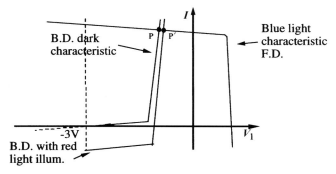

Fig. 7.9. Schematic current-voltage characteristics of the p-i-n-i-p device under -3 V bias voltage and blue light illumination

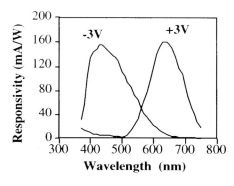

Fig. 7.10. Responsivity of p-i-n-i-p detector at $V_{\text{bias}} = +3$ V and -3 V [7.25]

(not sketched in figure), the reverse current in the front diode drops to the dark current value, and so the current in the structure is low.

For $V_{\text{bias}} = +3$ V role of the two diodes is exchanged, since the front diode is forward biased and the back diode is reverse biased. The device is now sensitive to red light, which penetrates deeper in the structure.

In Fig. 7.10 the responsivity at different wavelength in the visible is plotted, at $V_{\text{bias}} = +3$ V and -3 V. The device exhibits the following properties: a) a narrow spectral response, with full width at half maximum less than 130 nm for each peak; b) a high rejection ratio both at 430 nm with $R_{-3\,\text{V}}/R_{+3\,\text{V}} = 16$ (corresponding to 24 dB), and at 630 nm with $R_{+3\,\text{V}}/R_{-3\,\text{V}} = 23$ (27 dB). Along with these features, the device has good responsivity.

The response for a time-varying bias and cw red and blue illumination is shown in Fig. 7.11 together with the input waveform. The inset is a sketch of the test circuit. The time constants of the two outputs under illumination are due to the diffusion and junction capacitance in the forward biased and the reverse biased diode, respectively. With the chosen load resistor $\tau_{\text{red}} = 40\,\mu\text{s}$ and $\tau_{\text{blue}} = 110\,\mu\text{s}$ are the measured response times of red and blue illumination respectively.

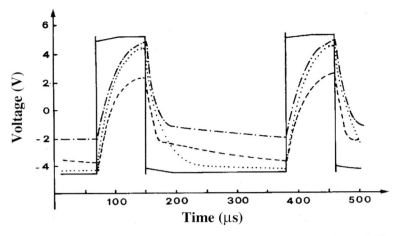

Fig. 7.11. Time response of p-i-n-i-p two color detector, under red (*dash-dot line*), blue (*dots*) illumination, and in dark (*dashes*). Voltage signal is also reported [7.27]

7.3.2 SPICE Model of the Two Color Detector

The device behavior under dark conditions can be simulated by SPICE, using an equivalent circuit based on two back-to-back diodes accounting for the two p-i-n structures, as shown in Fig. 7.8d [7.27], and two linearly decreasing shunt resistances for increasing reverse bias on each p-i-n cell simulating a soft breakdown.

For simulation of the device under illumination, Crandall's analytical relation between photocurrent and applied voltage for a p-i-n structure can be applied, as follows [7.28],

$$I_{\rm ph} = AqG(\mu_p\tau_p + \mu_n\tau_n)\frac{V_{\rm btin} - V}{d}$$
$$\times \left[1 - \exp\frac{-d^2}{(\mu_p\tau_p + \mu_n\tau_n)(V_{\rm btin} - V)}\right], \quad (7.12)$$

where A is the device area, G is the generation rate, d is the thickness of the i-layer and $V_{\rm btin}$ is the built-in potential of the p-i-n junction. This model leads to different slopes of $I_{\rm ph}(V)$ plot for different $\mu\tau$ products. The effect of blue and red illumination, can be reproduced by adding two nonlinear current sources calculated following (7.12). Figure 7.8 shows a fit to the data using this model.

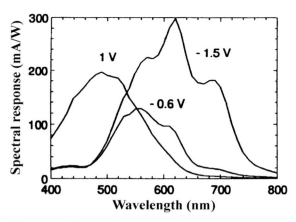

Fig. 7.12. Spectral response of a n-i-p-i-n three color detector for different bias voltages [7.30]

7.4 Three-Color Sensors

7.4.1 Three Color Discrimination with Two Electrical Terminals

Two color detection is based on spatially separated absorption of different wavelengths. Once we seek the additional information required for three color detectors, the discrimination between the different absorption regions becomes more complex.

The various different approaches taken to the problem of obtaining information about the third color in a device with only two terminals, can be divided in two main classes. In the first group are devices with essentially one depletion region which is modified by the external bias voltage. The modulated junction is generally included in a more complex stacked structure. In a device with one single modulated junction, blue light is detected if the depletion region is located at the front of the structure, green light is detected if it is at the center, and red light is detected if it is at the bottom portion of the structure. Generally, it is not possible simply to move the depletion region to the center, but rather it can be expanded, increasing the range of frequencies that are detected. In this case, the single color bands cannot be separately detected, but rather independent information on the illumination spectrum is obtained and analyzed externally.

Several devices belong to this class. Among these are the original n-i-p-i-n structure of Tsai and Lee [7.29] and the later development described by Stiebig et al. [7.30], where the depletion region of the second p-i-n junction is modulated by the bias voltage, tuning the detectivity from green toward red light. The front diode detects blue light as in the two color detector.

Figure 7.12 reports the spectral response of a n-i-p-i-n three color detector for different bias voltages [7.30]. Figure 7.13 shows a simulation of the

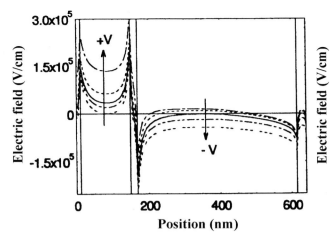

Fig. 7.13. Numerical simulation of electric field distribution in n-i-p-i-n three color detector for different bias voltages [7.30]

expected electric field in the structure. In the presence of a relatively high density of defects, and $V_{bias} = 0$ V, the electric field in the central portion of the back diode is low and can even be positive. In this condition red light absorbed in the rear portion is not detected. The electric field increases only under reverse bias, and in this condition, red light is detected and the spectral sensitivity of device is extended.

A similar process takes place in the n-i-p-i-n-i-n structure presented later by Rieve et al [7.31]. The intrinsic layer (i_2) of the central p-i-n junction is depleted at zero and at moderate bias. A positive bias corresponds to reverse polarization of the p-i_2-n cell. If larger reverse bias is applied, the central n-type layer is also depleted. The electric field penetrates in the rear intrinsic region, as schematically indicated in Fig. 7.14 and allows the detection of carriers photo-generated by red photons, which are absorbed in the last portion of the structure.

A modulated depletion region, obtained by a relatively large defect density, can be found in structures based on single p-i-n junction recently described [7.33, 34]. The intrinsic zone is made up of layers of different material quality (respectively, three layers in [7.33] and two in [7.34]). The single junction, a-SiC:H detector [7.35], can be included in this class, since the depletion region can be modulated by a reverse bias voltage due to the great number of defects in the material. Depending on the bias voltage, charge photogenerated in different portions of the absorber layer is extracted.

In all these structures, the presence of defects is necessary to control the extent of the depletion region. Defects may be induced by doping [7.31], alloying with germanium [7.36], or micro-crystallization of silicon [7.33]. The presence of defects, limited to a portion of the intrinsic absorber material,

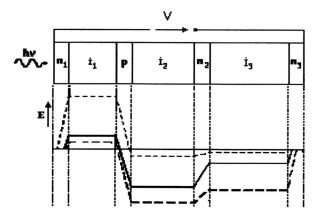

Fig. 7.14. The n-i-p-i-n-i-n structure and a sketch of the electric field distribution under different bias voltages [7.31]

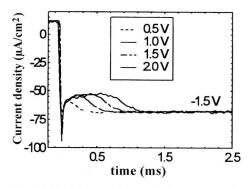

Fig. 7.15. Measured transient response when bias voltage is switched from different positive values to $v = -1.5\,\text{V}$ under illumination condition [7.36]

ensures that light absorbed in that defective region can be detected externally only under depletion conditions, which is obtained with a large reverse bias of the junction. This leads to the desired variation of quantum efficiency spectrum with applied bias voltage.

A shortcoming of these structures is that the presence of a large number of defects imposes a long response time in the device, due to the trapped charges which must be exchanged whenever the bias voltage is changed. An example of the transient response is reported in Fig. 7.15 [7.36]. Furthermore, the width of the depletion region may be modified by the presence of photogenerated charge, so that linearity is not ensured. An example is given in [7.31], where a sublinear behavior is found, and the photoconductivity exponent factor, γ, where $\sigma = I^\gamma$, is about 0.5.

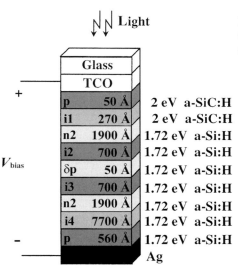

Fig. 7.16. Structure of the adjustable threshold three color detector (ATCD) as described in the text

7.4.2 Adjustable Threshold Three Color Detector (ATCD)

The second class of three color detectors is illustrated by the adjustable threshold three color detector (ATCD) [7.37, 38]. In this case, the depletion regions are fixed by a sequence of junctions, and each depletion region independently detects a portion of the spectrum depending on its thickness and position inside the structure. A simple two color detector discriminates color by choosing to reverse bias one of its two junctions, but a more sophisticated strategy is needed for the ATCD.

The ATCD structure is shown in Fig. 7.16 [7.32]. The front electrode of the photosensor is TCO. Both the front p-doped and i_1 layers are made of a-SiC:H ($E_g = 2\,\mathrm{eV}$). All the remaining layers are made of a-Si:H ($E_g = 1.7\,\mathrm{eV}$). The intermediate cell is n_1-i_2-δp-i_3-n_2, where δp is a thin layer lightly p doped. The rear cell is an n_2-i_4-p diode, and the back electrode is a $1\,\mu m$ thick evaporated silver film.

The ATCD operation can be understood by considering the photodetector as formed by three stacked devices: the p-i-n, the n-i-δp-i-n and the n-i-p cells. The thin, lightly doped p-layer in the n-i-δp-i-n cell is designed as the base of a bulk buried phototransistor [7.39], driven by the photocurrent generated by green light absorption. In the dark, this cell is in a high resistance state, and the current increases exponentially with bias voltage. The design adopted in the ATCD structure has a threshold voltage of approximately 2.5 V. Under illumination, the n-i-δp-i-n cell presents a nearly constant photocurrent at low voltage, growing rapidly as the bias voltage increases above threshold. Examples of the I–V characteristics of the bulk barrier phototransistor in the dark and under illumination, are shown in Fig. 7.17 [7.39]. The constant

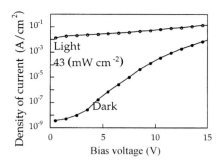

Fig. 7.17. I–V characteristics of bulk barrier a-Si:H phototransistor in dark (*full circles*) and under illumination (*open circles*) [7.39]

photocurrent is limited at 0 V by a sharp transition to the opposite polarity. The structure exhibits no photovoltage due to its symmetry.

Referring to a polarity in which V_{bias} is applied to the top contact in Fig. 7.16, the front diode is reverse biased when V_{bias} is negative and it behaves as a current source under blue illumination. Its load is constituted by the forward biased back junction and by the n-i-δp-i-n cell which is biased above threshold. As indicated in Fig. 7.18, the photocurrent of the front p-i-n cell determines the current in the whole structure (point A), irrespective of whether the n-i-δp-i-n cell is illuminated. Hence, in this bias condition the structure is sensitive to blue light, and without such illumination the current is low (point B). As shown later, the choice in the design of the structure, dictated by need to optimize the transient behavior, makes this bias condition sensitive to both blue and green radiation for the specific device that is being described.

With positive values of V_{bias}, the back diode is reverse biased and behaves like a current source. In this bias condition the structure is sensitive to red light.

At a bias of -0.8V, the front junction should be reverse biased, but its load is constituted by the n-i-δp-i-n cell in series with the forward biased back junction. Below threshold, the n-i-δp-i-n cell is a current generator. The thickness of the front cell is designed to give a larger photocurrent than the current of n-i-δp-i-n cell in the current generator regime, in particular under green illumination. The operating point of the front cell is thus driven into forward bias, as indicated in Fig. 7.18b (point C) and the current in the structure is then determined by green light absorbed in the central n-i-δp-i-n cell. In the absence of green radiation, this current drops to zero, since the inner cell is in a high resistance state (point D).

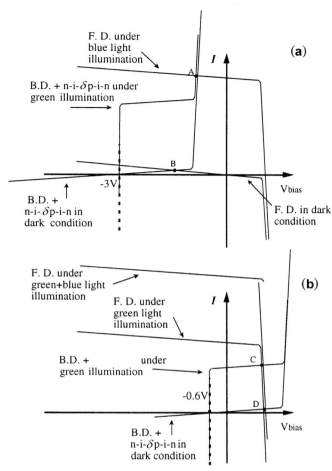

Fig. 7.18a,b. Load curves of the ATCD structure at, (a) a bias voltage of -3 V, (b) a bias voltage of -0.8 V

Figure 7.19 shows typical I–V characteristics for 1 cm² ATCD under blue (450 ± 5 nm), green (550 ± 5 nm) and red light illuminations (650 ± 5 nm). The illumination intensity in the three cases is 100 mW/cm² and as expected, the three colors are detected in three distinct regions of the external applied voltage. In the range -3 to -1 V, blue and green radiation have a very similar response due to the specific design choices made. In this case, external processing of information is necessary to separate the color spectra. Green radiation is independently detected in the range -0.8 to 0 V and red is detected in the range 0 to $+3$ V.

The response times of the ATCD are characteristic of a-Si:H p-i-n junctions, and thus are typically in the μs regime. The doped layer are never com-

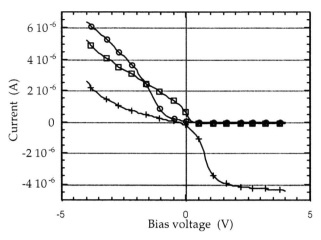

Fig. 7.19. I–V characteristics of the ATCD under blue illumination (*circles*), green illumination (*squares*) and red illumination (*crosses*)

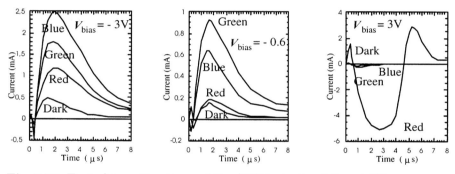

Fig. 7.20. Typical transition times of the ATCD at -3, -0.6, and 3 V

pletely depleted, including the δp inner layer in the phototransistor structure, so that either in forward or reverse biasing conditions, the response is rather fast. Figure 7.20 reports some typical response time for the ATCD device.

The device behavior may be described by an equivalent circuit model as reported in [7.38], which represents an extension of the model of the two-color sensor. Each p-i-n cell is modeled by an electrical parallel circuit comprising a diode, a current source to account for the photo-current, a shunt resistance and a capacitor which takes into account the junction capacitance. The inner n-i-δp-i-n cell is modeled as two diodes in a back-parallel configuration, a shunt resistance and a junction capacitance. This last parameter also allows the model to include the forward bias capacitance in addition to the geometrical one. The diode model accounts for the reverse current, the quality factor, the series resistance and the transit time.

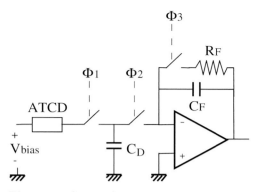

Fig. 7.21. Circuit for sensing a color device under transient conditions. The three phases necessary to complete the sensing cycle are shown

We complete this review of three color detector structures by summarizing the multi-terminal devices which can also be found in the literature. These devices use the filtering capability of a-Si:H layers but extract the electrical information by connecting independently to each junction. This approach requires additional device processing steps and greater interconnect complexity. An example is the device proposed by Topic et al. [7.40], which independently contacts a front p-i-n junction and a two color device underneath.

7.4.3 Three Color Detectors in the Time Integration Regime

The practical use of the color sensor in large area arrays requires the periodic readout of the photo-charge accumulated in the parasitic capacitance of the device, and is thus a transient technique of sensing. The integration time is the time of photo-charge accumulation, and determines the working rate of the array. This read out technique is referred to as the "time integration regime" and it is usually adopted for large area arrays (see the discussion of image sensors in Chap. 4).

No application in the time integration regime has yet been described for color detector of the "variable depletion region" class and so the following discussion focuses on the ATCD device. Nevertheless, general evaluation of the readout time can be given by considering the time response of the different types of device.

Figure 7.21 illustrates a possible circuit for sensing a color device under transient conditions. The device is connected to the readout circuit through a switch controlled by the signal Φ_1. When the switch is in the OFF-state (i.e., during the integration time t_i), the junctions forming the ATCD discharge due to the photocurrent. When the switch is in the ON-state (t_r), charge is restored to the structure by the external electronics. The column capacitance is in series and thus receives the charge exchanged by the detector. C_D holds this charge until it is connected to the sensing amplifier by the Φ_2 switch.

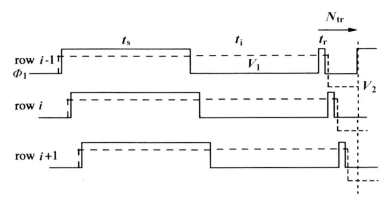

Fig. 7.22. The driving sequence of a three color detector. The *continuous line* is the switch signal Φ_1, while the *dashed line* is the bias voltage, which is assumed to have different values, V_1 and V_2, to detect different colors

For an RGB signal, the sensing cycle must be repeated for the three values of the bias voltage appropriate for sensing the different colors.

It is apparent from Fig. 7.15, that a settling time, t_s, of several milliseconds is required whenever a new bias voltage is applied to the n-i-p-i-n detector [7.36] For a matrix array with N rows and M columns, the voltage can be applied in sequence to the N rows. Once equilibrium is reached, each device is left in open circuit condition for the integration time, t_i, while the settling process proceeds to the following rows. After t_i, the photo-charge can be read out, applying again the bias voltage for a reading time, t_r. As soon as the reading process of one color is over, a new voltage is applied, starting a new settling time. Care must be taken in separating the readout time from the settling time. Two rows cannot be active at the same time, since charge accumulated in the column capacitance during the two processes would be superimposed. The settling times must start after all the reading pulses are complete. Figure 7.22 shows one possible driving sequence and the total time required to complete the whole cycle for three colors is

$$t_{\text{tot}} = 3\left(t_s + t_i + N\, t_r\right). \tag{7.13}$$

7.4.4 Mechanism of Autopolarization of the Stacked Cells

Biasing the stacked junctions in the transient integration regime is more complex than under dc conditions. The photo-charges are independently accumulated in the junctions (three for the ATCD) during the integration time when the device is under open circuit conditions. During this time a rearrangement of the internal charge takes place. In fact, the reverse biased capacitance discharges through the reverse current, whereas forward biased capacitance discharges through recombination of the excess carriers. As a result, in the same time interval the three capacitances in series discharge by different amounts.

The charge extracted from the device during the fast reading process is equal to the charge exchanged by each capacitance since the junctions are in series. It is important to note that at this stage the exchanged charge is not directly related to any one single junction.

This behavior is present in every device composed of stacked junctions, which may independently change their charge status during the integration time. The mechanism of *autopolarization* described below must therefore be taken into account, for example for p-i-n-i-p structures driven in the time integration regime, either for two and three color detection.

Repeating the sensing cycle at a fixed bias voltage, the readout charge varies, either in the dark or under illumination. However, experiments show that the device reaches an equilibrium condition after only a few readout cycles [7.42].

The time dependence of the readout charge and its final value at equilibrium can be understood by investigating the electrical evolution of stacked junctions. This is done with the aid of the graphs in Fig. 7.23, where the driver and load curves are plotted for a p-i-n-i-p device. In Fig. 7.23a, the initial condition is depicted with $V = -3\,\mathrm{V}$, and with the equilibrium point P. Figure 7.23b represents the situation under open circuit (during the integration time). In this condition, each junction independently reduces its voltage, discharging by different amounts.

The readout occurs after the time t_i. Figure 7.23c depicts the operation point soon after the voltage pulse has been applied to the structure. During this fast transient, the capacitance of each junction receives the same amount of charge, but this charge, in general, is different from the charge lost by each individual cell during the previous integration time. The net balance between the discharge and subsequent charging is non-zero, and thus the stacked cells undergo a shift of their internal voltage, V_1. The driver and load curves now have new operating points, P' and P'', which are different from the initial case (a). In Fig. 7.23c, the voltage drops satisfy Kirchoff law, but the currents are not the same throughout the structure, since this condition is out of equilibrium.

As a further consequence, the charge lost during the integration time of the following cycle is different in the forward biased junction (it is lower because the forward bias has decreased). The two junctions change their polarization, but while the reverse biased junction always maintains the same photocurrent, the forward biased junction reduces its forward polarization, thus reducing the discharge signal during the integration time.

Equilibrium is reached after a few cycles, when the charge lost in both junctions is equal, and thus equal to the charge restored during readout. At this stage the total charge exchanged in each cycle is fixed by the reverse bi-

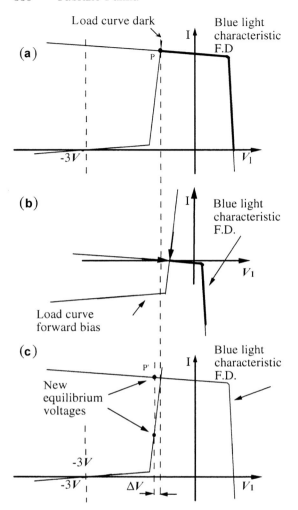

Fig. 7.23a–c. Load curves of a two-junction structure in time integration regime

ased junction, which maintains a constant photocurrent even if the operating point slightly changes due to auto-polarization.

The main consequences of this effect are: a) any stacked structure requires a number of cycles of integration, N_{cy}, to achieve correct sensing information, which leads to a longer frame time; b) the information obtained is unique since there exist one single reverse biased junction.

Autopolarization has been observed in the ATCD, and in particular it has been shown that information obtained in the time integration regime is the same as the information obtained from steady state measurements [7.38]. Referring again to Fig. 7.18, with a negative bias, the front junction is reverse

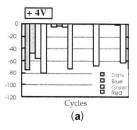 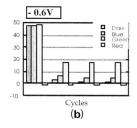 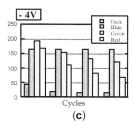

Fig. 7.24a–c. Readout signal for three bias voltages under different wavelengths with the same light intensity. Optimum color detection [Blue at +4 V (**a**), green at −0.6 V (**b**) and blue at −4 V] is obtained after four reading pulses

biased and behaves like a current source. The auto-polarization process causes the sensing of the photo-charge of the front n-i-p cell. With a positive bias, the rear junction is reverse biased and the auto-polarization process extracts the photo-charge from the rear n-i-p cell.

At −0.8 V, and due to the design of absorbing layers, the operating point of the front cell is driven into forward bias as indicated in Fig. 7.18b. The central n-i-δp-i-n cell works in the constant current region, proportional to the green radiation absorbed. In this condition, green photo-charge is detected after auto-polarization. From this discussion, it can be understood that the goal of the design is to achieve polarization in the constant current region of n-i-δp-i-n cell. Therefore a relatively thick absorber layer of the front cell is adopted; with this design, the front diode actually detects either blue and green radiation, and it has a short circuit current always greater than the n-i-δp-i-n cell in the constant current region. A reduced color selectivity is the trade-off of this choice, but it can be compensated by external processing of the information.

Figure 7.24 shows the readout charge versus the readout cycle number, under conditions of $V_{\text{bias}} = +4\,\text{V}$, -0.6, and $+4\text{V}$ respectively, with an integration time, $t_i = 2\,\mu\text{s}$. The best conditions for detection of the green and red light do not appreciably depend on the length of the integration time and quasi-equilibrium is achieved after only two cycles. In contrast, detection of the blue light requires four reading cycles.

The circuit sketched in Fig. 7.21 is also suitable for driving the ATCD. It is worth noting that the integration time depends on the radiation intensity, in the sense that lower intensities imply longer values of t_i. The dark current limits the minimum detectable radiation intensity and thus the maximum value of t_i. The reading pulse necessary to recharge the ATCD and achieve autopolarization is extremely short. Thus, periodic recharging of one row may be performed during the integration period of the others, as shown in Fig. 7.25. During each recharge pulse, the charge held in the column capacitance is modified. This implies that the capacitance C_D in Fig. 7.21 cannot

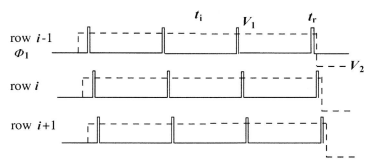

Fig. 7.25. The driving sequence of a three color detector. The *continuous line* is the switch signal Φ_1, while the *dashed line* is the bias voltage, which assumes different values, V_1 and V_2, to detect different colors

be used to store information from one single device while waiting for serial extraction of information along the data line.

The time required for a complete three color detection is approximately

$$t_{\text{tot}} = 3(N_{\text{cv}} t_{\text{j}} + N\, t_{\text{r}})\,. \tag{7.14}$$

A comparison of (7.13) and (7.14) indicates that the two type of devices can have similar frame rates.

In Fig. 7.26 linearity of photocharge of ATCD readout in time integration regime is presented. The photoconductivity exponent factor γ is about 0.98 ($\sigma = I^\gamma$). As described earlier, a high degree of linearity is ensured since one (and only one) junction is in the reverse bias condition, thus determining the charge exchanged by the structure after auto-polarization. The relatively small dynamic range (especially in green condition) is due to the present form of the design. The dynamic range can be enhanced once larger voltage thresholds are accepted.

7.5 a-Si:H Based UV Sensors

Although there is considerable interest in image sensing of ultraviolet radiation (UV), very few such devices sensitive in the far UV range (100–300 nm) are available. One example is the photon counting microchannel plate detector [7.43], whose photoresponse depends on the deposited photocathode. These devices are very sensitive and have extremely low noise, but need an ultra-high vacuum and a very high voltage (kV). UV-enhanced CCDs have been recently tested, [7.44, 45] and show very good performance. However, some significant problems need to be solved, such as the presence of hysteresis in the quantum efficiency due to the cooling of the devices, and the necessity to periodically treat them to maintain the efficiency.

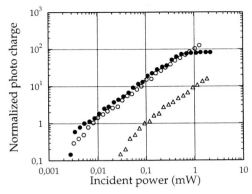

Fig. 7.26. Photo-charge of the ATCD readout in the time integration regime. Open circles represent blue radiation, triangles green radiation, full circles red radiation

Very recently, an innovative UV photodetector made with hydrogenated amorphous silicon and silicon carbide (a-Si/a-SiC:H) on glass has been described [7.46]. The device demonstrates excellent sensitivity to vacuum- and near-UV radiation.

Application of low cost single-channel UV photodetectors is of great interest in medical, laboratory, and chemical analysis, while large area arrays of solar-blind UV sensors could be valuable in astronomy and space applications. Furthermore, the possibility to tune the photoresponse in the visible range opens up some special applications, as in the detection of the Cherenkov radiation, which covers the spectrum from the far-UV to about 500 nm.

Another appealing application is for a-Si/SiC:H tunable photo-detectors with two or three sensitivity peaks. Including the UV sensor within a structure of a tunable detector, as those reported in [7.29, 34], the wide spectral range from the extreme UV to the red or the near infrared, respectively, can be scanned by varying the bias voltage over a few volts.

7.5.1 Structure and Operation of the UV Detector

The a-Si:H UV detector is fabricated in a sandwich configuration of the type, electrode/p-i-n/electrode on a glass substrate, as shown in Fig. 7.27. The amorphous semiconductor layers are deposited by PECVD. A thick a-Si:H n-doped layer is first deposited on TCO and then a thin intrinsic a-SiC:H layer and a very thin p-doped a-SiC:H layer are grown. A 1 μm-thick aluminum film is evaporated on top of the structure and subsequently patterned by photolitography. A grid pattern is used, with 50 μm spaced fingers.

Radiation with wavelength λ and intensity I_l penetrates through the grid and is absorbed at different depth, x, into the device, according to an exponential law

$$I_l(x) = I_l(0) \exp(-\alpha(\lambda)x). \tag{7.15}$$

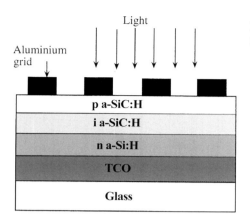

Fig. 7.27. The structure of the UV a-Si:H detector

Extrapolated values of the absorption coefficient of a-SiC:H indicate that far-UV radiation is absorbed within a few nanometers [7.47]. All the photogenerated carrier are thus in the p-doped layer. In order to detect the radiation, the minority carrier electrons must be able to reach the depletion region without recombination. In the depletion region the electric field pushes electrons toward the n-doped layer where they can be detected.

The electron extraction length in p-type a-Si:C:H is a few times the electron diffusion length, which has been investigated by several techniques [7.48], and found to be of the order of one nanometer. On the basis of these considerations, the thickness of the p-doped layer in the present device is fixed at 5 nm, since thinner layers offered serious problems of microshunts and of a reduction of the built-in junction potential. The p-layer thus achieves three goals: 1) it maximizes the absorption of the UV light, 2) it allows the collection of the photogenerated carriers, and 3) it enhances the transmission of the visible light.

The quantum efficiency at $V_{\text{bias}} = 0$ is reported in Fig. 7.28. A very high efficiency (around 60%) is obtained at 50 nm. The curve shows an exponential decreases at longer wavelength with very good rejection of the visible radiation. The high efficiency in the UV range may be related to multiple generation and hot carrier relaxation in the p-layer due to the high photon energy [7.47]. Measurements do not show saturation toward 50 nm, thus suggesting a high sensitivity also in the soft X-ray region of the spectrum.

7.6 a-Si:H Based IR Sensors

Wavelengths of 1.3 µm and 1.5 µm are commonly used in optical fiber transmission, while 3–5 µm wave-lengths are common for night vision and fire alarms. A-Si:H has an optical gap of 1.7 eV which sets its absorption edge

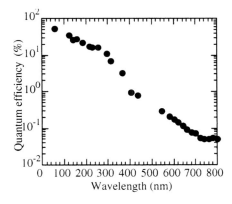

Fig. 7.28. The quantum efficiency of the a-Si/a-SiC:H UV photodetector at $V_{bias} = 0$

well inside the visible, apparently not suitable for IR detection. Several attempts have been made to extend the absorption spectrum of a-Si:H to longer wavelength. Indeed the possibility for low temperature deposition of this semiconductor as an infrared sensor is extremely attractive. In integrated optics, a-Si:H technology is compatible with most waveguides materials. In particular an IR sensor would complete the technological requirements of a-Si:H/a-SiC:H waveguides and modulators for a low cost opto-electronics technology [7.49]. Moreover a-Si:H technology is compatible with integrated circuit technology since it can be deposited above metal layers and could be easily included in vision systems.

The use of a-Si:Ge alloy to reduce the energy gap has been extensively studied. In particular a-Si:Ge has been introduced in solar cells as absorber layer in stacked cells [7.50] (see Chap. 6). A-Si:Ge is also used for photodetection up to 900 nm [7.51, 52] Unfortunately, deterioration of the electronic quality of the material with increasing Ge content reduces the detector efficiency as soon as a low band gap is achieved.

Attempts have been made to use a-Si:H directly for IR detection. Wind and Müller [7.53] report the infrared response of a-Si:H p-i-n diodes up to 2400 nm under forward bias at low temperature (198 K). Direct IR photodetection using a gain enhanced a-Si:H photodiode with a reach-through structure was presented by Okamura and Suzuki [7.54]. The device showed sensitivity in the 1.3–1.55 μm range, even though avalanche multiplication could induce noise in the current-voltage characteristic of high reverse biased devices.

Recently, an IR sensor based on the measurement of photocapacitance in an a-Si:H structure has been demonstrated, and is an example of a novel de-

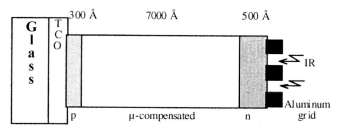

Fig. 7.29. The structure of the IR a-Si:H photo-capacitor described in the text

tection approach [7.55]. Measurements described next show that this method is sensitive to radiation up to 4 µm.

7.6.1 IR Detection by Differential Photo-Capacitance

The differential capacitance detection method overcomes the limitation of the wide band gap of a-Si:H to the detection of IR. Absorbed IR photons excite electron transitions between valence band extended states and localized states in the forbidden gap. After the transition, the electrons remain trapped until they fall back to the valence band. At equilibrium, no photo-current flows due to the absorbed radiation, rather the trapped charge and the electric field distributions in the junction are modified. These changes can be detected as changes of the capacitance.

Figure 7.29 shows a cross-section view of this a-Si:H IR photodetector. The whole structure is deposited from SiH_4 by PECVD [7.56], and the device is similar to a p-i-n structure, except that micro-doping of the internal layer is performed. This structure is referred as p-c-n, since the central compensated layer is deposited from silane with 0.25 ppm of PH_3 and 18 ppm of B_2H_6 in the gas mixture. The thickness of the micro-doped, p, and n layers are respectively 7000 Å, 300 Å and 500 Å and the substrate is TCO-coated glass. The front electrode is aluminum deposited by evaporation, and then patterned as a thin finger grating, with spacing of 50 µm.

Figure 7.30 shows a typical capacitance versus frequency measurement on a p-c-n device, with zero bias voltage under illumination of a IR light in the 3.25–4.3 µm range, obtained by filtering a halogen lamp. The total incident power is 50 µW cm^{-2}. Capacitance measurements performed on the same device in the dark, are reported for comparison. IR radiation leads to an increase of capacitance. As expected, the capacitance variation decreases with frequency due to a reduction of the trap-emission process. At frequencies higher than 2 kHz, there is no difference between dark conditions and under IR illumination, since the capacitance reduces to the geometrical value for the structure.

Figure 7.31 shows the spectral response at room temperature. In this case the capacitance variations are detected by including the device in the

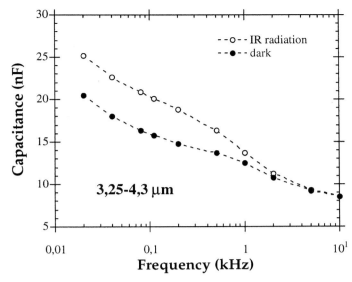

Fig. 7.30. Capacitance versus frequency measurement on a p-c-n device, with zero bias voltage under 50 mW illumination of a IR light in the 3.25–4.3 µm range and in dark

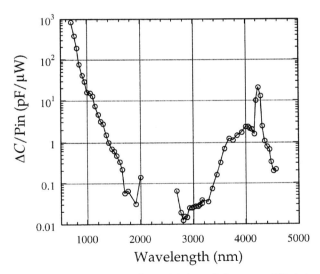

Fig. 7.31. The spectral sensitivity of the p-c-n IR detector at room temperature. Spectral sensitivity is defined as the ratio between the frequency variation of the detector and the incident power

feedback loop of an astable oscillator for which the capacitance determines the oscillation frequency. The spectral sensitivity is defined as the ratio between the frequency variation and the incident power. A relatively high responsivity in the range 800–1400 nm occurs, and a further peak between 3 and 5 µm is observed. The sensitivity at 3–5 µm requires further studies to understand the mechanism.

References

7.1 R. A. Street, X. D. Wu, R. Weisfield, P. Nylen, "Color document imaging with amorphous silicon sensor array", Mater. Res. Soc. Symp. Proc. **336**, 873 (1994).

7.2 L. E. Antonuk, Y. El-Mohri, W. Huang, J. Siewerdsen, J. Yorkston, "A large area, high resolution a-Si:H array for X-ray imaging", Mat. Res. Soc. Symp. Proc. **336**, 855 (1994).

7.3 M. Bohm, F. Blecher, A. Eckhardt, K. Seibel, B. Schneider, J. Sterzel, S. Benthien, H. Keller, T. Lule', P. Rieve, M. Sommer, B. Van Uffel, F. Librecht, R. C. Lind, L. Humm, U. Efron, E. Roth, "Image sensor in TFA technology – status and future trendes", MRS Symp. Proc. **507**, 327 (1998).

7.4 J Wyatt, J. Rizzo, "Ocular implant for the blind", IEEE Spectrum, May (1996), p. 47;
M. B. Schubert, A. Hierzenberger, V. Baumung, H. N. Wanka, W. Nisch, M. Stelzle, E. Zrenner, "Amorphous Silicon Photodiodes for replacing degenerated photoperceptron in the uman eye", Mat. Res. Soc. Symp. Proc. **467**, 913 (1997).

7.5 "Digital Television", Edited by C. P. Sanbdbank, John Wiley & Sons (1990)

7.6 A. G. Gaydon, "The spectroscopy of Flames", Chapman and Hall LTD., London (1957).

7.7 D. Caputo, F. Irrera, F. Palma, S. R. Rachele, M. Tucci, "Amorphous silicon optical spectrum analyzer for the visible spectrum", **198–200**, 1172 (1996).

7.8 I. Shimizu, "Enhancement of long wavelenght sensitivity", Amorphous Semiconductor Technologies & Devices, JARECT **16**, 300 (1984).

7.9 D. S. Shenj. P. Conde, V. Chu, S. Aljishi, S. Wagner, "Effect of material properties on the performance of a-Si, Ge:H,F photodetectors", Mat. Res. Soc. Symp. Proc. **118**, 457 (1988).

7.10 L. Ley, "Photoemission and optical properties", in "The physics of hydrogenated amorphous silicon", edr. J. D. Joannopoulos, G. Lucovsky, Topics in Applied Physics **56**, 61 (1884).

7.11 J. Ristein, G. Weiser, "Influence of doping on the optical properties in plasma deposited amorphous silicon", Solar Energy Mat. **12**, 221 (1985).

7.12 G. D. Cody, B. Abeles, C. Wronski, C. R. Stephens, B Brooks, "Optical characterization of silicon-hydride films", Solar Cells **2**, 227 (1980).

7.13 J. Tauc, Amorphous and liquid semiconductors, Plenum Press, New York (1974).

7.14 H. Fritsche, "Characterization of glow-discharge deposited a-Si:H (Review)", Solar Energy Mat. **3**, 447 (1980).

7.15 B. Von Roedern, L. Ley, M. Cardona, "Photoelectron spectra of hydrogenated amorphous silicon", Phys. Rev. Lett. **39**, 1576 (1977).

7.16 S. B. Zhang, W. B. Jackson, D. J. Chadi, " Diatomic-Hydrogen-Complex dissociation: a microscopic model for metastable defect generation in Si", Phys. Rev. Letters **65**, 2575 (1990).

7.17 X Xu, S. Wagner, "Physics and electronic properties of amorphous and microcrystalline silicon alloys", in Amorphous and Microcrystalline Semiconductors Devices, edr. J. Kanicki, Artech House (1992) p. 89.

7.18 A. Desalvo, F. Giorgis, C. F. Pirri, E. Tresso, P. Rava, R. Galloni, R. Rizzoli, C. Summonte, "Optoelectronic properties, defect structure and composition of a-Si:H films grown in undiluted and H_2 diluted silane-metane", Journal Appl. Phys. **81** (12), 7973 (1997).

7.19 S. Wagner, V. Chu, J. P. Conde, J. Z. Liu, "The optoelectronic properties of a-Si, Ge:H(F) alloys", J. Non-Cryst. Solids, **114–166**, 453 (1998).

7.20 S. Aljishi, Z.E. Smith, S. Wagner, "Optoelectronic properties and the gap state distribution in a-Si,Ge alloys", Amorphous silicon and related material, edr. H. Fritzsche, World Scientific, Singapore, (1989) p. 887.

7.21 L. P. Landau, E. M. Lifshitz, Electronics of continuous media, Cambridge Univ. press, Cambridge.

7.22 S. Adachi, "Calculation model for the optical constant of amorphous semiconductors", J. Appl. Phys. **70**, 2304 (1991).

7.23 H. K. Tsai, S. C. Lee and W. L. Lin "An amorphous SiC/Si two-color detector", IEEE-Electron Device Lett. **EDL-8**, 365–67 (1987).

7.24 H. Stiebig and M. Bohm, "Optimization criteria for a-Si:H nipin color sensors", J. Non-Cryst. Solids **164–166**, 785 (1993).

7.25 G. de Cesare, F. Irrera, F. Lemmi, F. Palma, "Tunable photo-detectors based on amorphous Si/SiC heterostructures", IEEE Transaction on Electron Devices **42** (5), 835 (1995).

7.26 G. de Cesare, F. Galluzzi, F. Irrera, D. Lauta, F. Ferrazza, M. Tucci, " Variable spectral response photodetector based on crystalline/amorphous silicon heterostructure", J. of Non-crystalline Solids **198–200**, 1189–1192 (1996).

7.27 G. de Cesare, F. Irrera, F. Lemmi, F. Palma, "a-Si:H/a-SiC:H heterostructure for bias-controlled photodetectors", Mat. Res. Soc. Symp. Proc. **336**, 885 (1994).

7.28 R. S. Crandall, J. Appl. Phys. **53** (4), 3350 (1982).

7.29 H.K. Tsai and S.C. Lee "Amorphous SiC/Si three-color detector", Appl. Phys. Lett. **52** (4), 275 (1988).

7.30 H. Stiebig, J. Gield, D. Knipp, P. Rieve, M. Bohm, "Amorphous silicon three color detector", MRS Symp. Proc. **337**, 815 (1995).

7.31 P. Rieve, J. Giehl, Q. Zhu, M. Bohm, "a-Si:H photo diode with variable spectral sensitivity", Mat. Res. Soc. Symp. Proc. **420**, 135 (1996).

7.32 F. Irrera, F. Lemmi, F. Palma, "Transient behaviour of Adjustable Threshold a-Si:H/a-SiC:H Three-Color Detector", IEEE Transaction on Electron Devices **44** (9), 1410 (1997).

7.33 H. Stiebig, D. Knipp, P. Hapke, F. Finger, "Three color piiin-detector using microcrystalline silicon", J. Non-cryst. Solids **227–230**, 1330 (1998).

7.34 T. Neidlinger, R. Bruggemann, H. Brummach, M. B. Schubert, "Color separation in a-Si:H based p-i-i-n sensor: temperature and intensity dependance", J. Non-cryst. Solids **227–230**, 1335 (1998).

7.35 M. C. Rossi, R. Vincenzoni, F. Galluzzi, "n+-SnO_2/a-SiC:H/Metal thin film photodetectors with voltage controlled spectral sensitivity", IEEE Transaction on Electron Devices **42**, 153 (1996).

7.36 H. Stiebig, D. Knipp, J. Folsch, F. Finger and H. Wagner, "Optimized three-color detector based on a-SiGe:H heterojunctions", Mat. Res. Soc. Symp. Proc. **420**, 153 (1996).

7.37 G. de Cesare, F. Irrera, F. Lemmi, F. Palma, "Amorphous silicon/silicon carbide three color photodetector with adjustable threshold", Appl. Phys. Lett. **66** (10), 1178 (1996).

7.38 G. de Cesare, F. Irrera, F. Lemmi, F. Palma, "Adjustable threshold a-Si:H color detectors", MRS Symp. Proc. **337**, 785 (1995). Italian Patent No. RM94A000294; US Patent Sep. 17,1996, Patent Number 5,557,1333.

7.39 G. Masini, G. de Cesare and F. Palma, "Current induced degradation in boron-doped hydrogenated amorphous silicon: a novel investigation technique", J. Appl. Phys. **77** (3), 1133 (1995).

7.40 M. Topic, F. Smole, J Furlan, W. Kusian, "Stacked a-SiC:H/a- Si:H heterostructure for bias controlled three-colour detector", Journal of Non-Crystalline Solids **198–200**, 1180 (1996).

7.41 R. L. Weisfield, "High performance input scanning array using amorphous silicon photodiodes and thin film transistors", Mat. Res. Soc. Symp. Proc. **258**, 1105 (1992).

7.42 F. Irrera, F. Lemmi, F. Palma, "Driving of a-Si:H/a-SiC:H adjustable threshold three color detectors for video rate applications", J. Non-cryst. Solids **227–230**, 1340 (1998).

7.43 J. L. Wiza, "Microchannel plated detector", Nucl. Instr. Met. **162**, 587–601 (1979).

7.44 J. Janesick, T. Elliot, G. Fraschetti, S. Collins, M. M. Blouke, B. Corrie, "Charge-coupled device pinning technologies", Proc. SPIE **1071**, 153–169 (1989).

7.45 R. A. Stern, R. C. Catura, R. Kimble, M. Wienzenread, M.M. Blouke, R. Hiyes, D. M. Walton, J. L. Culhane, "Ultraviolet and extreme ultraviolet response of CCD detectors", Opt. Eng. **26** (10), 972–980 (1987).

7.46 G. de Cesare, F. Irrera, F. Palma, M. Tucci, E. Jannitti, P. Naletto, and P. Nicolosi, "Amorphous silicon-silicon carbide photodiodes with excellent sensitivity and selectivity in the vacuum ultraviolet spectrum", Appl. Phys. Lett. **67** (3), 335 (1995). Italian patent No. RM95A000073. Italian patent No. RM95A000073; E.C. patent 96830052.5-2203.

7.47 D. Caputo, G. de Cesare, F. Irrera, F. Palma, "Solar blind UV photodetector for large area application", IEEE Trans. on Electron Devices, **43**, 1351 (1996).

7.48 H. Weinert, M. Petrauskas, J. Kolenda, A. Galecka, F. Wang, and R. Schwarz, "Ambipolar diffusion coefficient in a-SiC:H alloys in steady state and transient grating measurements" Mat. Res. Soc. Symp. Proc. **297**, 497 (1993).

7.49 G. Cocorullo, F. G. Della Corte, R. De Rosa, I. Rendina, A. Rubino, E. Terzini, "a-Si:H/a-SiC:H waveguides and modulators for low-cost silicon integrated optoelectronics", J. Non-Cryst. Solids **227–230**, 1118 (1998).

7.50 S. Guha, J. Yang, A. Banerjee, T. Glatfelter, K. Hoffman, S. R. Ovshinsky, M. Izu, H. C. Ovshinsky, X. Deng, Mat. Res. Soc. Symp. Proc. **336**, 645 (1994).

7.51 P.P. Deimel, B.Heimhofer, G. Krötz, H.J. Lilenhof, J. Wind, G. Müller, E. Voegs, IEEE Photon. Technol. Lett. **2**, 499 (1990).

7.52 Y.K.Fang, S.B. Hwang, K.H. Chen, C.R. Liu, L.C. Kuo, IEEE Trans. Electron Devices **39**, 1350 (1992).

7.53 J. Wind, G. Müller, Appl. Phys. Lett. **59**, 956 (1991).

7.54 Masamichi Okamura, Satoru Suzuki, IEEE Photon. Technol. Lett. **6**, 412 (1994).
7.55 D. Caputo, G. de Cesare, A. Nascetti, F. Palma, M. Petri, "Infrared photodetection at room temperature using photocapacitance in amorphous silicon structures" Appl. Phys. Lett. **72**, 1229 (1998).
7.56 Italian patent n. RM97A03341 (1997).

8 Thin Film Position Sensitive Detectors: From 1D to 3D Applications

Rodrigo Martins and Elvira Fortunato

Dep. de Ciência dos Materiais da Faculdade de Ciências e Tecnologia da Universidade Nova de Lisboa (FCT-UNL) e Centro de Excelência de Microelectrónica e Optoelectrónica de Processos (CEMOP) do Instituto de Desenvolvimento de Novas Tecnologias, Lisboa, Portugal
E-mail:rm@uninova.pt; emf@mail.unl.pt

Abstract. This chapter is a review of 1D, 2D and 3D thin film position sensitive detectors (TFPSD) made with hydrogenated amorphous silicon. Examples of different types of TFPSD are given, including the technological implications of the fabrication process and the physics ruling their behaviors. A detailed numerical analysis and electro-optical characterization is made for large area thin film position sensitive detectors and linear array devices, including the peripherals to be used in measurement/inspection systems.

8.1 Introduction and Historical Background

When a light beam is projected onto the surface of a p-n junction, a photovoltage is produced on each plane of the junction. The photocurrent generated on the surface flows laterally toward the electrodes due to the potential gradient in the lateral direction. This phenomenon is usually called the *lateral photo-effect*. Schottky [8.1] in 1930, was the first who observed the lateral photo-effect. At that time Schottky investigated the position dependence of the output photocurrent of a $Cu-Cu_2O$ diode and tried to correlate the optical behavior of this diode with its electrical characteristics. For this experiment he used a rectangular metal-semiconductor junction with a few cm^2 area, with a single contact on the semiconductor, in parallel with one of the detector sides (Fig 8.1a). The contact was short-circuited to the metal face, which behaves as an equipotential layer due to its low resistivity. By shining a light line on its surface and parallel to the collecting contact, Schottky observed that the photocurrent collected decreased as the distance between the contact and the light line position increased. Schottky explained this phenomenon by the aid of a leaking transmission line with a constant longitudinal resistance and a constant transverse conductance.

The work performed by Schottky remained unnoticed until Wallmark [8.2] observed it again in 1957. Wallmark used a Ge–In p^+-n junction with floating p^+ layer and point contacts on the n-layer (Fig 8.1b). The diode had two opposite point contacts separated by approximately 1 mm. Under these conditions external currents are not involved and Wallmark correlated this

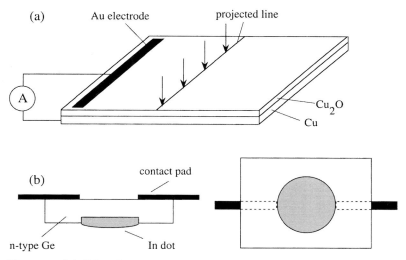

Fig. 8.1. (a) Schottky position sensitive Cu-CuO$_2$ diode [8.1]. (b) Cross section and top view of the Wallmark position sensitive detector [8.2]

photovoltage to the feed-in/feed-out effect described by Moore and Webster [8.3]. This means that the photo-induced carriers diffuse along both junction faces and recombine in the p-n barrier at a distance faraway from the position where they were separated. This effect was commonly called *surface recombination* or *barrier reinjection*.

After Wallmark rediscovery, the R&D work in the field of position sensitive detectors (PSD) has increased, aiming to apply it in various areas of optical measurements such as surface inspection and control [8.4–14].

To the best of our knowledge the most common semiconductor used to produce PSD devices is crystalline silicon (c-Si) [8.15–21]. Nevertheless, its detection area is small (around 1 cm^2 due to constrictions related to the detection principle [8.15]), which implies the use of expensive and complex optical magnification systems to support its application for large area inspection systems.

8.1.1 Why Use Amorphous Silicon to Produce Position Sensitive Detectors?

Hydrogenated amorphous silicon (a-Si:H) technology is now well established as a viable low-cost technology for a variety of large area applications such as solar cells [8.22, 23], image sensors [8.24], flat panel displays [8.25] etc. These devices take advantage of certain a-Si:H properties such as: low temperature processing capability; high photosensitivity; short response time; thermal stability and high production yield, besides allowing the production of devices in a monolhitic manner, either in rigid or flexible substrates [8.26].

These attributes now make a-Si:H the most common thin film semiconductor for large size optical sensor applications [8.27–32]. This material is often discussed in the popular press [8.33] in terms of its potential for *spinning it on*, into rug-like soft display screens. The a-Si:H base devices have also the unique property of not requiring an electrically common substrate, as happens, for instance with c-Si. In simplified terms, we can *paint* the "stuff" onto large surface areas.

It has taken more than 26 years of efforts since Chittick et al. [8.34] first deposited a-Si:H from SiH_4 in a plasma. To understand the distinctive role of the performances of the a-Si:H in device applications, in Table 8.1 a comparison is made with other materials also used as thin film photoconductors, aiming their use in large area optical sensors.

Table 8.1. Comparison of thin film photoconductors [8.35]

	Chalcogenides	CdS-CdSe	Poly-Si	a-Si:H
Photosensitivity	good (field dependent)	good (controlled by grain boundary)	low (limited by grain boundary)	high ($\mu\tau \approx 10^{-6}\,cm^2/V$)
Sensitive to visible light	possible	yes	maximum in the red	yes
Response speed	slow ($\approx$ ms)	slow (5 ms)		fast ($\approx \mu s$)
Carrier diffusion (crosstalk)	$< 1\,\mu m$	$\approx 1\,\mu m$	100 nm (crystallite size)	$< 1\,\mu m$
Non injecting contacts	possible	possible	yes	yes
Homogeneity	amorphous	granular ($\approx \mu m$)	granular (100 nm)	amorphous
Simplicity of manufacturing (low cost substrate)	yes	no (annealing at 500°C)	no (deposition at 630°C)	yes
Compatible with cryst. Semic. technology	no	no	yes	yes

Although Table 8.1 lists only some of the most important properties, the superiority of a-Si:H is clearly seen. For example, one of the most promising large area applications of a-Si:H is on LCD (Liquid Crystal Displays) and image sensors. Image sensors have been made in the past of chalcogenides [8.36] and CdS/Se [8.37, 38], but by now, they are no longer so popular, since a-Si:H was generally recognized as a suitable, material around 1980! The main arguments in favor of a-Si:H are the simplicity of manufacturing very uniform thin films in large areas, the relatively low dark currents, the high sensitivity and its suitable response speed (not too high due to the low electron mobility, but enough to these applications). Here, the uniformity of the material is an essential requirement for large area optical sensor applications. Heterogeneities, as they occur for example in polycrystalline material, may be avoided by using proper deposition parameters and systems [8.39–42]. Although the electron $\mu\tau$ product of a-Si:H ($\approx 10^{-6}\,\mathrm{cm^2/V}$) is much lower than that of c-Si, a photoconductivity gain of one is easily obtained. This is because the transport path of the generated photocarriers to the collecting electrodes run towards the film thickness ($d \leq 1\mu\mathrm{m}$), which is much smaller than the Schubweg ($\mu\tau E$) [8.43], where the electric field $E \geq 10^4\,\mathrm{V/cm}$.

An overview of the significant advantages and some of the disadvantages of a-Si:H are given in Table 8.2. The great interest in a-Si:H is due to its unique combination of a single fabrication process (the elements can be mixed and matched in arbitrary proportions, and different layers can be deposited on each other without the constraints of a crystalline structure) with an almost complete set of the most important semiconductor properties. The deposition (and doping, if necessary) of a-Si:H thin films is performed by a single step process at a temperature between 200°C and 250°C in a plasma reactor, not requiring any subsequent treatment. These performances made that several researchers started also to use the a-Si:H technology to produce thin film position sensitive detectors (TFPSD), either for small or large area applications. The results achieved up to now are promising regarding the application of these sensors to a wide variety of optical inspection systems like: machine tool alignment and control; angle measuring; rotation monitoring; surface profiling; medical instrumentation; targeting; remote optical alignment; guidance systems; etc., to which automated inspection control is needed, at moderate sampling time rates.

Table 8.2. Advantages and disadvantages of a-Si:H [8.44]

Advantages	Disadvantages
Large-area fabrication	Low deposition rate
Low temperature process	Small drift mobility
Multilayer structures	Schubweg smaller than the diffusion length in c-Si
Compatible with microelectronics technology	Light degradation behaviour of the absolute current.
Photoconduction (similar to c-Si)	
Dopability (similar to c-Si)	
Interfaces blocking or ohmic (similar to c-Si)	
Light sensitivity in the visible region	
Optical absorption greater than the one of c-Si	
Lateral carrier diffusion smaller than c-Si	
Intrinsic material highly photosensitive	

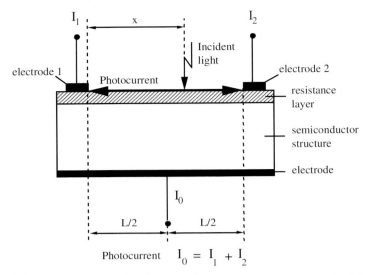

Fig. 8.2. Schematic illustration of 1D TFPSD operation principle

8.2 Principles of Operation of 1D and 2D PSD

Most one-dimensional (1D) and two-dimensional (2D) measurements require automated processes, accomplished with high resolution and linearity, able to be achieved through optical methods in conjunction with the required optical detector. Among them are the PSD devices whose working principle is based on the lateral photo-effect. These devices differ from quadrant cells and linear or matrix arrays, because they use only a single element with continuous detection capability, eliminating *dead* regions that otherwise exist between cells.

The easiest way of 1D position sensing is to measure the light spot using a linear image sensor. However, the size of the light spot should be smaller than the length of the sensor and the correlation between the position and the center of the light spot is achieved through proper hardware and software manipulations. Each electrode of the sensor element is bonded to an amplifier, so that the photogenerated charge can be read one after another, under control of a shift register (SR). Using this image sensing system we can obtain information about dark and illuminated points and their position, according to the resolution of the sensor, limited only by sensor size and pitch.

When a light spot falls on the PSD, an electric charge proportional to the light energy is generated at the incident position. This electric charge is driven through the resistive layer and the photocurrent collected by an electrode is inversely proportional to the distance between the incident position and the electrode. So, it is possible to obtain the photocurrents I_1 and I_2 collected by the two electrodes, as a function of the electrodes interdistance (L) and of

8 Thin Film Position Sensitive Detectors: From 1D to 3D Applications 347

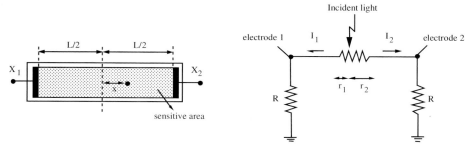

Fig. 8.3. Top view and equivalent electric circuit for 1D PSD

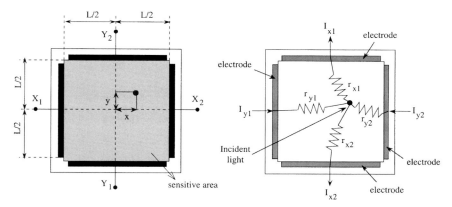

Fig. 8.4. Top view scheme of operation of a 2D PSD, indicating also the equivalent distributed resistances between the position of a spotlight and the contacts

the total photocurrent (I_0) that flows towards both directions, as it is shown in Fig. 8.2.

Figure 8.3 shows the top view of 1D PSD and the equivalent electric circuit where r_1 and r_2 are the distributed resistances of the photocurrent divider and R the load resistance. From the output currents, I_1 and I_2 the position (P) is obtained by:

$$I_1 = I_0 \frac{L-x}{L} \quad I_2 = I_0 \frac{x}{L} \tag{8.1}$$

$$P = \frac{I_2 - I_1}{I_2 + I_1} = \frac{2x - L}{L}. \tag{8.2}$$

Similarly, Fig. 8.4 shows the top view and the equivalent electric circuit for a 2D PSD. The position information for each direction (P_x and P_y) is given by

$$P_x = \frac{I_{x_2} - I_{x_1}}{I_{x_2} + I_{x_1}} \quad \text{and} \quad P_y = \frac{I_{y_2} - I_{y_1}}{I_{y_2} + I_{y_1}}. \tag{8.3}$$

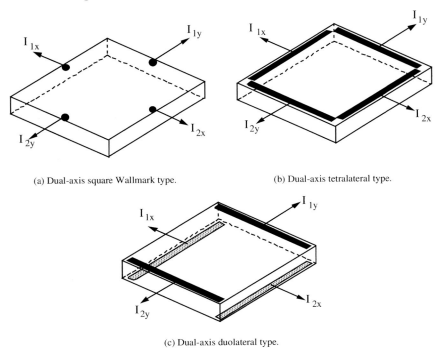

Fig. 8.5a–c. Types of 2D PSD: (**a**) Wallmark type, (**b**) tetralateral type and (**c**) duolateral type

Here, the "positions" are relative. In order to get the absolute position of the laser spot, the dimensions of the sensor must be taken into account.

The directional photo-induced current flowing through the external contacts depends on the relative location and intensity of the light beam over the active area of the sensor, where the reference corresponds to the geometrical center of the device active area. Thus, what is then obtained is the average or the centroid of the image spot and since there is no *dead* region on the device, no defocusing would necessarily be required.

8.2.1 The Different Types of PSD Devices That Can Be Produced

PSD's are classified in one-dimensional (1D) and two-dimensional (2D). The 2D are further divided into Fig. 8.5: (a) Wallmark type, (b) tetralateral type and (c) duolateral type.

The Wallmark type consists of one resistive layer and four dot electrodes. The tetralateral type may be regarded as an improved version of the Wallmark type consisting of four extended electrodes instead of dot electrodes. The dual-axis duolateral consists of a pair of resistive layers at the top and bottom and two parallel extended lateral electrodes at opposite sides, on each surface. These resistive layers function as dividers of generated photocurrent so that an incident light position can be detected using the output currents collected at each electrode.

The Wallmark type is ascribed with severe nonlinearity and shows other drawbacks due to carrier reinjection, e.g. poor sensitivity, background illumination dependency, temperature dependency and low speed. The tetralateral PSD does not show the drawbacks related to carrier reinjection, since a reverse-bias voltage suppresses carrier reinjection. However, a considerable nonlinearity is still present. The dual-axis duolateral PSD shows high position linearity and has no drawbacks due to carrier reinjection. Besides that, the dual-axis duolateral PSD has sensitivity twice that exhibited by the tetralateral one. Therefore, the dual-axis duolateral PSD is the most promising device for practical applications [8.45]. Beyond the different configurations, the PSD are also classified in terms how they operate: photovoltaic mode (without bias polarization) or photodiode mode (with reverse bias polarization).

8.2.2 Different Types of a-Si:H TFPSD and the Production Processes Used

MIS Type TFPSD. The first 1D and 2D (duolateral type) TFPSD were developed in 1983 by Arimoto et al. [8.46, 48], in Japan. The devices had sizes of 3 mm × 26 mm and 10 mm × 10 mm, respectively and consisted on a heterojunction based in a metal insulator structure (MIS), produced by anodic oxidation process. The reason to use this process is that it offers high breakdown voltages, high speed, noise immunity and design flexibility [8.49]. A cross sectional view of the proposed 1D and 2D TFPSD are shown schematically in Figs. 8.6 and 8.7, respectively.

The fabrication steps for the MIS 1D and 2D TFPSD devices consists in depositing by vacuum evaporation two parallel extended lateral Al electrodes at opposite sides and a thin Au–Cr film (which is the bottom resistive layer of 2D TFPSD), as metal electrodes. This is followed by the deposition of an indium tin oxide (ITO) transparent and conductive film with a sheet resistivity high enough so that there is only negligible effects on the sheet resistance of the resistive layer. For the 1D TFPSD, the ITO film is deposited directly on the substrate to form the ohmic contacts to the a-Si:H n^+ layer grown on top of them. besides this, the other fabrication steps to produce 1D (3 mm × 26 mm) and 2D (10 mm × 10 mm) TFPSD are the same, consisting in the deposition of the n^+- and i-layers followed by the oxidation step and the deposition of the front metal contacts, respectively.

The thicknesses of the i- and the n^+-layers are 6000 Å and 400 Å for both 1D and 2D TFPSD, respectively. The anodic oxidation is performed in two steps. The first one is an oxidation to passivate material defects and the second is to grow a thin oxide layer of 60 Å to form the MIS structure.

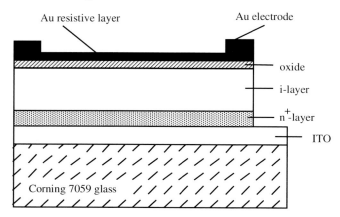

Fig. 8.6. Cross-sectional view of 1D TFPSD, made from anodic oxidation [8.46]

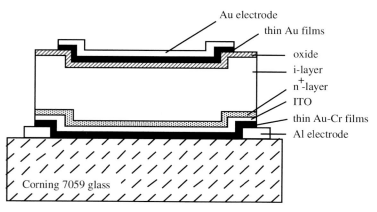

Fig. 8.7. Cross-sectional view of 2D duolateral TFPSD, made from anodic oxidation [8.48]

The thin Au film will act simultaneously as a metal barrier and a resistive layer. These devices presented good linearity with a correlation of 0.996, for the 1D TFPSD. The maximum position error ascribed to the 2D TFPSD devices is of about 20%, with an average error of 10%. Two years later, the same authors [8.48] improved substantially the performances of the 2D duolateral TFPSD with an area of 30 mm × 30 mm, presenting now a maximum detection error of 2%. Since these structures are semi-transparent, it is possible to realize a detectable angle TFPSD by using two superimposed TFPSD. This system has a variety of potential applications such as sensors for manufacturing robots.

Schottky Based TFPSD. In 1988, Okumura et al. [8.50] from Xerox Corporation, USA, proposed a new type of 1D TFPSD structure based in a

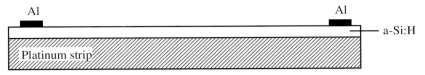

Fig. 8.8. Cross-sectional structure of a Schottky-barrier a-Si:H 1D TFPSD [8.50]

Schottky-barrier with a length of 16 cm. For his experiments, he used a platinum strip of about 20 cm length and deposited an n-type a-Si:H film by plasma enhanced chemical vapor deposition (PECVD) with a thickness between 0.5 µm and 1 µm. A pair of Al electrodes were vacuum deposited near the ends of the sample strip, as indicated in Fig. 8.8.

Besides the observation of the photo-effect in this type of device, the position dependence was non-linear, as it was in the other cases described previously.

One year later, Yamaguchi et al. [8.51] from Kanegafuchi, Japan, presented a TFPSD with a similar structure of that one presented by the Takeda's group. In their paper, they present a comparison between a 2D TFPSD (analog detection) and a linear image sensor (digital detection) both of a-Si:H where the main advantages and disadvantages of these devices are presented in Table 8.3. Although the amount of information obtained from a simple TFPSD system is limited, the analog TFPSD has some advantages in its simplicity and high response speed. The analog TFPSD is applicable to a digitizer (the authors call it a "light tablet") which takes the role of a man–machine (computer) interface. The large area TFPSD seems to be useful as a drawing tool on a cathode ray tube (CRT) or other display devices with the aid of a microcomputer.

P-I-N Based TFPSD. In 1988, Takeda et al. [8.52, 53] from Komatsu Research Division, Japan, presented a new type of 2D duolateral TFPSD with an area of 10 mm × 10 mm based on an a-Si:H p-i-n structure, as shown in Fig. 8.9. The structure consisted of a glass substrate, two resistance layers made of transparent conductive films (deposited by sputtering) and a semiconductor a-Si:H p-i-n structure (deposited by PECVD), incorporated between both resistance layers (four lateral extended Al or Au electrodes were deposited at opposite sides on both resistance layers). The aim of this device was to be used in a telephone terminal (Kom-Tel) that can send and receive different formats of information such as voice, image and characters.

In the same year Dutta et al. (1988) [8.54], presented a new type of 2D a-Si:H p-i-n tetralateral TFPSD with a total and effective areas of 12 mm × 12 mm and 10 mm × 10 mm, respectively with modified electrodes (top view of Fig. 8.10).

The shape of the electrodes was designed with the aid of numerical analysis by computer to compensate device non-linearity close to the edges [8.55].

Table 8.3. Comparison between analog and digital a-Si:H TFPSD

	Advantages	Disadvantages
Analog detection 1D and 2D duolateral a-Si:H TFPSD with a p-i-n structure	• Information about the center of light spot • High response speed* due to the simple mechanism to get the center position	• Resolution is limited by the uniformity of the constituting layers
Digital detection Linear image sensor (a-Si:H Schottky-barrier)	• Information about light intensity distribution • The amount of information is higher than in analog detection	• Resolution is limited by cell size and pitch

* For large area analog TFPSD, the resistance and capacitance of the device limit the response time so care must be taken when selecting and designing the device.

Besides that, they used pre-mounted electrodes on the glass substrate to minimize the leakage currents (bottom view of Fig. 8.10). The maximum detection deviation, within 80% of the active-area, was 2.5%, much better than the calculated deviation (6.25%) of a conventional TFPSD. Furthermore, this TFPSD uses an n$^+$ a-Si:H layer as a resistive dividing layer ($\rho \approx 10^2 \Omega\,\text{cm}$) deposited directly on the p-i-n structure. These results still show some position non-linearity that the authors attribute to the inter-electrodes separation at the edges of the two electrodes of opposite axis and to some difficulties related to the process fabrication.

P-I-N Based TFPSD with Strip Metal Electrodes. In 1990 [8.56] and 1992 [8.57] Yamamoto et al. presented an improvement on the 2D tetralateral TFPSD also with an a-Si:H p-i-n structure produced by PECVD. The new TFPSD has striped metal electrodes (length 13 mm, width 100 μm and spacing 100 μm) on its photosensitive surface as shown in Fig. 8.11. The device fabrication steps used are the following: 1) Cr layer sputtered onto a glass substrate; 2) Al layer vacuum evaporated on the Cr layer and patterned with standard photolithographic techniques for the bonding pads (after the patterning of the Al layer, the Cr layer is patterned for the striped electrodes); 3) a-Si:H p, i and n layers consecutively grown by PECVD; 4) Al electrode layers vacuum evaporated, using a metal shadow mask.

The experimental data show that these devices can be used to determine the position of two or more incident light spots, simultaneously. This type of TFPSD is more effective to solve the position non-linearity than the one with the modified four electrodes, as described earlier [8.53]. The operation princi-

8 Thin Film Position Sensitive Detectors: From 1D to 3D Applications   353

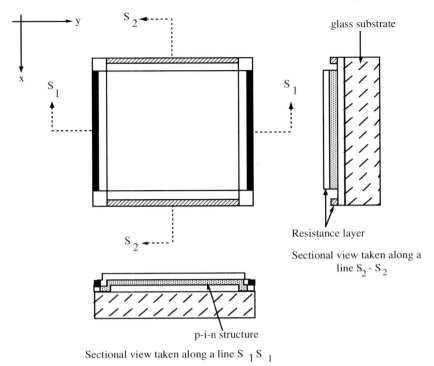

Fig. 8.9. Structure of an a-Si:H p-i-n 2D duolateral TFPSD [8.53]

ple is similar to that one of 1D TFPSD since each stripe acts as a 1D TFPSD. The only difference deals with the fact that to obtain 2D information it is necessary to use a scanning technique to find which electrode stripe is illuminated by the use of an array of metal-oxide semiconductor (MOS) switches. From these results, the authors conclude that this TFPSD with striped metal electrodes is more effective to solve 2D-position non-linearity than the 2D tetralateral TFPS. Another advantage of this device is its capability of simultaneously detecting two illuminated positions, without interference from adjacent electrode stripes.

Multilayer Systems Based TFPSD (MLS). In 1993, Panckow et al. [8.57] at Magdeburg, Germany, presented a new 1D TFPSD structure (with a length of 18 mm) based on a multilayer systems (MLS) as shown in Fig. 8.12. The MLS structure consists of ten periods of alternate a-Si:H and very thin Ti or Mo sub-layers. The results obtained lead to non linear curves and were fitted using the equation derived by combining Ohm's law and the continuity equation, assuming ideal Schottky contacts [8.58]. The authors attributed the linearity achieved to the continuous thin metal sub-layers that guarantee

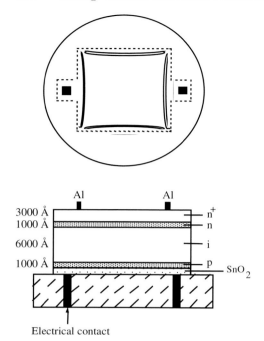

Fig. 8.10. Structure of 2D tetralaterala-Si:H TFPSD with modified electrodes to compensate the edge effects of the device on its performances, top view (*above*) and cross-sectional view (*below*) [8.54]

a charge carrier generation and separation, without considerable inner short circuit currents.

Large Area 1D and 2D p-i-n Based TFPSD. In 1993 and in 1994, E. Fortunato et al. [8.59–65] at FCT-UNL/CEMOP in Portugal, presented large area 1D (5 mm × 80 mm) and 2D (80 mm × 80 mm) duolateral TFPSD with an a-Si:H p-i-n structure produced by PECVD. This structure was deposited on ITO substrate with a sheet resistance of 20 Ω/□ and a transmittance of 80%. A temperature of about 210°C and a pressure of about 600 mTorr were used for the deposition of the p-i-n structure. The main characteristics of the layers that constitute the structure are shown in Table 8.4.

The ohmic contacts are formed by the ITO layer underneath the p-layer and by a thin resistive Al layer on top of the n-layer that form the required device equipotentials for carriers collection. Finally, a co-evaporation of Al and Ag is made to produce the edge contacts, located at both ends of the ITO layer for the 1D or alternatively, at the four edges (two in the ITO and the other two orthogonally placed on the thin Al electrode) for the 2D TFPSD. The fabrication of these TFPSD require the use of 3 or 4 photolithography patterning steps, depending on the type of TFPSD considered (1D or 2D). As

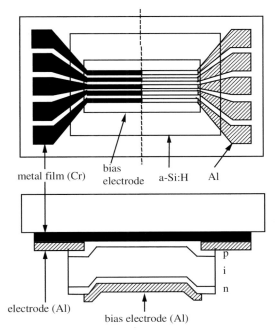

Fig. 8.11. Structure of a-Si:H 2D tetralateral TFPSD with striped metal electrodes: plane view (*above*) and cross-sectional view (*below*) [8.56]

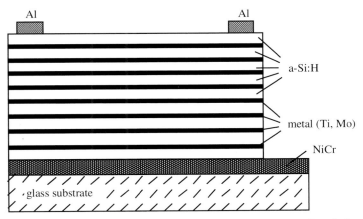

Fig. 8.12. Structure of an a-Si:H MLS TFPSD on a metallized glass substrate [8.57]

an example, Fig. 8.13 depicts the top view (above) and cross section (below) for an 1D TFPSD.

The devices developed by this group [8.59–65] exhibit excellent linearity, up to device sizes of 80 mm opening a new field of applications where large area TFPSD are required, for un-manned control processes. This makes this

Table 8.4. Main electrical and optical characteristics of the p, i and n a-Si:H large area 1D and 2D TFPSD constituting layers

Properties	p-layer	i-layer	n-layer
Dark conductivity [S cm^{-1}]	5×10^{-5}	1.4×10^{-9}	5×10^{-3}
Optical gap [eV]	1.71	1.75	1.71
Activation Energy [eV]	0.40	0.90	0.18
Photosensitivity [100 mW/cm^2]	–	5.0×10^5	–
Hydrogen content [at %]	20	13	17
Density of defects [cm^{-3}]	–	1×10^{16}	–

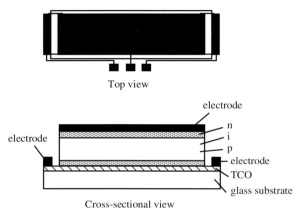

Fig. 8.13. Example of the structure of the 1D TFPSD, showing the top and cross-sectional views

Table 8.5. Characteristics presented by a-Si:H TFPSD and c-Si PSD

Characteristics	Amorphous silicon TFPSD	Crystalline PSD
Detecting range [cm]	10×10	1×1
Detecting accuracy	±0.5% F. S.	±0.5% F. S.
Wavelength at peak sensitivity [nm]	550	850
Sensitivity [A/W]	0.3	0.6
Frequency response (kHz)	> 1	500
Transparency	More than 50%	non transparent

type of TFPSD competitive when compared with the conventional c-Si PSD. In Table 8.5, we present some of the main characteristics exhibited by a-Si:H TFPSD and c-Si PSD devices used in optical position sensing applications.

For illustration, Fig. 8.14 shows a collection of some experimental a-Si:H 1D and 2D TFPSD fabricated at FCT-UNL/CEMOP, Portugal.

8 Thin Film Position Sensitive Detectors: From 1D to 3D Applications   357

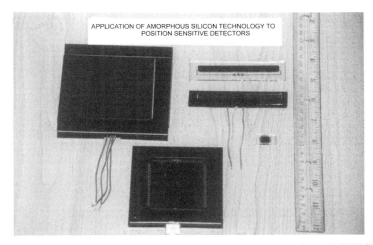

Fig. 8.14. Photograph of some a-Si:H 1D and 2D duolateral TFPSD with different areas, produced at FCT-UNL/CEMOP, Portugal

Linear Thin Film Position Sensitive Detector (LTFPSD). In 1995 Martins and Fortunato [8.64] proposed a new linear array LTFPSD based on p-i-n structures with 128 integrated elements. The aim is to use them in 3D inspections/measurements as a substitute of the conventional charge-coupled devices (CCD) used in optical cameras. This idea was stimulated by the desire to perform high-speed 3D profile measurements for industrial applications. Each element consists of 1D TFPSD (15 mm × 100 μm, with a separation of 100 μm to 10 μm), based in a p-i-n diode produced in a conventional PECVD system. By proper incorporation of the LTFPSD in an optical camera it

Table 8.6. Main process conditions for the fabrication of a typical LTFPSD [8.65]

N°	Process step	Conditions
1	Glass substrate	chemical and physical cleaning
2	ITO deposition (vacuum evaporation)	$d = 1000\,\text{Å}$
3	ITO patterning	Mask # 1
4	p-i-n semiconductor deposition (PECVD system)	p-layer: $d = 200\,\text{Å}$ i-layer: $d = 5000\,\text{Å}$ n-layer: $d = 600\,\text{Å}$
5	p-i-n patterning	Mask # 2
6	Metallization (electron gun)	Aluminium $d = 5000\,\text{Å}$
7	Patterning of thin resistive layer and contacts	Mask # 3
8	Pad protection	Mask # 4

top view through the glass substrate

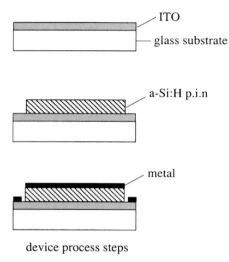

device process steps

Fig. 8.15. Simplified fabrication process steps of an LTFPSD

will be possible to acquire information about an object/surface, through the classical optical cross-section method [8.66, 67]. The main advantages of this system when compared with the conventional CCD's are the low complexity of hardware and software used and that the information can be continuously processed (analog detection). Figure 8.15 shows the fabrication steps of the LTFPSD. Table 8.6 shows the main process conditions for its fabrication that requires the use of 4 photolithographic masks.

8.3 Physical Model for the Lateral Photo-effect in a-Si:H p-i-n 1D and 2D TFPSD

8.3.1 Introduction

The photovoltage (ΔV) effect is a well-known phenomenon in a-Si:H devices under uniform illumination but less known when the device is under non-uniform illumination, as it is the case of the lateral photo-effect. This effect was discovered by Schottky in 1930 [8.1] and rediscovered by Wallmark in 1957 [8.2], being further developed by other researchers such as Lucovsky [8.68], Noorlag [8.69], Connors [8.70] and Woltring [8.71].

The lateral photo-effect although not well known in a-Si:H p-i-n devices was used to determine the ambipolar diffusion length [8.72–74] and later proposed to develop TFPSD. To explain this effect in a-Si:H p-i-n devices, an analytical model able to determine the device spatial detection limits under reverse bias and steady state conditions, was proposed by Martins et al. [8.74]. This model is able to interpret the lateral photo-effect behavior observed in a-Si:H 1D TFPSD and like the Lucovsky equation [8.68], a fall-off parameter (α) was deduced but now dependent on the peculiarities exhibited by the amorphous/disorder devices.

In the following we present a model of the 1D TFPSD incorporating the effects of the collecting resistance and the terminating resistance loads, either when the device works in the photovoltaic mode or in the photodiode mode [8.75]. As a first approach, we assume that the device width (Δ) (in the zz' direction) is much smaller than the device length (L). We also consider that the light beam reaching the device leads to a steady-state photocurrent generation within a line of length $2a$ and width equal to that one of the device, during the time interval between two consecutive measurements. Based on that, the same model will be extended to the two dimensional case, to interpret the behavior of the 2D TFPSD.

To derive the device response time and its behavior under transient conditions an equivalent electric circuit is constructed based on the results of the developed model, where the input light signal is approximated to a step unit function. Finally, the model is checked against the experimental results obtained in an 1D and 2D TFPSD.

8.3.2 General Description of the 1D Theoretical Model

In this section we present the general equations in which the 1D model is based, assuming that the a-Si:H p-i-n device when under reversed bias and non uniform illumination leads to a carrier's gradient between the irradiated and non irradiated regions, dependent on carrier's flow distribution, carrier's transit time, device relaxation time, conductivity of the collecting doped layer and on the recombination losses in the interfaces. This leads to the appearance of a lateral photo-effect dependent on the lateral diffusion rate of the carriers generated within the i-layer, towards the non-illuminated region [8.2, 68–70, 76]. Here the carriers laterally collected are also influenced by losses due to the collecting doped layer.

To analyze the device behavior, we consider that the carriers are collected in the p-layer by a thin collecting resistive layer (TCRL). The reference equipotential is also a thin resistive layer deposited on the top of the back doped layer and that the reverse bias is used to enhance the device response time as well as to decrease the role of the temperature on device performances

[8.61]. This will allow defining the boundary conditions, under the following assumptions:

- conduction is mainly dependent on carriers accumulated in the i-layer edges;
- the doped layers behave as sources of carrier's losses (lateral and transverse);
- the current collected by the TCRL depends on the carriers storage time in the interface, the Schubweg [8.78] and the diffusion length of the excess of carriers;
- the net flow involves both carriers and hence the ambipolar transport mechanism;
- the rate of lateral diffusion from the irradiated area depends on the average diffusion velocity of the carriers (v_d), their spatial distribution and reverse bias.

These conditions lead to the build up of an electric field E along the junction plane that helps the ambipolar transport away from the irradiated region, as sketched in Figs. 8.16a and b, without and with the reverse bias, respectively.

E is obtained by combining the Poisson, continuity and current density equations, including or not the recombination losses (R_l), taking into account that p_1 and n_1 are the excess of electron–hole pairs generated by the photon flux f_λ and that $n' = n_0$ outside the irradiated area.

The losses are given by $R_l \approx n_a/\tau_a$ where τ_a is the average mean time that the carriers generated in the i-layer take to reach the TCRL, given by $\tau_a \approx \varepsilon\varepsilon_0/2\pi\sigma_a$, where σ_a is an equivalent conductivity, dependent on the ones of the TCRL and the doping collecting layer [8.74]

$$J_y = q\mu_a n' E + \varepsilon\varepsilon_0 v_d \nabla E + C_l \frac{dE}{dt}. \tag{8.4}$$

In (8.4) C_l is the capacitance per unit length of the device and J_y is the lateral current density that depends on the excess of ambipolar carriers generated within the illuminated region to which are ascribed two main components

$$J_y = J_{yd} + J_{yl}, \tag{8.5}$$

where $J_{yd} = \sigma_r E_y$ (σ_r is the conductivity of the TCRL) and $J_{yl} = J_{sy}$ (J_{sy} is the saturation current density of the device reverse biased, $J_{sy} \approx \sigma_0 E_y$, where σ_0 is the conductivity of the i-layer, outside the illuminated region) are respectively the drift and the lateral leakage current density components. In addition to J_y we also consider the role of the transverse current density, J_x, on the device performances. Within the illuminated region J_x is given by $J_x \approx J_s - J_{ph}$, where J_{ph} is the generated photocurrent density, dependent on the excesses of generated electron–hole pairs (G_{na}) and R_l, through the

8 Thin Film Position Sensitive Detectors: From 1D to 3D Applications 361

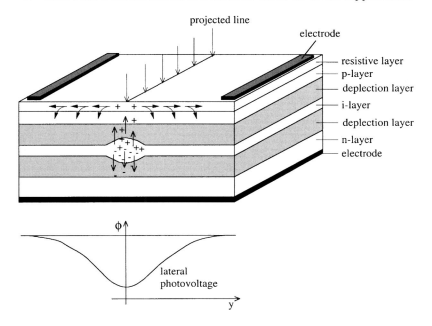

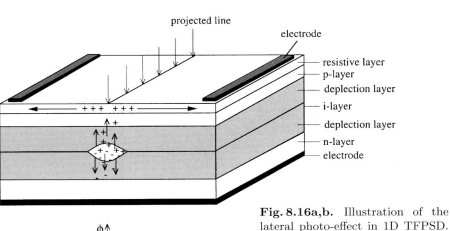

Fig. 8.16a,b. Illustration of the lateral photo-effect in 1D TFPSD. The + and − signs represent free holes and electrons generated by the light beam: (**a**) operating in the photovoltaic mode (without reverse bias); (**b**) operating in the photodiode mode (with reverse bias)

relation $1/q(\mathrm{d}J_x/\mathrm{d}x) = G_{\mathrm{na}} + R_\mathrm{l}$. Outside the illuminated region, $J_{\mathrm{ph}} \sim 0$, and so, $J_x \sim -J_{\mathrm{s}}$ (Fig 8.16). The correlation between J_x and J_y is ruled by the principle of charge conservation, leading to (for the 1D case) [8.68]:

$$\frac{\mathrm{d}J_y}{\mathrm{d}y} = \frac{J_x}{W} \Rightarrow J_y = \int_{-a}^{a} \frac{J_x}{W} \mathrm{d}y, \tag{8.6}$$

where W is the thickness of the i-layer and a is the half length of the irradiated line on the top of the device (Fig 8.16), with a negligible width z. If the device is reversed biased, J_x is equal to the contribution of the photocurrent density generated by the irradiated light, J_{ph} and the reverse current density (J_s) $\{J_x = J_\mathrm{s}[\exp(qV/kT) - 1] - J_{\mathrm{ph}}$. As $|V| \gg kT/q$, this leads to $J_x \approx -J_\mathrm{s} - J_{\mathrm{ph}}\}$, and so:

$$J_y = -\frac{2a}{W}(J_\mathrm{s} + J_{\mathrm{ph}}). \tag{8.7}$$

E_y is determined assuming that the carriers collected by the TCRL flow through it towards the end terminals, leading at each point, inside or outside the illuminated region, to an ohmic dependence of J_y on E_y.

Substituting (8.7) in (8.4) and taking into account the boundary conditions and that ΔV is conditioned by σ_0, σ_r and σ_d, the dc (0) and ac (1) components, of E_y are obtained through the equations:

$$\frac{\mathrm{d}E_0(y,t)}{\mathrm{d}y} + \alpha \frac{4a^2}{WW_\mathrm{r}} E_0(y,t) + \frac{C_1}{\sigma_\mathrm{r}} \frac{\mathrm{d}E_0(y,t)}{\mathrm{d}t} = \frac{-J_{\mathrm{ph}_0}(y,t)}{\sigma_\mathrm{r}} \frac{2a}{W}, \tag{8.8}$$

$$\frac{\mathrm{d}E_1(y,t)}{\mathrm{d}y} + \alpha E_1(y,t) + \frac{C_1}{\sigma_\mathrm{r}} \frac{\mathrm{d}E_1(y,t)}{\mathrm{d}t} = -\frac{J_{ph1}(y,t)}{\sigma_\mathrm{r}} \frac{2a}{W}, \tag{8.9}$$

where W_r is the thickness of the TCRL, ρ_s is the sheet resistance of the TCRL ($\rho_\mathrm{s} = 1\sigma_\mathrm{r} W_\mathrm{r}$ and J_{ph} has a dc (0) and an ac (1) components: $J_{\mathrm{ph}} = J_{\mathrm{ph0}} + J_{\mathrm{ph1}}$, respectively, outside and within the irradiated area. In the above relations α is the lateral fall-off parameter, given by, $\alpha = 4\rho_\mathrm{s}\sigma_0 a^2/WW_r$, when R_l is neglected.

8.3.3 Role of the Recombination Losses for the Fall-Off Parameter

The role of R_l for the device performances depends on σ_a determined through the equivalent electric circuit of Fig. 8.17. There, the illuminated and the non-illuminated regions are substituted by two equivalent current sources, respectively equal to the photocurrent I_{ph} and the saturation current I_s. These current sources are connected to the corresponding distributed resistances due to the TCRL and to the doping collecting layer, within an incremental distance Δ. To the resistance of the doped collecting layer are associated two components, a transverse one (R_{td}) and a longitudinal one (R_{ld}).

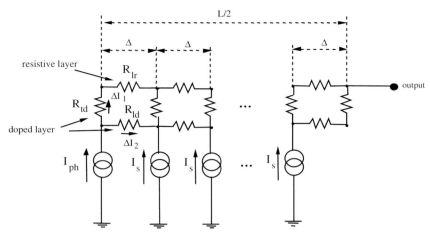

Fig. 8.17. Equivalent incremental electric circuit of the static 1D TFPSD. R_td and R_ld are the transverse and longitudinal resistances of the doped collecting layer, within a spatial increment Δ. R_lr is the longitudinal discrete resistance of the TCRL within the same Δ. I_ph and I_s are the equivalent current sources, in regions with and without illumination

For the TCRL we only consider the longitudinal component, (R_lr), through which the carriers flow towards the end terminals, as also shown in Fig. 8.17. In this discrete model ΔI_1 is the incremental current that transversely reaches TCRL and flows through it towards the terminals located on the edges of the TCRL. The ΔI_2 is the incremental current that flows laterally through the doped collecting layer and can be considered as a leakage current.

Under these conditions, ΔI_1 and ΔI_2 are respectively given by:

$$\Delta I_1 = \frac{\Delta V_y}{R_\mathrm{td} + R\mathrm{lr}} \text{ and } \Delta I_2 = \frac{\Delta V_y}{R_\mathrm{td} + R_\mathrm{ld}}, \tag{8.10}$$

where ΔV_y is the potential voltage ascribed to Δ and related to the total potential voltage V_y by the relation $V_y = \sum \Delta V_y$.

The total current supplied by the current sources in the boundary of the illuminated region is given by $I_\mathrm{ph} + I_\mathrm{S} = \Delta I_1 + \Delta I_2$, while outside the illuminated region we have $I_\mathrm{S} = \Delta I_1 + \Delta I_2$. The equivalent conductivity σ_a is then determined through (8.10) taking into account the Kirchoff's rules. Under these conditions the value of α in (8.8) replaced by an α' parameter such that:

$$\alpha' = \frac{\alpha}{1+\gamma}, \text{ with } \gamma = \frac{1}{4\pi}\rho_\mathrm{sd}\sigma_0 \frac{a^2}{W_0}\rho_\mathrm{s}\sigma_0 \frac{1}{W_\mathrm{d}}, \tag{8.11}$$

where ρ_sd and W_d are, respectively, the sheet resistance $[\rho_\mathrm{sd} = (\sigma_\mathrm{d} W_\mathrm{d})^{-1}]$; and the thickness of the doped collecting layer and $W_0 = W + W_\mathrm{d} + W_\mathrm{r}$. Here, it is also important to notice that $W_\mathrm{d} < W_\mathrm{r}$, to decrease the optical and the ohmic losses due to the doping collecting layer.

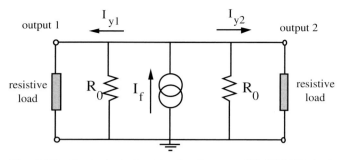

Fig. 8.18. Equivalent electric circuit to an 1D TFPSD device, showing the corresponding current source, output resistance and the equivalent load resistance

8.3.4 Static Behaviour of E_y and ϕ_y

To solve (8.8) we place the origin of the co-ordinates in the center of the device. That is, $-L/2 < y < L/2$; E vanishes when $|y| \to \infty$ and the generated carriers depend now on their distribution within the irradiated region. Under these conditions the general solution of (8.8), for a semi-infinite slab is $E_y = \pm E_0\{1 - \exp[\alpha'(a \pm y)]\}$, with $E_0 = 2a(J_{\text{ph}} + J_s)\rho_s/W\alpha'$. The static potential spatial distribution function $[\phi(y)]$ is obtained taking into account that $E = -d\phi(y)/dy$

$$\phi_y = \pm \frac{E_0}{\alpha'} \frac{1 - \exp[-\alpha'(a \pm y)]}{\tanh(\alpha'L/2)} + c \tag{8.12}$$

or after mathematical manipulations, where $\phi(y)$ is assumed finite at $y = \pm L/2$ [8.81]

$$\phi_y = \pm \frac{E_0}{\alpha'} \frac{\sinh(\alpha'y)}{\sinh(\alpha'L/2) + \sinh(\alpha'a)} + c'; \tag{8.13}$$

c and c' are constants related to the field distortions at the device edges [82]. Here $\Delta V = \phi(y) - \phi(L-y)$. The spatial range up to which a linear correlation exists between ΔV and y depends on α'. So, to enhance the linearity, spatial resolution and the signal to noise ratio, $\alpha' \ll 1$ cm^{-1}.

Once defined ΔV the device can be substituted by an equivalent electric circuit based on a current source, I_f and on an output resistance R_0,

$$I_f \approx J_{\text{ph}}(2a\Delta) \frac{\sinh(\alpha'y)}{\sinh(\alpha'L/2) + \sinh(\alpha'a)} \Rightarrow \tag{8.14}$$

$$R_0 \approx \rho_s \frac{L \tanh(\alpha'y)}{2\Delta\alpha'L/2}, \tag{8.15}$$

where R_0 depends on ρ_s and on the spatial position of the light beam impinging on the surface of the TFPSD, as shown in Fig. 8.18.

8.3.5 Role of ρ_s and ρ_{sd} for the Device Detection Limits, Linearity, and Spatial Resolution

The role of ρ_s and ρ_{sd} on the device performances depends on the electro-optical characteristics of the layers of the 1D TFPSD. For an ITO/p-i-n/metal structure where the ITO layer is TCRL the device performances are enhanced if the losses lead to very low α'. This condition is reached when the device is reverse biased, with σ_0 as low as possible (to decrease the lateral losses from the region where carriers are generated and simultaneously to sustain them during the time required for their collection by the end collecting terminals). This means that $\sigma_0 < 5 \times 10^{-8} \Omega^{-1}$ cm^{-1} and that the i-layer has a low density of defects. In addition: $0.05\,\mu\text{m} < W_r < 0.25\,\mu\text{m}$, to allow a high light transmittance and a low ρ_s; $W_d = 0.025\,\mu\text{m}$, to decrease the optical and carrier losses, but thick enough to sustain the needed built-in potential; and $0.15\,\mu\text{m} < W < 1.2\,\mu\text{m}$, to match the spectral response of the light source and to be completely depleted under reverse bias. The back doped layer might have also the enough conductivity and thickness that leads to the establishment of a good ohmic contact with the back metal electrode that behaves as a true equipotential. Thus, for a particular 1D TFPSD we can put $\sigma_0 \approx 10^{-9}\,\Omega^{-1}$ cm^{-1}, $W=0.5\,\mu\text{m}$, $W_d=0.02\,\mu\text{m}$ and $W_r = 0.1\,\mu\text{m}$ and so, to determine the influence of ρ_s and ρ_{sd} on α', through (8.11), leading to:

$$\alpha' \approx \frac{0.08\rho_s}{1 + 2.7 \times 10^{-8}\rho_s\rho_{sd}} . \tag{8.16}$$

Figure 8.19 shows the predicted behavior of α' as a function of ρ_{sd} using ρ_s as a parameter. The data show that low α' are obtained when $\rho_{sd} > 5 \times 10^6\,\Omega/\square$, almost independent of ρ_s while for $\rho_{sd} < 5 \times 10^6\,\Omega/\square$, $\rho_s < 20\,\Omega/\square$, to get $\alpha' < 1$ cm^{-1}. Thus, to reach high spatial detection limits with a high linearity and spatial resolution, $\sigma_d > 10^{-2}\,\Omega^{-1}$ cm^{-1}, to avoid a high lateral leakage current through the doped collecting layer. Besides that $\rho_s > 10\,\Omega/\square$ for the TCRL not to be considered as an equipotential and $\rho_s < \rho_{sd}$ at least two orders of magnitude, to allow that the current that flows laterally through the doped collecting resistive layer is much smaller than the current that flows through the TCRL.

8.3.6 Static Distribution of the Lateral Current

Once known $\phi(y)$ the lateral photocurrent, $I_L(y)$ is derived through the conductance $G_0(y)$ of the "channel" formed close to the collecting interface and dependent on the charge accumulated in the interface, through $I_L = G_0(y)\phi(y)$, where $G_0(y)$ depends on the number of ambipolar carriers existing per unit volume in the "channel" (Q_a) and on the ambipolar mobility of the carriers (μ_a):

$$G = \int_{-L/2}^{L/2} \mu_a Q_a(y) dy = \int_{-L/2}^{-a} \mu_a Q_a(y) dy + \int_{a}^{L/2} \mu_a Q_a(y) dy . \tag{8.17}$$

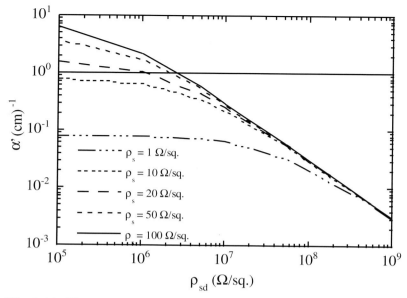

Fig. 8.19. Theoretical dependence of α' on ρ_{sd}, using ρ_s as a parameter (8.16). The *horizontal line* helps to find the conditions for which $\alpha \geq 1$ cm^{-1}

From (8.6) we can identify a distributed shunt conductance (similarly to a transmission line) G_0 as $G_0 = \varepsilon\varepsilon_0 v_d a^2$. This leads to:

$$R_0 \approx \frac{2}{\varepsilon\varepsilon_0 v_d (\alpha' W)(\alpha' L)} . \tag{8.18}$$

Once known R_0, the corresponding $I_L(y)$ is given by:

$$I_L(y) \approx -[J_s + J_{ph}(y,t)] \Delta a \frac{a}{W} \frac{\sinh(\alpha y)}{\sinh(\alpha L/2)} , \tag{8.19}$$

where Δa is the illuminated area. The correlation of $I_L(y)$ with the spatial position is given by $I_{L1}(y) = G(y)\phi(y)$ or $I_{L2}(L-y) = G(L-y)\phi(L-y)$, where the subscripts 1 and 2 refer to the terminal where the current is measured. Considering now that the terminals are connected to a load of value R_L and if $R_L \ll R_0$, R_0 can be neglected, leading to $V(y) = R_L I_L(y)$. That is, the signal detected varies as a sinh function with y, implying that good response linearity is only achieved when $\alpha L \ll 1$ [8.74].

An important point to be considered is the role of the temperature (T) on device performances. In the discussed model we consider that the device is under reverse bias and so, the diode current density was approached to the reverse saturation current which has a small dependence on T. Otherwise, the diode current density presents an exponential dependence on T [$J_s \propto J_0 \exp(-\Delta E/\eta kT)$], which strongly affects device linearity behavior.

8.3.7 Extension of the Theoretical 1D Model to the 2D Case

For the 2D sensors the starting point is (8.8) [8.62]

$$\nabla^2 \Phi(x,y) = -\frac{\rho_s}{W}(J_{\text{ph}} + J_s) \ , \qquad (8.20)$$

where Φ is the potential in the collecting resistive sheet. To solve (8.20) we consider the homogeneous boundary conditions where it is assumed that the field strength component normal to the boundary is zero ($\delta\Phi/\delta n$) along both axis and that the principle of separation of variables is applied. We also assume that $\delta_B^2(x,y)/\delta x^2 = 0$ and $\Phi_B(0,y) = \Phi_B(x,L)$ (the structure does not lead to any type of anisotropy as observed previously in crystalline devices [8.82]. This leads to orthogonal solutions for the current of the type:

$$I_{j,i}(x,y) = I_{0j,i}(x,y)\sin\left(j\pi\frac{x}{L}\right)\cos\left(i\pi\frac{y}{L}\right) \ . \qquad (8.21)$$

This means that each individual solution for one specific direction is always modulated by a fix amount of the other direction. That is

$$I_{j,i}(x,y) = I_j(x)I_i(y), \text{ where} \qquad (8.22)$$

$$I_j(x,) = I_{0j}\sin\left(j\pi\frac{x}{L}\right) \text{ and } I_i(y,) = I_{0i}\cos\left(i\pi\frac{y}{L}\right) \ , \qquad (8.23)$$

and the summation (or integration) over the entire collecting surface leads to the determination of the x,y position. Neglecting R_l and assuming a detection resolution (α^{-1}) of $8\,\text{cm}^{-1}$, in both directions for a $8\,\text{cm} \times 8\,\text{cm}$ size 2D detector we have $\alpha L = 1$. This leads to a low resolution and linearity, similarly as it happens for the 1D case. If $\alpha^{-1} \gg L$, in both x and y directions, we get:

$$\begin{aligned} I_{j,i}(x,y) &\approx I_{0j,i}(x,y)(1-\alpha_x x)(1-\alpha_y y) \\ &\approx I_{0j,i}(x,y)\left(1-\frac{x}{L}\right)\left(1-\frac{y}{L}\right) \ , \end{aligned} \qquad (8.24)$$

under the condition that the sampling rate is compatible with the elementary time constant per unit area $\beta = C_1 \rho_s$, of the device. The values of $I_{0j,j}$ depend on J_{ph} and on the current losses ascribed to the device. That is:

$$I_{0j,i} \approx \sqrt{I_{0,j}^2 + I_{0,i}^2} \ , \quad \text{with} \qquad (8.25)$$

$$I_{o,j} \approx \frac{\int J_{\text{ph}_j}\,dxdy}{\alpha L} - I_{\text{leak}} \text{ and } I_{o,i} \approx \frac{\int J_{\text{ph}_i}\,dxdy}{\alpha L} - I_{\text{leak}} \ , \qquad (8.26)$$

where the detector is assumed to be square, I_{leak} is the current leak due to the reverse saturation current and possible carrier's shunt paths, including also the ohmic losses ascribed to the contact resistances and the external R_L used.

Here, we must also consider the noise current in the collecting resistive layer due to the Johnson noise [8.81] given by:

$$I_\text{n} = \left(4k_\text{B}TB\frac{w_\text{d}}{\rho_\text{d}}\right)^{1/2} \quad \text{where} \quad B = \pi/(2\beta L^2) .\tag{8.27}$$

For large area 2D devices $\beta L^2 \approx 1\,\text{ms}$. Considering that $w_\text{d} \approx 0.2\,\mu\text{m}$, $\rho_\text{d} \approx 50\,\Omega/\square$ and $T \approx 300\,\text{K}$, we get for the bandwidth $B \approx 374\,\text{kHz}$ and $I_\text{n} \approx 748\,\text{pA}$. Thus, the minimum value of the current detected has to be at least a factor of five larger than the total value of $I_\text{n} + I_\text{s} + I_\text{leak}$.

As already stated for the 1D case, the point of the detector where the current is null corresponds to the center of gravity whose x coordinate (x_G) is given by:

$$x_\text{G} = \frac{\iint x_0 J_\text{ph}(x_0, y_0)\mathrm{d}x_0\mathrm{d}y_0}{\iint J_\text{ph}(x_0, y_0)\mathrm{d}x_0\mathrm{d}y_0} .\tag{8.28}$$

The value of x_G is highly influenced by the distribution of the sheet resistance of the collecting layer and in the constancy of $J\text{ph}$ and so, on the uniformity of the i-layer. Changes in such parameters lead to an offset of the origin, proportional to percentage of variation of the corresponding collecting resistance and of the photoconductivity of the i-layer (highly influenced by the background illumination). For the TCRL, the variations observed are mainly attributed to changes in the shape of the electrodes and in non- uniformities of the layer along the surface. Indeed, most of the non-linearities seen are due to the effect of the edge shape of the electrodes on the induced potential. This can be compensated by proper design of the electrodes. Now, if the contacts at $x = 0$ and $x = L$ are terminated with impedances R_0 and R_L, we get;

$$\Phi_{0,y} = R_0 I_{x-} \quad \text{and} \quad \Phi_{L,y} = R_\text{L} I_{x+} = R_0 \left(I_\text{ph} - I_{x-}\right)\tag{8.29}$$

and

$$I_{x-} = \frac{\rho_\text{s}/w + R_\text{L}}{\rho_\text{s}/w + R_\text{L} + R_0} I_\text{ph} - \frac{\rho_\text{s}/w}{\rho_\text{s}/w + R_\text{L} + R_0} I_\text{ph} \frac{x_\text{G}}{L} .\tag{8.30}$$

That is, the position dependence remains linear with the use of finite terminating impedances, with a possible offset function of x_G. Thus, during the design of the lateral electrodes especial care has to be put to balance both collectig resistors, to guarantee that $R_y = R_x$. The main parameters to play with are the width and thickness of the TCRL and the design of the soldering contacts.

8.3.8 Determination of the Transient Response Time of the TFPSD

The transient response of the device can be determined through an equivalent electric. For the device under reverse bias we can assume that the irradiated

8 Thin Film Position Sensitive Detectors: From 1D to 3D Applications 369

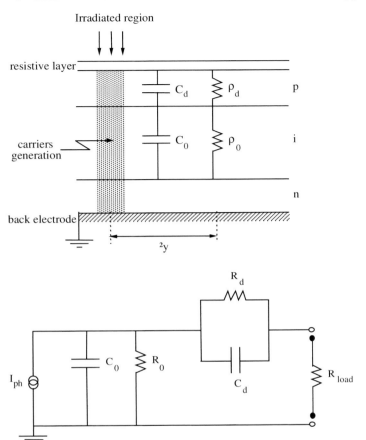

Fig. 8.20. (a) Elementary cross section of a p-i-n junction where C_0 and C_d refer to the capacitance per unit area of the i- and p-layers, and ρ_0 and $\rho_d d$ represent the resistivity of the i- and p-layers, respectively. (b) Simplified equivalent circuit of the 1D TFPSD where I_{ph} represents the current generated by the incident light

region behaves as a current source (the output current of the device is taken as its short circuit current) connected in parallel to R_0 and to a capacitance $C_0 (C_0 = C_1 L/2)$, as shown in Fig. 8.20a, b.

There we also show the equivalent load resistance R_L, connected to the collecting terminals. Using the parameters of the equivalent electric circuit together with the Laplace transform principles [8.79]

$$V(y,t) = \int V(y,s) e^{st} ds \qquad (8.31)$$

$$J_{\text{ph}_1}(y,t) = \int J_L(y,s) e^{st} ds \Rightarrow I_L(y,t) = \int I_L(y,s) e^{st} ds . \qquad (8.32)$$

The transient solution can be derived, considering that $V(y, s) = -I_L(y, s)R_L$. Assuming now that a step function is applied to the device $[J_L(s) = J_{ph1,0}/s]$ and that $R_0 \gg R_L$, the solution of (8.9) (under Laplace transformation [8.80]) is:

$$I_L(s) = \frac{I_L(y, s)}{s} \Rightarrow I_L(s) = I_0 \frac{1}{s} \frac{\sinh(\alpha_1 y)}{\sinh(\alpha_1 L/2)}, \quad \text{where} \tag{8.33}$$

$$\alpha_1 = \alpha + \frac{C_1}{W_r \sigma_r} \cdot s \quad \text{and} \quad I_0 = \frac{2a^2}{W} \frac{J_s + J_{ph\,1,0}}{(2\Delta a)^{-1}}. \tag{8.34}$$

As α_1 can be written in the form of $\alpha_1 = (\alpha\omega_0 + s)/\omega_0$ with $\omega_0 = W_r\sigma_r/C_1$, its substitution in (8.33) leads to:

$$I_L(s) = \frac{I_L(y, s)}{s} \Rightarrow I_L(s) = I_0 \left\{ \frac{1}{s} \frac{\sinh\left[\frac{y}{\omega_0}(\alpha'\omega_0 + s)\right]}{\sinh\left[\frac{L}{2\omega_0}(\alpha'\omega_0 + s)\right]} \right\} \tag{8.35}$$

and so, $i(t)$ is obtained through the inverse of the Laplace transform [8.81–84]

$$i(t) = I_0 \mathbf{L}^{-1} \left\{ \frac{1}{s} \frac{\sinh\left[\frac{y}{\omega_0}(\alpha'\omega_0 + s)\right]}{\sinh\left[\frac{L}{2\omega_0}(\alpha'\omega_0 + s)\right]} \right\} \tag{8.36}$$

that can be simplified, using $s' = s + \alpha'\omega_0$ [8.21]

$$i(t) = I_0 \exp(-\alpha'\omega_0 t)\mathbf{L}^{-1}\left[\frac{1}{s' - \alpha'\omega_0} \frac{\sinh\left(\frac{y}{\omega_0}s'\right)}{\sinh\left(\frac{L}{2\omega_0}s'\right)}\right]. \tag{8.37}$$

From the table of Laplace transforms [8.85] and after some mathematical manipulations we get [8.84]

$$i(y, t) \approx I_0 \frac{\sinh(\alpha' y)}{\sinh(\alpha' L/2)} \left\{ 1 - \exp\left[-2\frac{\omega_0}{L}t\left(\alpha'\frac{L}{2}\right)\right] \exp\left(-\pi\frac{2\omega_0 t}{L}\right) \right\} \tag{8.38}$$

if we consider that G_0 is time independent. Otherwise, the second exponential term of (8.38) should be multiplied by a time dependent factor, leading to

$$2\frac{\omega_0}{L}t \exp\left\{-2\frac{\omega_0 t}{L}[\alpha'(L/2 + a)]\right\}. \tag{8.39}$$

These solutions show that when a modulated light pulse is used the output signal is not disturbed by the presence of a background illumination and it is only valid if $R_0 \gg R_L$. Here the limit of the frequency response depends on $2\omega_0/L$ (where $\omega_0 \approx 1/R_L C_1$), decreasing as the device length increases. The current rise time is dependent on the position of the incident radiation, increasing as the distance from the contact $(y - L/2)$ increases. Also, the output current achieves its steady-state value for t such that $\omega'_0 t L(\alpha'L + 2\pi)/4$.

8 Thin Film Position Sensitive Detectors: From 1D to 3D Applications

We can also determine the overall attenuation factor a_n of the device as being:

$$a_n \approx \left[\alpha^2 L^2 + (2n-1)^2 \pi^2\right]^{1/2}, \tag{8.40}$$

and so, to a spatial modulating term given by the exponential factor

$$\exp\left[-\left(a_n^2 + m^2\pi^2\right)xy\alpha_x\alpha_y\right], \tag{8.41}$$

where for the 1D case $y\alpha_y$ (or $x\alpha_x$) are considered to be constant, including the possible device offset ascribed to possible changes in x_G.

8.4 Static and Dynamic Detection Limits

8.4.1 Static Detection Limits of 1D TFPSD

The device detection limits is the maximum distance that a light spot from the collecting electrodes can be detected with a linear correlation between the spatial position and the lateral photocurrent measured, within an allowable detection time, τ_d such that $\tau_d \geq L[(\alpha' L) + 2\pi]/4\omega_0$, considered as a dynamic fall-off parameter. On the other hand, τ_d has to be smaller than the device relaxation time ($\tau_r = \varepsilon\varepsilon_0/\sigma_0$), that is the time required for the device to store the information associated with the photocarriers generated within the illuminated area.

The static detection limits ($s_d \sim \alpha'^{-1}$) is the maximum distance that a light spot from the collecting electrodes can be detected with a linear correlation between the spatial position and ΔV. It depends on α and on its correlation with R_l. For a-Si:H diodes with $10^{-10}\,(\Omega\,\text{cm})^{-1} < \sigma_0 < 10^{-8}\,(\Omega\,\text{cm})^{-1}$, $\sigma_d \approx 10^3\,(\Omega\,\text{cm})^{-1}$, $10\,\Omega/\square < \rho_s < 10^3\,\Omega/\square$, $100\,\text{cm}^{-1} < \alpha < 10^{-3}\,\text{cm}^{-1}$, or $10\,\text{cm}^{-1} < \alpha' < 10^{-4}\,\text{cm}^{-1}$, neglecting or not R_l. This means that high s_d are obtained using devices with small σ_0 and moderate ρ_s, for which α' is close to zero.

Figure 8.21 shows the normalized predicted ϕ_y curves $[\phi_y/(E_0/\alpha')]$ at different α', (respectively, 1.04, 5.9×10^3 and 2.9×10^3 cm^{-1}), assuming $L = 8$ cm, following (8.13), neglecting or not the losses. The theoretical curves depicted were plotted using the data given in Table 8.7 and they agree well with the previous assumptions about the role of α'. Nevertheless, still remains to demonstrate that these curves are able to match the experimental behavior observed in devices. Thus, in the same figure we superimposed (dark circles) the experimental normalized voltages as a function of y, where the different layers of the device have the characteristics listed in Table 8.7 and another (open circles) device where σ_0 was of about $5\times10^{-8}\Omega^{-1}\,\text{cm}^{-1}$. The data depicted clear show that the first device exhibits a good linearity along L while the second device only shows a relative linearity for values of $y \ll L$. In addition, we also observe a good fitting of the experimental points by the predicted curves, revealing so a good agreement of the model with the experimental values.

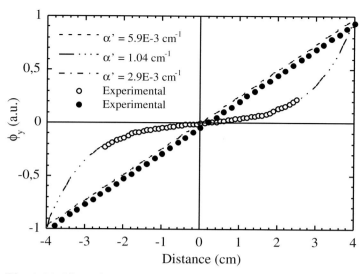

Fig. 8.21. Normalized predicted behavior of experimental data of ϕ_y, (*full* and *open circles*, respectively for α' of $0.24\,\mathrm{cm}^{-1}$ and $1.58\,\mathrm{cm}^{-1}$), and the predicted curves (*dashed lines*) for the conditions labeled in the figure

Table 8.7. Main electrical and optical characteristics of the p, i and n 1D TFPSD constituting layers

Properties	ITO layer	p-layer	i-layer	n-layer
• Conductivity $(\Omega\,\mathrm{cm})^{-1}$	4×10^3	5×10^{-5}	$1.4 \times 10^{-9} / 1.0 \times 10^{-8}$	5×10^{-3}
• Optical gap (eV)	3.5	1.71	1.75	1.71
• Activation Energy (eV)	–	0.40	0.90	0.18
• Photosensitivity $(100\,\mathrm{mW/cm^2})$	–	–	$5.0 \times 10^5 / 2.0 \times 10^4$	–
• Film thickness (µm)	0.12	0.02	0.7	0.07

8.4.2 Linearity and Spatial Resolution of 1D TFPSD

The device nonlinearity (position detection error) (δ), is obtained taking into account that [8.93]: $\delta = 2\sigma/F$ where σ is the *rms* (root mean square) deviation from the regression line data and F is the measured full scale. In Fig. 8.21 the linearity exhibited by the experimental data of the good TFPSD (dark circles) is better than 99% while the second one shows a linearity below 5% ($\alpha'L \geq 1$).

The nonlinearity dependence on light intensity was performed at different monochromatic light intensities. Figure 8.22 shows that as the light intensity

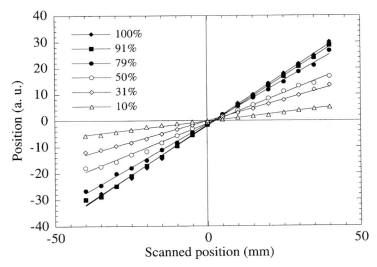

Fig. 8.22. Nonlinearity measurements performed in 1D large area TFPSD for different light intensities. The light intensity was changed from $2.4 \times 10^{-4}\,\text{W/cm}^2$ to $2.4 \times 10^{-5}\,\text{W/cm}^2$, which corresponds to 100% and 10%, respectively on the figure's labels

decreases from $2.4 \times 10^{-4}\,\text{W/cm}^2$ to $2.4 \times 10^{-5}\,\text{W/cm}^2$, δ varies from 2% to 5%. Below, the error increases reaching a value of 14% at $0.24\,\mu\text{W/cm}^2$.

The spatial resolution (s_r), is the minimum distance that can be clearly measured when the light spot is moved from one position to another. To determine s_r a special designed XY table was used [8.74]. The measured values are shown in Table 8.8, together with the deviations in linearity observed, for the device whose layer's characteristics are shown in Table 8.7.

The data in Table 8.8 show that $10\,\mu\text{m} < s_\text{r} < 20\,\mu\text{m}$, much smaller than the width of the projected laser beam (1 mm). This shows that the sensor response is mainly a function of the center of the light distribution on the device surface.

Table 8.8. Position detection resolution

Step (μm)	Distance scanned (μm)	Standard deviation (σ)	Nonlinearity (%)
80	4000	0.297	0.98
40	2000	0.173	0.98
20	1000	0.144	2.60
10	500	1.249	15.01

That is, good s_r are obtained when the characteristics of the different layers of the TFPSD obey to the following conditions: $\sigma_0 < 10^{-8}(\Omega\,\mathrm{cm})^{-1}$, $\sigma_d < 10^{-2}\,(\Omega\,\mathrm{cm})^{-1}$ and $10\,\Omega/\square < \rho_s < 10^3\,\Omega/\square$. These conditions allow us to define the discrete parameters of the electric equivalent circuit shown in Fig 8.18. This leads to $R_0 > 100\,\mathrm{k\Omega}$, for the device exhibiting the highest linearity while for the device having $\sigma_0 \sim 5 \times 10^{-8}\,(\Omega\,\mathrm{cm})^{-1}$, $R_0 < 2.5\,\mathrm{k\Omega}$, exhibiting so the highest leakage current, as predicted by the theoretical model.

The distributed resistance associated with the "channel" can be also determined experimentally. This is done through the dependence of the current and voltage measured on the spatial position, respectively for the short circuit and open circuit voltage conditions. The data obtained were fitted by straight lines and are depicted in Figs. 8.23a and b, leading to: $-I_y = 0.54118 - 0.7924y$ (µA), with a resolution better than 0.999, and $-V_y = 0.49412 - 1.4121y$ (mV), with a resolution better than 0.999. These values lead to $R_L \approx 900\,\Omega$ ($\ll R_0$), consistent with the measured ρ_s of the ITO layer and the device width used. This shows that the carriers' flow detected at the end terminals drift mainly through the TCRL being little influenced by the doped collecting layer, when $\rho_s \ll \rho_{sd}$. These data also show that the response frequency decreases as the device length increases, reaching values in the range of few kHz as $L \geq 10\,\mathrm{cm}$. That is, large area TFPSD can be used in continuous measurement processes where the time rate of measurements is mainly limited by the mechanical parts. The systems based on CCD's make it by a discrete process. This is the relevancy and superiority of the TFPSD's or other similar devices that make them quite suitable to be used in inspection/control systems directly, simplifying the optical components required and allowing a control in real time, either in alignments (1D), surfaces (2D) or objects (3D).

8.4.3 Position Response to Multiple Light Beams

One of the main advantages of the TFPSD over the CCD's is that they do not require precise focusing of the incident light beam and scanning of the pixels. besides that, different beams of different intensities or wavelengths can irradiate the same detector. When different light beams with the same wavelength impinge the detector the way that the response is affected is quite important. In most applications these systems are used indoors with different backlights, some of them could be of the same wavelength as the optical source used with the detector.

Figure 8.24 shows the situation where three different lasers impinge the TFPSD, leading to 2^3 possible combinations of light sources (3 responses due to one single source; 1 response due to the 3 light sources and 4 responses due to combinations of two sources). Assuming that the intensity of the three sources is the same; the number of the points able to be taken in each case is of about 409.

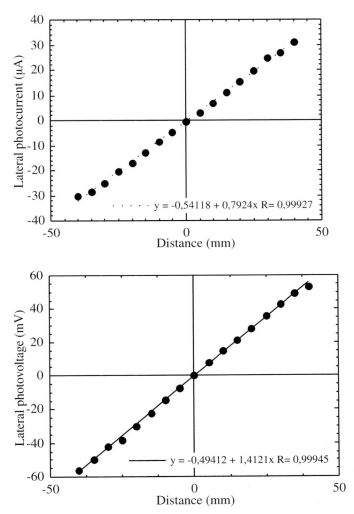

Fig. 8.23a,b. Experimental measured lateral photocurrent (**a**) and photovoltage (**b**), for a device with $L \approx 8\,\text{cm}$, where the layers' characteristics are the ones listed in Table 8.7

Under this condition $P(x)$ and $P(y)$ are the result of the relative weight of the currents detected. This leads to a maximum error for $P(x)$ and $P(y)$ below 0.16% when the positions are taken after weighting the mean value of the intensities supplied by the different sources or, about 0.40%, if we consider that the output current collected by each electrode of the detector is the result of the sum of the currents generated by each source. This means that the position output for multiple beams using the outputs for single beams give identical values within an average error of about 0.2% of the total scale.

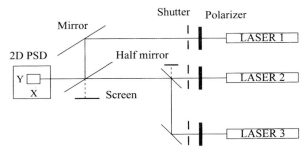

Fig. 8.24. Set up used for the detection of multiple beams

Dependent on the number of illuminated elements it is also possible to obtain information about the light intensity distribution from the variation of the photocurrent at each sensor element with the aid of a simple calculation circuit.

8.4.4 Static Predicted and Experimental Performance of the 2D TFPSD Device

Equations (8.21) and (8.29) rule the predicted behavior for the 2D devices where it is of extreme importance the value of the αL product in both directions, to determine the device detection limits, its linearity and spatial resolution.

The edge effects that can occur and the possible offset from the centre of gravity of the mean value of the currents recorded depend on the electrode's configuration chosen (Fig. 8.5) and on how they were designed, if the proper TCRL and i-layer were chosen. Figures 8.25a and b show the predicted contours for a device of size of $8\,\text{cm} \times 8\,\text{cm}$, for $\alpha L \approx 1$ and $\alpha L < 1$. Figure 8.25a shows that for high αL the predicted behavior leads to devices with a very poor resolution, more pronounced close to the device edges. Figure 8.25b reveals the behavior expected for a device with low αL. Under these conditions the device linearity improves substantially.

Figure 8.26 shows the experimental values obtained in a 2D TFPSD with the same size as above, with a duolateral configuration [8.71], using specific control hardware and the corresponding software for data acquisition and treatment.

8.5 Dynamic Performance of the 1D and 2D TFPSD

The behavior of the normalized photocurrent $i(t)/I_0$ is simulated through (8.38), for both type of devices under analysis. Figure 8.27 shows that behavior for $\alpha = 0.244$ (dashed lines) and $\alpha' = 0.032$ (solid lines)], following (8.38) and using as parameter $2\omega_0 t/L$. There, the y values are measured from

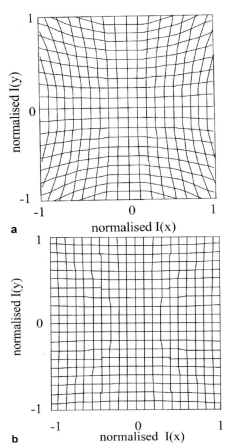

Fig. 8.25a,b. The theoretical geometrical distribution of currents in a 2D TF-PSD for (**a**) $\alpha L \approx 1$ and a tetralateral configuration and (**b**) $\alpha L \approx 0.16$ and a duolateral configuration

the center of gravity of the device for one of the collecting terminals, assumed to be symmetric. The curves depicted show a distortion close to the collecting electrode, more pronounced as $2\omega_0 t/L$ ratio is close to 1. Besides that, as $2\omega_0 t/L$ ratio starts decreasing (by changing the sweep time), $i(t)/I_0$ also decreases, not reaching its steady state value meaning that τ_d is out of the device's detection time range and so, the output signal is attenuated. There still exists a linear correlation between the spatial position of the light line and the detected current up to a value of $2\omega_0 t/L \approx 0.02$.

The experimental data were taken only for $2\omega_0 t/L = 1$ and fitted to the predicted curves of Fig. 8.27. The results show that experimental data are well fitted by (8.38). The data also show that good device linearity is only obtained when $\alpha L \ll 1$. Indeed, the predicted behavior depicted in Fig. 8.27

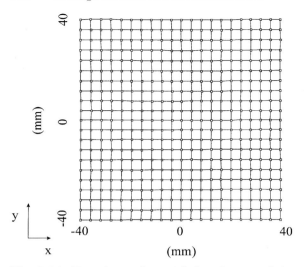

Fig. 8.26. Experimental spatial distribution of the currents in a 2D duolateral TFPSD where $\alpha L \approx 0.016$

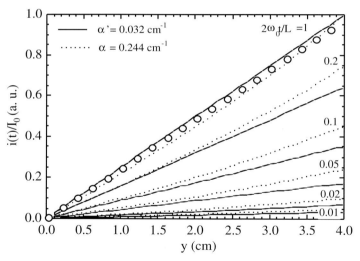

Fig. 8.27. Short circuit current distribution versus the signal position, using the normalized time after a step input, as a parameter following (8.38) for $\alpha = 0.244\,\mathrm{cm}^{-1}$ (dashed lines) and $\alpha' = 0.032\,\mathrm{cm}^{-1}$ (solid lines)

shows that when $\alpha \approx 0.244$ ($R_l = 0$), the device linearity over an extension of $L/2 \approx 4\,\mathrm{cm}$ is poor.

Figure 8.28 shows the simulation of $i(t)/I_0$ on $2\omega_0 t/L$, for $\alpha = 0.244\,\mathrm{cm}^{-1}$ (dashed lines) and $\alpha' = 0.032\,\mathrm{cm}^{-1}$ (solid lines), following (8.38) and using y as parameter. The data show that the shape of the curves is preserved in

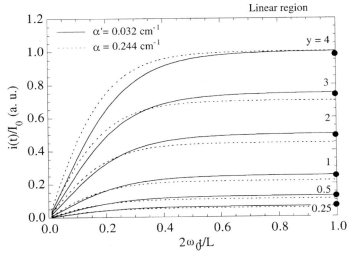

Fig. 8.28. Predicted normalized output current response to a step input of radiation using y as a parameter, following (8.38). The *solid* and *dashed lines* correspond to the predicted behavior when $R_l \neq 0$ and $R_l = 0$, respectively

both cases and that the main role of the losses is in improving the device linearity and to slow down the device response time in about 5%.

The curves depicted also reveal a dependence of the detection time on the recorded steady-state value, which is a function of the light line position on the device surface. Thus, $\omega_0 t/L$ decreases as y approaches the center of gravity of the device. In general, we observe that the output current achieves its steady-state value for $\omega_0 t/L > 0.5$ to which corresponds $\tau_d \geq 2\pi L/\omega_0$ (considering negligible the product of $\alpha' L$). This means that if a modulated light is applied to the device, to obtain a coherent output when consecutive impulse steps are applied, the delay time (τ) between pulses must be at least within the order of $\tau \approx 2\pi L/\omega_0$, equivalent to the expected value for the device response time.

In Figure 8.28 we also depicted the experimental points for a time interval between pulses in the range of 100 µs, corresponding to different detected response times, normalized to the device junction response time. The shadow region in Fig 8.28 corresponds to output signal values not influenced by τ (steady state device behavior) considered as the "active" region of the device.

8.5.1 Response Time of the TFPSD

The response time is a measure of the response speed of a TFPSD to a square-wave input incident light. If the sensor is instantaneously illuminated, it takes a certain time for the photogenerated current to flow in the external circuit. This is called the rise time (t_r) to a light pulse. If the light beam

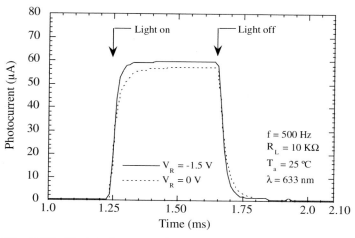

Fig. 8.29. Electrical output photoresponse to a square-wave optical input, of a p-i-n TFPSD (dashed line without polarization, solid line with reverse bias applied), under a fixed R_L

is instantaneously turned off, it takes a certain time for the photogenerated current to stop flowing, function of the load resistance used and on the device size, that determines the total device capacitance. This is called the decay time (t_d) to a light pulse. Figure 8.29 shows the photoresponse speed of a TFPSD with and without bias (solid and dashed lines) respectively. The experimental conditions are labeled inside the figure.

The data show similar t_r either when the reverse bias is applied or not ($\approx 50\,\mu\mathrm{s}$). The same does not happen with t_d since the capacitance associated with the depletion layer decreases when the reverse bias is applied. t_d decreases from 3.4×10^{-4} s to 1.2×10^{-4} s and the steady state output signal is enhanced by almost 4% when the voltage changes from 0 V to -1 V. These results show that large area 1D TFPSD are frequency limited up to values of the about 10 kHz. To improve this, R_L has to decrease and optimized the device design.

The role of R_L on the response speed was also analyzed for two TFPSD in the same run but with different areas (0.09 cm^2 and 0.49 cm^2). The data obtained are depicted in Fig. 8.30a and b. Table 8.9 shows the calculated t_r and t_d for two TFPSD with different areas, as a function of R_L, revealing an optimum load to which t_d presents its minimum. Besides that, the data also show that the output signal is distorted when $R_\mathrm{L} \geq R_0$, as predicted by the model already discussed.

This means that, if fast response is required R_L has to be low, in opposition of what is usually used for crystalline-based sensors [8.62].

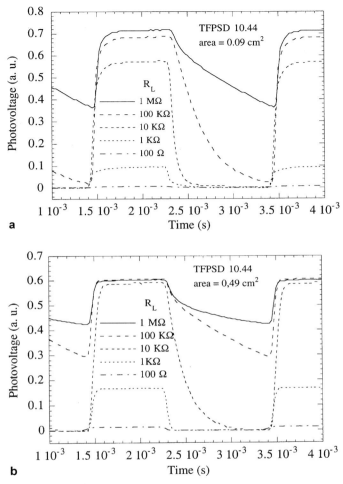

Fig. 8.30a,b. Photoresponse speed of a TFPSD for different R_L. (**a**) with $A = 0.09\,\text{cm}^2$ and (**b**) $A = 0.49\,\text{cm}^2$

8.5.2 Detection of Light Signals with Different Wavelengths

Another interesting situation for the TFPSD is when light sources of different wavelengths (λ) are used. If the TFPSD is based on a single p-i-n junction, the use of light beams with different λ will give rise to different types of output photocurrents, function of the sensitivity of the device to the particular λ. This will influence the S/N ratio and so the level of currents able to be detected, reducing its static analysis to a situation similar to one where light beams with different intensities are used. Nevertheless if we intend to perform a multiple detection where the different outputs could be identify with the λ of the beam used, the following procedure should be followed:

Table 8.9. Calculated rise and decay times for two TFPSD with different areas, as a function of the load resistance

	Area = 0.09 cm²		Area = 0.49 cm²	
$R_L (\Omega)$	t_r (s)	t_d (s)	T_r (s)	t_d (s)
100	–	–	–	–
1 K	1.7×10^{-4}	1.9×10^{-4}	1.8×10^{-4}	2.0×10^{-4}
10 K	1.1×10^{-4}	1.2×10^{-4}	1.1×10^{-4}	3.4×10^{-4}
100 K	1.0×10^{-4}	6.2×10^{-4}	–	–
1 M	–	–	–	–

a) to modulate the different beams with a frequency f_m in such away that the maximum number N of detectable light beams should obey to the relation [8.87]:

$$N \leq \frac{1}{T_s f_m} - 1 \qquad (8.42)$$

where T_s is the sampling time of the sample/hold device plus the peripherals.

b) f_m should be at least a factor of 5 below the cut off frequency of the detector.

c) The maximum power permitted by each light beam should such that

$$P_{max} \leq \frac{P_s}{N} . \qquad (8.43)$$

d) I_{ph} due to each individual light beam has to be at least a factor of 5 above the equivalent current noise, plus the leakage and saturation currents of the device.

e) The frequency response is independent of the beam diameter used. A non-focused beam leads to a maximum error below 0.4% [8.87].

f) As the cut off frequency of each device depends on its size, we obtain modulation frequencies that vary from about 500 Hz up to 200 kHz, as the size of the devices varies from 80 mm×80 mm to 5 mm×5 mm. These ranges are highly suitable for most of the applications involving electromechanical parts. Apart from that, if each frequency is identify with a λ, a color matrix can be achieved, after proper image treatment using the adequate software.

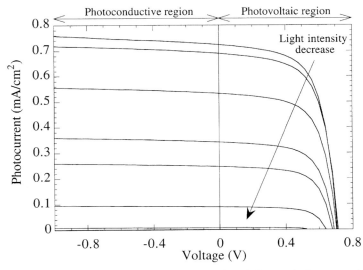

Fig. 8.31. $J-V$ curves of a TFPSD, for different light intensities (from 2.6×10^{-4} W/cm^2 to 2.6×10^{-7} W/cm^2)

8.6 Characteristics of the a-Si:H p-i-n Structures Used to Produce the TFPSD

In the following, it is presented a set of electrical and optical results related to the p-i-n a-Si:H structures and TFPSD's developed.

8.6.1 $J-V$ Curves

Figure 8.31 shows the current density-voltage ($J-V$) curves of a-Si:H 1D TFPSD under different light intensities (monochromatic light, $\lambda = 632.8$ nm). The data show that TFPSD are high light sensitive even for intensities in the range of 10^{-6} W/cm^2 to which S/N ≈ 10 dB. The data also show that the devices have a shunt resistance in the range of 10^5 Ω cm^2 while the series resistance is in the range of $40\,\Omega$ cm^2. Under reverse bias the photocurrent is found to be nearly independent of the negative applied voltage. This is a direct consequence of the non-injecting behavior of the contacts, which leads to a saturation photocurrent, when the Schubweg exceeds the transverse electrode's distance [8.88]. Another feature of the devices is the almost linear dependence of the short circuit photocurrent on the light intensity.

This value is usual referred to monomolecular recombination kinetics in intrinsic a-Si:H samples with the Fermi level near the midgap [8.89]. The slope of the power law dependence of the light intensity and photocurrent was found to be 0.988. As we can see, the correlation is good up to values of light intensity of about 2×10^{-6} W/cm^2. Below this value the detected

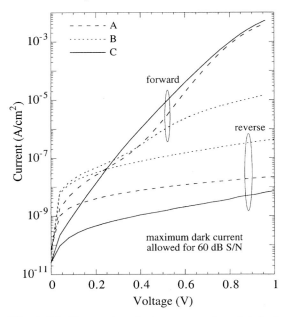

Fig. 8.32. Comparison between experimental dark J–V characteristics for TFPSD under reverse and forward bias polarization, for two structures with different p-layer thicknesses (devices A and B) and a TFPSD presenting a good rectification ratio with the semiconductor edges etched away (device C)

photocurrent starts being limited by the background reverse dark current, responsible for the non-linearity behavior observed, related to the limitation imposed by the noise equivalent power (NEP), as will be explained later.

Figure 8.32 shows the dark J–V characteristics of two TFPSD, A and B, prepared with different p-layer thicknesses (≈ 260 Å and ≈ 120 Å, respectively). The main differences concerns the diode quality factor (η), J_s and the rectification ratio (R_r).

For structure A, $\eta \approx 1.7$, $J_\text{s} \approx 2 \times 10^{-8}\,\text{mA/cm}^2$ and $R_\text{r} \approx 10^5$ while for structure B, $\eta > 2$, $J_\text{s} \approx 4 \times 10^{-7}\,\text{mA/cm}^2$ and $R_\text{r} \approx 10^2$. These results show the role of the p-layer thickness on the dark J–V characteristics. Nevertheless, for small bias voltage, devices A and B present η values not easily measurable. This behavior is explained by some macro-structural defects such as the incomplete junction formation due to pinholes propagating away through the film [8.90, 91]. Another cause for this beavior deals with the fact that the collecting metal electrode does not cover entirely the doped collecting layer, promoting so, the appearance of a lateral leakage current [8.90]. This gives rise to an abnormal beavior of the J–V characteristics, in the low forward bias regime. This evidence agrees well with the model presented by Martins et al. [8.92] since this abnormal beavior disappears in devices where the surrounding semiconductor region of the collecting elec-

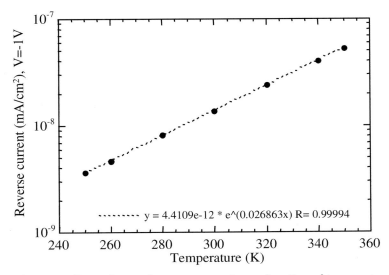

Fig. 8.33. Dependence of reverse current as a function of temperature

trode is etched away. To prove this effect, we present also in Fig. 8.32 a p-i-n a-Si:H TFPSD (labeled C) where the surrounding region of the collecting metal electrode was etched away. There, we obtain a well defined $\eta = 1.8$, $J_s = 7.3 \times 10^{-9}$ mA/cm^2 and $R_r = 7.5 \times 10^5$, and the S-like shape observed in the other J–V curves disappears, confirming so the model proposed by Martins et al. The photo-to-dark current ratios [under illumination intensity of 2.6×10^{-4} W/cm^2 ($\lambda = 632.8$ nm)] for devices C, A and B depicted in Fig. 8.32 are higher than 10^5, 10^4 and 10^3, respectively. This result is in line with the previous results obtained in the three structures.

8.6.2 Dependence of the Saturation Current of the Device on T

Changes in temperature (T) greatly affect dark current of the sensor and so, J_s. This beavior is caused by the increased probability of having more excited free steady-state carriers as the device temperature increases. As the output recorded signal is dependent on J_s and on the reverse current density (J_R), it is important to notice how affects J_R to determine the range of T in which the device response is used. Figure 8.33 shows the dependence of J_R at a reverse bias equal to -1 V, on T. There, the experimental points are fitted by the equation $J_R = J_{R_0} \exp(-\alpha_T T)$, with α_T the temperature coefficient, presenting a correlation better than 0.9999. From that we calculate the dJ_R/dT variation. The data show that J_R almost doubles every 26°C, indicating that T has to be of about 90°C to enhance J_R by one order of magnitude. This means that under low power illumination and close to the device detection limits (below $2\,\mu$W/cm^2), the maximum working T is in the

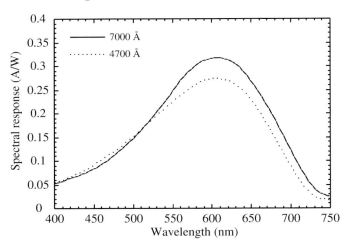

Fig. 8.34. Spectral response of two TFPSD with different i-layer thicknesses. All the other device parameters have been kept constant

range of 120°C. Above this value the NEP dominates leading to a poor S/N ratio.

8.6.3 Spectral Response and Detectivity

The spectral response (R_λ in A/W) of the device is a measure of its sensitivity to light, being defined as the ratio of the current generated by the device to the amount of light falling on it. Figure 8.34 shows a typical spectral response for a thin and thick i-layer p-i-n devices, respectively.

The data show, that both devices match well with the spectral range of the visible light and peaks at a wavelength (λ) of about 610 nm which almost corresponds to the maximum quantum efficiency of the i-layer. Overall, the data show that R_λ decreases as λ increases beyond 610 nm, due to an enhancement of the optical losses in the red region of the spectrum. On the other hand as λ decreases towards the blue region, R_λ decreases due to reflection losses and small absorption on the TCO layer and at the p-layer, as well as possible back diffusion of carriers generated near the top of the i-layer which are not collected [8.93].

The device detectivity (D^*) is a measure of its detecting ability and it is limited by the shot noise (related to the dark current) and the thermal noise (if a load is connected across the device) [8.94]:

$$D^* = \frac{\sqrt{AB}}{\text{NEP}}, \tag{8.44}$$

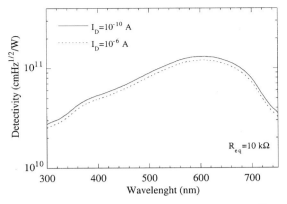

Fig. 8.35. Dependence of D^* for a typical TFPSD at room temperature. The *dashed lines* represent a simulation for a dark current of 10^{-6} A

with B the bandwidth (usually 1 Hz), A the area of the device and NEP, given by:

$$\mathrm{NEP} \approx \frac{1}{R_\lambda}\sqrt{2qI_\mathrm{D} + \frac{4kT}{R_\mathrm{eq}}}, \qquad (8.45)$$

where I_D is the dark current and R_eq the equivalent resistance of the sensor. The NEP, by definition, is the amount of light falling on a device, which produces a signal equal to the noise generated internally by the device.

Figure 8.35 shows $D^*(\lambda)$ for a typical TFPSD, indicating a "good" D^* in the range of 450 nm to 700 nm, with $R_\mathrm{L} = 10\,\mathrm{k\Omega}$. The observed peak on the $D^*(\lambda)$ plot coincides with the one of R_λ.

We have also simulated D^* to determine I_DR that leads to a decrease on D^*. The simulated data show that when $I_\mathrm{DR} \geq 10^{-6}\,\mathrm{A}$, D^* is reduced (Fig. 8.35), as expected.

8.7 Peripherals for 1D and 2D TFPSD Signal Processing

8.7.1 Optical Methods

In the following we refer to the optoelectronic schemes for application in dimensional measurement and inspection of surfaces. These schemes can be divided into three main operation units [8.95]. The first unit concerns the sources of light illuminating the workpiece or the scene. The second operation unit is the optoelectronic transducer, which will determine the resolution by which the viewed object is detected and if the detection is analog or digital. Finally, we have the system through which the signal detected is processed. There, the analysis of the signals results in a decision to intervene in the

Table 8.10. Components of optoelectronical techniques

Light sources	Methods of illumination	Auxiliaries	Detectors
White light lamp	Front light	Fiber optics	Photodiodes
Fluorescent strip lamp	Back light	Scanner	PSD
Flashing lamp	Dark-field illumination	Filters	various (+CCD)
Light emitting diode	Luminescence	Polarizer	Linear image sensor
Laser diode	Luminescence	Dynam. Focusing	Area image sensor
Laser (HeNe, A)	Structured illumination (light section method, Moiré contouring)	Holography	Area image sensor

production process. The most common components ascribed to these units are listed in Table 8.10. Methods of illumination and special auxiliaries are also mentioned for clearness.

Typically, as light sources are preferred Lasers or LED. The first light sources are usually used in scanning light modes, allowing so static and dynamic measurements. Usually, the scanning rate is beween 45 to 600 times per second. The scanning width, the range of measurements is, according to the apparatus, 1.2 mm up to 450 mm. The typical resolution varies from 0.1 μm to 5 μm, with a repeatability from 0.5 mm to 10 μm. Averaging the measurements of a certain number of scans is required to improve the accuracy and the reliability of the measurement (type of environment, presence of vapors, etc.).

One of the most common measurement processes for surface inspection is the laser (or LED) triangulation method (see Fig. 8.36). Their, the light emitted is narrowed by a proper lens and applied to an object. In an inclined angle to the direction of incidence the diffused reflection is focused through a reception lens and the spot image is formed on a light detector, like a PSD or TFPSD. The position on the spot depends on the distance beween the sensor and the workpiece. A displacement Δx of the object shifts the spot on the detector by Δy.

The relation beween displacement and the shift is non-linear. The reference distance of such systems is 1 mm up to several meters. It depends on the length of the base line beween the light source and the detector and on the angle beween the direction of incidence and the detection. The available resolution range from 0.1% to 0.01% of the measuring range. Pulsing the emitted light minimizes the sensitivity against the ambient light.

This technique can be applied to determine distances, thicknesses, surface inspection and roughnesses. The principle of measurement of defects in a surface bases in the fact that the defects cause alterations in intensity of the light

8 Thin Film Position Sensitive Detectors: From 1D to 3D Applications   389

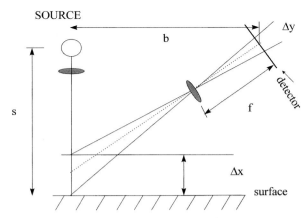

Fig. 8.36. Sketch of the system used for measurements based on the laser triangulation method

scanning over it. Knowing the incident angle and the time of measurement it can localize the recognized defect. Typically the scanning frequency is beween 1 kHz to 5 kHz, with resolutions from 0.05 mm to 3 mm. The typically recognized defects are spots, holes, folds, scratch tears and inclusions. This sort of inspection can be realized in the production of paper, textiles, plastics and glasses, cork plackets, metalworking, etc. It is also important to notice that these detectors can be used to monitor the position and motion of an object in space, such as a robot arm, part of the human anatomy, etc.

8.7.2 Peripherals for Signal Processing

The basic electronics blocks for monitoring the displacement of the position of a light spot on a 1D or 2D position sensors are displayed in Fig. 8.37. It is important to notice that the circuitry and configuration depends on the envisaged final application (1D or 2D). Figure 8.37 represents a configuration that can be used in stand alone systems (the XY position is executed digitally by the processor), where a microprocessor selects which signal to read (a sample and hold gives a clean signal for an analog to digital conversion of the values measured).

The first step is to convert I_{ph} into a proportional voltage by using a conventional electronic scheme, as the one depicted in Fig. 8.38. There, the quality and selection of the components are very important since there are some precautions to be taken such as to use precision low noise high impedances OPAMP, in the input stages. Also the gain selection resistor (R_1) should be a precision one. The bypass capacitor (C_1) must have a very low leakage current. If extra precision is desired a zero offset null trimmer might be added (not shown in the figure). With this circuit variations in light conditions or

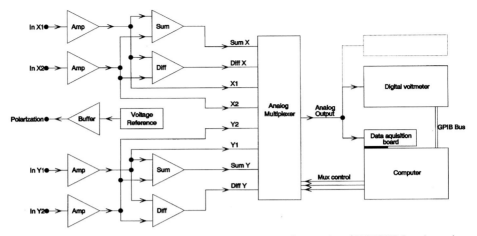

Fig. 8.37. Basic steps for displacement monitor electronics (SENSIT hardware): complete circuit schematic for 1D and 2D applications

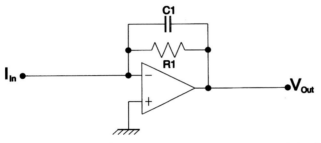

Fig. 8.38. Current/voltage converter circuit stage of the SENSIT hardware

sensor currents can be easily corrected by simply changing the gain resistor so that the output never saturates. The transfer function of this circuit is:

$$V_{\text{Out}} = R_1 I_{\text{In}} \qquad (8.46)$$

The resistor (R_1) is not fixed. In fact, a slot with four locations exists so that the resistors can be easily changed and by so, the gain of the circuit can be adjusted to several kinds of sensors. For the TFPSD developed the SENSIT [8.96] hardware works well with values from $10\,\text{k}\Omega$ to $1\,\text{M}\Omega$. In the circuit the capacitor C_1 is used to reduce the gain at high frequencies and hence the noise. The type of amplifier used in the circuit was an AD712. These amplifiers present a fast response, low noise, low output offset voltage and most important, FET inputs. This gives to the input very high impedance (about $10^{12}\,\Omega$) which minimizes errors due to input bias currents. The output is a voltage proportional to the input current.

The next stages of the SENSIT hardware (Fig. 8.37) are a summing and a difference circuits. These stages produce the summing and the difference

8 Thin Film Position Sensitive Detectors: From 1D to 3D Applications   391

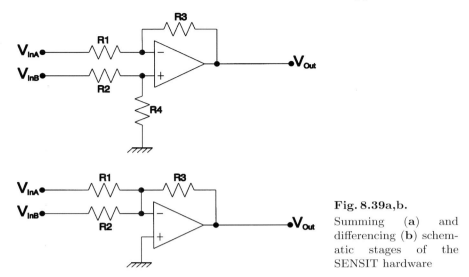

Fig. 8.39a,b. Summing (**a**) and differencing (**b**) schematic stages of the SENSIT hardware

voltages derived from the outputs of the first stages. The basic schematics are shown in Fig. 8.39. Here, some precautions must be taken. In the summing stage the value of the resistors should be the same so that the result equals the sum of the inputs. That is, the sum done will have the same weight for both inputs. The transfer function of the summing circuit is:

$$V_{\text{Out}} = \left[-V_{\text{In}A}\left(\frac{R_3}{R_1}\right)\right] + \left[-V_{\text{In}B}\left(\frac{R_3}{R_2}\right)\right], \tag{8.47}$$

where $R_{\text{in}}/R_1 = 1$. We used all resistors equal to $10\,\text{k}\Omega \pm 1\%$.

To know the difference of the two components we use the circuit of Fig. 8.39b. There, the input signals are subtracted and the result appears at the outputs as a function of $R_{1,2}$ and R_3. Here, the gain of the amplifier and the coefficient resistors should be paired so that the output as the maximum possible accuracy. If a different range is desired, resistors R_3 can be changed to a suitable value. The transfer function of the difference circuit is (with $R_1 = R_2$ and $R_3 = R_4$):

$$V_{\text{Out}} = (V_{\text{In}A} - V_{\text{In}B})\left(\frac{R_3}{R_1}\right). \tag{8.48}$$

To obtain the difference without any coefficients, all resistors are equal and with a value of $10\,\text{k}\Omega \pm 1\%$. If extra precision is needed, an offset null trimmer can be used in the OPAMP. Dependent on the final application, there can be several configurations of how to use the available signals.

When a bias voltage is required for a specific application of the sensors, a circuit similar to the one shown in Fig. 8.40 is used to allow that option. This circuit picks a user-selected voltage by means of a trimmer and buffers, with

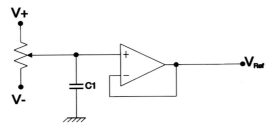

Fig. 8.40. Bias voltage circuit schematic

an unity gain voltage follower (TL081) (polarization in Fig. 8.37), where C_1 is used to filter any noise. This circuit can produce voltages of both polarities and has a current capability of up to 20 mA, enough for the range of sensors we deal with.

8.8 Simulated and Experimental Data in 2D Optical Inspection Systems with TFPSD Detector

In Fig. 8.26 we display the result of a real 2D TFPSD with 80 mm × 80 mm in size, with a linearity better than 1% and a resolution of about 10 mm. The results obtained are highly accurate and precise, when compared with the ideal situation.

Figure 8.41a shows the response of a 2D TFPSD where the table that contains the sensor was under an excessive acceleration, causing sensor movements (vibrations) in relation to the XY table. Figure 8.41b shows the measurements achieved without averaging the number of measurements per each

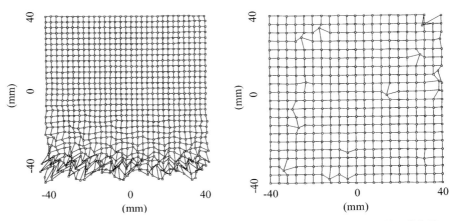

Fig. 8.41. (a) Results from a loose sensor with respect to the XY table. (b) Untreated measures showing errors produced by dust particles

8 Thin Film Position Sensitive Detectors: From 1D to 3D Applications   393

point read. Indeed, to get the best results, each point has to be read an arbitrary number of times (defined by the user in the used software). This allows to determine an average reading value and so a range within which the values can be considered or eliminated, depending on the percentage of the error allowable. If, for instance, after 8 trials there are still "bad" points, the process stops and the best points from the total read are taken to produce the final average value. One should note that this feature is only active if the number of reads per point is at least 3. Measures taken without this feature produce results as the one shown in Fig. 8.41b. Here, we can see that the sample is fairly linear in response but some points seem to be out of the allowable range. These points were affected by errors in the acquisition process and if corrected (using 3 or more reads per point) will produce results as the one shown in Fig. 8.26.

8.9 Linear Array of Thin Film Position Sensitive Detector (LTFPSD)

1D TFPSD can also be used in arrays with n equal elements aligned side by side. One possible implementation is to use 128 of this small 1D sensor as depicted in Fig 8.42, to form an array. In this array each sensor works in the same way as a single 1D TFPSD but now supplying information about the x–y plane. A typical connection is with all pads on one of the sides of the array connected to a common point (signal ground) while the other pads work as outputs giving a current proportional to the position of the laser along the active area.

Here, each elemental 1D TFPSD has the end-terminals connected to an analog shift register (SR) to process the information recorded in both end-contacts, as shown in Fig. 8.43. Per stripe, we get the information concerning the "point" of the line projected on the array as:

$$P_n(y_n) = \frac{L(I_{0,n} - I_{1,n})}{(I_{0,n} + I_{1,n})}, \tag{8.49}$$

where L is the length of each stripe, the subscripts 0 and 1 refer to the position of each end-contact, n is an integer that varies from 1 to 127 and the reference is considered to be the center of gravity of each elemental 1D LTFPSD.

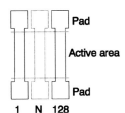

Fig. 8.42. Array of multiple 1D sensors

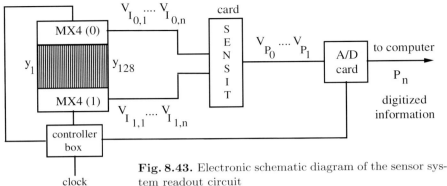

Fig. 8.43. Electronic schematic diagram of the sensor system readout circuit

The recorded values are processed in sequential series ($[I_{0.1} \ldots 1I_{0.n}]$ and $[I_{1.1} \ldots I_{1.n}]$) by SR [MX4(0) and MX4(1)] whose output correspond to a signal voltage that is fed to the SENSIT [96] card. From that card, a sequential voltage information is obtained for P_n as given by (8.49). After that, the information is supplied to an A/D card to digitize the signals (P_n) prior to be to the computer. The SR's and the A/D converters (12 bits) are driven by a controller box that determines the need time delays and the number of scans allowed per second beween information packets, using the same clock drive 1 MHz.

The position of an image line projected in the z–y plane is determined by the y_n data obtained by the 128 stripes and related to the currents detected by both SR. The third variable (x) is the angle beween the light source (Laser) and the object plane to be inspected and by so, the determination of the depth profile related to the image received by array is performed.

8.9.1 Principles of the Optical Methods Used

The principle of operation of a detector array is that an image line projected in the array induces photocurrents in the illuminated elements. All elements are then scanned to determine the position of the image line. By focusing a laser line on the surface to be inspected under a certain angle ϕ (Fig 8.44) the relation beween the movement of the light line (d_d) and the object movement (d_0) is:

$$\phi = \sin^{-1} \frac{d_0(\min)}{d_\mathrm{d}(\min)}. \tag{8.50}$$

For instance, for a detector resolution of 10 µm and an object resolution of 5 µm (allowable displacement of the object) implies that $\phi = 30°$. Similarly, the angle of incidence and the maximum movement of the test object determine the minimum active length of the detector:

$$d_\mathrm{d}(\max) \geq \frac{d_0(\max)}{\cos \phi}, \tag{8.51}$$

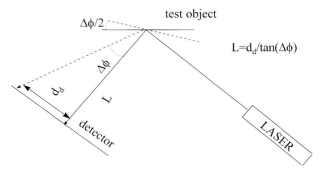

Fig. 8.44. Setup used to measure linear displacements and the rotation of a test object

meaning, for instance, that for a movement of 100 µm and an incident angle of 41° the detector must have a length of at least of 133 µm. If, for instance, the rotation of a test object about a fixed point is intended to be known (Fig 8.44), and as $\tan(\Delta\phi) = d_\mathrm{d}/L$, we get $L \leq d_\mathrm{d}/\tan(\Delta\phi)$. The error can occur if there is any linear displacement when an object is rotated or if there is any rotation when it is positioned linearly. Many applications require that this error be corrected in the measurement. A combined linear displacement and rotation can also be measured, resulting in the superposition of the conditions described by Fig. 8.44. One way to avoid errors related to displacements is by using an autocollimator.

8.9.2 Positional Resolution of the Array

The positional resolution depends on the active length of each sensing element of the array and of the signal to noise ratio (S/N):

$$dP \approx \frac{L}{2\,S/N}. \tag{8.52}$$

Taking into account that at each current read there are an uncertainty related to the noise (I_n ascribed to the measurement, the measure position is given by:

$$P = \frac{(I_1 \pm I_{n_1}) - (I_2 \pm I_{n_2})}{I_1 \pm I_{n_1}) + (I_2 \pm I_{n_2})} \cdot \frac{L}{2} \tag{8.53}$$

The maximum measurable position occur when both noise signals are negative and approximately equal in value:

$$P_\mathrm{max} = \frac{I_1 - I_2}{I_1 + I_2 - 2I_\mathrm{n}} \cdot \frac{L}{2} \tag{8.54}$$

Since the S/N ratio is:

$$S/N = \frac{I_1 + I_2}{2I_\mathrm{n}}, \tag{8.55}$$

assuming that S/N ≫ 1, the proper manipulation of the above equations lead to (8.52). If we consider the array such that $L = 1\,\text{cm}$, $I = 50\,\text{mA}$ and $I_n = 10\,\text{nA}$, we get $dP \approx 1\,\mu\text{m}$. This value is much smaller than the pixel size or its separation, factors that limit the final device resolution.

8.9.3 Hardware to Control Arrays of Multiple 1D Sensors

To control arrays of multiple 1D sensors there are specialized shift registers as the one shown in Fig. 8.45. This SR processes the analog information, through 128 separate charge amplifiers and the need logic to deal with required electronics to process the information recorded by the array.

In this circuit the first stage (Fig. 8.46) is a current (charge)/voltage amplifier. Without any input signal (Laser beam off), the *Reset* is activated. This discharges the integrating capacitor and puts the circuit in an initial known state (quiescent state) with a stable voltage at the output. This voltage is sampled by the *sampling*. After an event (Laser beam on) and after all the charge from the sensor strip has been collected, a new stable voltage is developed at the output. Again, the *sampling* circuit samples this voltage. Pulses from *Cal* are used to calibrate/ test the circuit. Since the value of the coupling capacitor and the amplitude of the pulse voltage are known, charge transferred to the circuit is also known.

The *sampling* stage has two sampling and hold circuits and a differential output stage, controlled by a digital strobe line, as shown in Fig. 8.47. Activating S_1 when the input amplifier is in its quiescent state turns T_1 on.

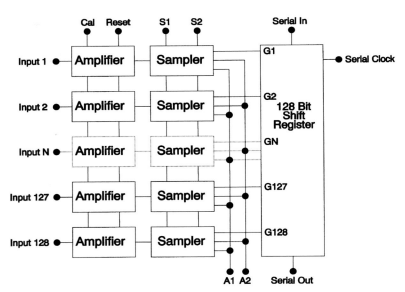

Fig. 8.45. Generic schematic of the MX4 chip used in the hardware to control the LTFPSD

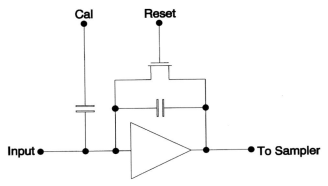

Fig. 8.46. Schematic of the charge amplifier that integrates the MX4

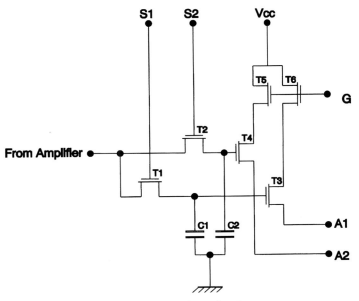

Fig. 8.47. Schematic of the sampling circuit

This charges C_1 with the quiescent voltage of the amplifier. After charging C_1 T_1 is then deactivated leaving the quiescent voltage on hold on C_1. After an event S is activated and a new voltage is then sampled and put on hold by T_2/C_2 Now, the two voltages (quiescent and event) are stored and can later be read. This process occurs in all the 128 amplifiers/samplers at the same time.

When the gating signal G is activated, T_5 and T_6 are turned on. This applies power to T_3 and T_4 a proportional current is then put on the differential analog bus lines A_1 and A_2. Through this scheme common mode signals and low frequency noise are removed and the result is a voltage that is propor-

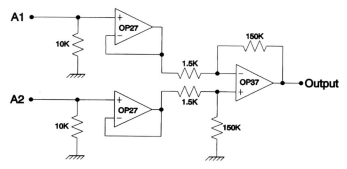

Fig. 8.48. Schematic of the differential to single ended amplifier

tional to the input charge from the sensor. The G comes from a 128-stage SR, with a *Serial Input* and a *Serial Output* that can be used to cascade up to 5 chips.

The differential outputs A_1 and A_2 can be converted to an amplified single ended signal by a circuit as the one shown in Fig. 8.48. The input resistors convert the output current in a proportional voltage that is buffered by the OP27 amplifiers. Then, a differential amplifier (OP37) is used to remove the common mode voltage and, at the same time, to amplify the signal 100 times.

This signal is then ready to be processed by a computer through a data acquisition card or visualized in an oscilloscope.

8.9.4 Bandwidth Requirements for the Preamplifiers Used in the Hardware Control Unit of the LTFPSD

Due to the presence of a junction capacitor, C_{TFPSD} ascribed to each 1D TFPSD that constitute the array, a compensation capacitor C_F in parallel to a resistor R_F is required to prevent the occurrence of a strong resonance in the step response. Thus, representing the high frequency stability of the preamplifiers by the angle of the roots of the output signals p_1 and p_2 [$p \approx t_t/(t_1 + t_2)$] in the s-plane [8.79], then for $f < 60°$, the value of C_F should satisfy to the condition:

$$C_\text{F} > \sqrt{\frac{C_{\text{TFPSD}}}{R_\text{F}\omega}}, \tag{8.56}$$

where ω is the cut-off angular frequency of the preamplifiers. For instance, if $C_{\text{TFPSD}} = 5000\,\text{pF}$, $R_\text{F} = 1\,\text{M}\Omega$, and $\omega = 5 \times 10^5\,\text{rad/s}$, we find $C_\text{F} > 100\,\text{pF}$.

As the proposed detection process is dependent on the charge detected, we must take into account the response time constant of each sensing element, given by

$$\tau = \frac{2R_\text{F}}{\omega R_{\text{TFPSD}}} \tag{8.57}$$

and error in the period ratio Δp, given by:

$$\Delta p \approx \frac{RF}{\omega F_{\text{TFPSD}}} f_s (1-x), \tag{8.58}$$

where f_s is the frequency of the clock and x is the linear position of the beam upon each array element. For a clock such that $f_s = 5\,\text{kHz}$; a resistance for the collecting layer of $10\,\text{M}\Omega$ and that x lays beween 10% and 90% of the total length of the element, we get $\tau = 0.4\,\mu\text{s}$ and a systematic error of $\Delta p < 4 \times 10^{-4}$. Considering the 128 elements that constitute the array, this implies an overall error of 5%, in the array response time. That is, the use of the LTFPSD is limited to applications where a substantial delay time is allowed beween pulses and where no very fast processing times are required. To this condition obey most of the known control processes for inspection and measurement purposes.

8.10 Summary and Future Outlook

In this chapter we have described the technologies and physics of a-Si:H TFPSD and LTFPSD. The described sensors can be used in automated inspection systems, improving the quality and the reliability of manufacturing products in a wide range of industries. The devices have a non-linearity of $\pm 2\%$ (for long-range detection processes) and a position resolution better than $20\,\mu\text{m}$.

It is also important to notice that good signal to noise ratio (above $5\,\text{dB}$) were obtained at light intensities below $2\mu\text{W}/\text{cm}^2$, not disturbed by the background illumination or changes in the temperature, in the range of $\pm 10°\text{C}$, when the device is reversed bias.

The devices show a good response speed (from $2\,\text{kHz}$ up to $1\,\text{MHz}$, respectively for TFPSD and LTFPSD, function of the device size and load resistance), making them attractive to control mechanical systems or for surface inspection. The LTFPSD can substitute the classical CCD's used in laser triangulation measurement processes since they do not use integrated cells and so, the resolution is not limited by cell size. As the detection process in the TFPSD or LTFPSD is analog, the recorded information is continuos while the one of the CCD is discrete (digital detection process).

In conclusion, the authors are extremely encouraged by the promising large area TFPSD and LTFPSD associated to a low cost technology due to the fact that these types of sensors could have a profound impact in the development of several industrial applications where control in real time is needed. However a sustained research effort will be required to develop the sensor's material and design as well as the electronic readout system associated to them.

Acknowledgements. The authors are indebted to many people for the opportunity to perform this work. To their colleagues A. Maçarico, G. Lavareda and C.N. Carvalho for the production of the p-i-n structures and ITO depositions, F. Soares, L. Fernandes and C. Machado for the specific hardware and software developments, A. Bicho, I.Ferreira and F. Barbosa for the inestimable electric and optical characterisation of the devices, T. Chee for the dry etching experiments and M. Vieira for her fruitful discussions.

The authors also wish to express their gratitude to FCT-UNL (Faculdade de Ciências e Tecnologia da Universidade Nova de Lisboa) and UNINOVA (Instituto para o Desenvolvimento de Novas Tecnologias) for the working conditions that make possible the present work. Thanks are also due to organisations such as: NATO (North Atlantic Treaty Organisation through the Scientific and Environmental Affairs); PEDIP (Programa Específico para o Desenvolvimento da Indústria Portuguesa); DG-XII (Directorate Generale); JNICT (Junta Nacional de Investigação Científica e Tecnológica de Portugal); Fundação para a Ciência e Tecnologia; programa PRAXIS XXI; FLAD (Fundação Luso Americana para o Desenvolvimento); Fundação Calouste Gulbenkian, that sponsored or are still sponsoring projects or travel missions in the field related to the present overview, allowing the authors to stay up dated on the R&D field of a-Si:H based devices.

Special thanks are due to L. Ferreira, for the smart industrial applied ideas, concerning TFPSD applications. All the photographs presented in this work, were taken from experimental devices fabricated in this laboratory, by A. Cerdeira and B. Sequeira to whom the authors express their thanks.

Finally the authors would like to dedicate this work to his sons (R. S. Martins and C. F. Martins) and E. Fortunato would like to remember and to dedicate this work, to her father A. Fortunato (1939–1995) for his advice stimulation and support during his life.

References

8.1 W. Schottky, Phys. Z **31**, 913 (1930).
8.2 J.T. Wallmark, Proc. IRE **45**, 474 (1957).
8.3 A.R. Moore and W.M. Webster, Proc. IRE **43**, 427 (1955).
8.4 B.O. Kelly and R.I. Nemhauser: in *Proceedings of the 21st Annual IEEE Machine Toll Conference* (Hartford, 1973).
8.5 L.D. Hutcheson, Optical Engineering **15**, 61 (1976).
8.6 B.O. Kelly, Laser Focus **12**, 38 (1976).
8.7 S. Middelhoek and D.J.W. Noorlag: *Silicon microtransducers: a new generation of measuring elements*, Modern electronic measuring systems (Delft University Press, 1978), Chap. 1.
8.8 W. Light, UDT Application Note, 1982 (unpublished).
8.9 R.A. Foulds and P.W. Demasco: in *Proceedings of the 5th Annual Conference on Rehabilitation Engineering* (Houston, 1982), p. 81.
8.10 E. Feige and T.B. Cleeg, Am. J. Phys. **51**, 954 (1983).
8.11 F.J. Schuda, Rev. Sci. Instrum. **54**, 1648 (1983).
8.12 G.C.M. Meijer and R. Chrier, Sensors & Actuators **A21–A23**, 538 (1990).
8.13 R. Ohba: *Intelligent Sensor Technology* (Wiley, New York, 1992).

8.14 H. Walcher: *Position Sensing – Angle and Distance Measurement for Engineers* (Butterworth-Heinemann, Oxford, 1994).
8.15 S. Middelhoek and S.A. Audet: *Microelectronics and Signal Processing – Silicon Sensors* (Academic Press, London, 1989).
8.16 H. Muro and P.J. French, Sensors & Actuators **A21–A23**, 544 (1990).
8.17 G.C.M. Meijer and R. Schrier, Sensors & Actuators **A21–A23**, 538 (1990).
8.18 Y. Morikawa and K. Kawamura, Sensors & Actuators **A34**, 123 (1992).
8.19 A. Kawasaki, M. Goto, H. Yashiro and H. Ozaki, Sensors & Actuators **A21–A23**, 529 (1990).
8.20 F.R. Riedijk, T. Smith and J.H. Huijsing, Sensors & Actuators **A40**, 237 (1994).
8.21 J. Geits: in *Planar Silicon Photosensors*, edited by L. Ristic, Sensor Technology and Devices (Artech House, Boston, 1994) Chap. 9.
8.22 Y. Hamakawa: in *Proceedings of Technical Digest of the 6th International PV Conference* (N. Delhi, 1992), p. 3.
8.23 C.R. Wronski: in *Proceedings of the 1st World Conference on Photovoltaic Energy Conversion* (Waikoloa, 1994) (in press).
8.24 K. Rosan: in *Amorphous semiconductor image sensors: physics, properties and performances*, edited by J. Kanicki, Amorphous & Microcrystalline Semiconductor Devices (Artech House, Boston, 1991), Vol. 1, Chap. 7.
8.25 K. Suzuki: in *Flat panel displays using amorphous and microcrystalline semiconductor devices*, edited by J. Kanicki, Amorphous & Microcrystalline Semiconductor Devices (Artech House, Boston, 1991), Vol. 1, Chap. 3.
8.26 H. Shinohara, M. Abe, K. Nishi and Y. Arai: in *Proceedings of the 1st World Conference on PV En. Conversion* and 24th IEEE PV Specialist Conference (Waikoloa, 1994), vol. I pp. 1216.
8.27 R. Street, Physics World, 54, April (1993).
8.28 T. Kagawa, N. Matsumoto and K. Kumabe, Jpn. J. App. Phys. **21**, Suppl. 21-1, 251 (1981).
8.29 L.E. Antonuk, J. Boudry, W. Huang, D. L. McShan, E. J. Morton, J. Yorkston, M. J. Longo and R. A. Street, Med. Phys. **19**, 1455 (1992).
8.30 C. Van berkel, N.C. Bird, C.J. Curling and I.D. French, Mat. Res. Symp. Proc. **297**, 939 (1993).
8.31 R.A. Street, X.D. Wu, R. Weisfield, S. Nelson and P. Nylen, Mat. Res. Symp. Proc. **336**, 873 (1994).
8.32 H. Miyake, K. Sakai, T. Abe, Y. Sakai, H. Hotta, H. Sugino, H. Ito and T. Ozawa, SPIE Proc. **1448**, 150 (1991).
8.33 D. Lake, Advanced Imaging, 72 November 1994.
8.34 R.C. Chittick, J.H Alexander and H.F. Sterling, J. Electrochem. Soc. **116**, 77 (1968).
8.35 K. Kempter, Adv. Sol. Stat. Phys. (Festkörperprobleme) **27**, 279 (1987).
8.36 H. Yamamoto, M. Matsui, T. Tsukada, Y. Eto, T. Hirai and E. Maruyama, Jpn. J. Appl. Phys. **17**, 135 (1978).
8.37 S. Boronkay, P. Gustin and D. Rossier, Acta Electronica **21**, 55 (1978).
8.38 K. Komiya, M. Kanzaki and T. Yamashita: in *IEEE Tech. Dig. Int. Electron Devices Meeting* (1981), p. 309.
8.39 R. Martins, I. Ferreira, N. Carvalho and L. Guimarães: in *Proceedings of the Technical Digest of the 6th International Photovoltaic Conference* (New Delhi, 1992), p. 655.

8.40 R. Martins, I. Ferreira, N. Carvalho and L. Guimarães, J. Non-Cryst. Sol. **137–138**, 757 (1993).

8.41 J.P.M. Schmitt, Mat. Res. Symp. Proc. **219**, 631 (1991).

8.42 G. Strasser, J.P.M. Schmitt and M.E. Bader, Solid State Technology **37**, 83 (1994).

8.43 A. Shah, J. Hubin, E. Sauvain, P. Pipoz, N. Beck and N. Wyrsch, J. Non-Cryst. Solids **164**, 485 (1993).

8.44 K. Kempter, SPIE Proc. **617**, 1220 (1986).

8.45 D.J.W. Noorlag, Ph.D thesis, Delft University, 1982.

8.46 S. Arimoto, H. Yamamoto, H. Ohno and H. Hasegawa, Electron. Lett. **19**, 628 (1983).

8.47 S. Arimoto, H. Yamamoto, H. Ohno and H. Hasegawa: in *Proceedings of the 15th Conference on Solid State Devices and Materials* (Tokyo, 1983), p. 197.

8.48 S. Arimoto, H. Yamamoto, H. Ohno and H. Hasegawa, J. Appl. Phys. **57**, 4778 (1985).

8.49 H. Yamamoto, M. Moniwa and H. Hasegawa, Jpn. J. Appl. Phys. **21**, Suppl. 21-2, 53 (1982).

8.50 K. Okumura, Mat. Res. Symp. Proc. **118**, 605 (1988).

8.51 M. Yamaguchi, S. Murakami, S. Todo and Y. Tawada, Mat. Res. Symp. Proc. **149**, 631 (1989).

8.52 T. Takeda and S. Sano, Mat. Res. Symp. Proc. **118**, 399 (1988).

8.53 T. Takeda: in *Applications of amorphous silicon position-sensitive detectors*, edited by J. Kanicki, Amorphous & Microcrystalline Semiconductor Devices (Artech House, Boston, 1991), Vol. 1, Chap. 9.

8.54 A.K. Dutta, K. Murakami, Y. Hatanaka and T. Yamamoto: in *Proceedings of the Technical Digest of the 7th Sensor Symposium* (Tokyo, 1988), p. 233.

8.55 T. Yamamoto, K. Murakami, S. Takayama and Y. Ono: in *Proceedings of the Technical Digest of the 9th Sensor Symposium* (Tokyo, 1990), p. 217.

8.56 T. Yamamoto, K. Murakami, S. Takayama & Y. Ono, Sensors & Actuators **A30**, 193 (1992).

8.57 A.N. Panckow, J. Bläsing and T. P. Drüsedau, J. Non-Cryst. Sol. **164–165**, 845 (1993).

8.58 B.F. Levine, R.H. Willens, C.G. Behea and D. Brasen, Appl. Phys. Lett. **49**, 1537 (1986).

8.59 E. Fortunato, M. Vieira, L. Ferreira, C.N. Carvalho, G. Lavareda and R. Martins, Mat. Res. Symp. Proc. **297**, 981 (1993).

8.60 E. Fortunato, M. Vieira, G. Lavareda, L. Ferreira and R. Martins, J. Non-Cryst. Solids **164–166**, 797 (1993).

8.61 E. Fortunato, G. Lavareda, M. Vieira, R. Martins and L. Ferreira, Vacuum **45**, 1151 (1994).

8.62 E. Fortunato, G. Lavareda, M. Vieira and R. Martins, Rev. Sci. Instrum. **65**, 3784 (1994).

8.63 R. Martins and E. Fortunato, IEEE Trans. on Electron Devices **43**, 2143–2152 (1996).

8.64 R. Martins, G. Lavareda, E. Fortunato, F. Soares, L. Fernandes and L. Ferreira, SPIE Proc. **2415**, 148 (1995).

8.65 E. Fortunato, F. Soares, M. Fernandes, G. Lavareda and R. Martins, Mat. Res. Symp. Proc. **420**, 165 (1996).

8.66 A. Kawasaki, M. Goto, H. Yashiro and H. Ozaki, Sensors and Actuators **A21–A23**, 529 (1990).

8.67 P.W. Verbeek, F.C.A. Groen, G.K. Steenvoorden and M. Stuivinga: in *Proceedings of the IASTED Int. Symp.* (Lugano, 1985), p. 67.
8.68 G. Lucovsky, J. Appl. Phys. **31**, 1088 (1960).
8.69 D.J.W. Noorlag and S. Middelhoek, Solid-State Electron Devices **3**, 75 (1979).
8.70 W.P. Connors, IEEE Trans. Electron Devices **ED-18**, 591 (1971).
8.71 H.J. Woltring, IEEE Trans. Electron Devices **ED-22**, 581 (1975).
8.72 R. Martins, M. Vieira, F. Soares and L. Guimarães: in *Proceedings of the Technical Digest of the 5th International Photovoltaic Conference* (Kyoto, 1990), p. 975.
8.73 M. Vieira, R, Martins, E. Fortunato, F. Soares and L. Guimarães, J. Non-Cryst. Sol. **137–138**, 479 (1991).
8.74 R. Martins and E. Fortunato, Rev. Sci. Instrum. **66** (4) 2927 (1995).
8.75 C. Narayanan, A.B. Buckman, I.B. Vishniac and W. Wang, IEEE Trans. Electron Devices **40**, (1993).
8.76 C.A. Klein and R.W. Bierig, IEEE Trans. Electron Devices **ED-21**, 532 (1974).
8.77 I. Chen, J. Appl. Phys. **64**, 2224 (1988).
8.78 R.S. Crandall, J. Appl. Phys. **55**, 4418 (1984).
8.79 J.L. Buchanan and P.R. Tunner: *Numerical Methods and Analysis* (McGraw-Hill, New York, 1992).
8.80 G.A. Korn and T.M. Korn: *Mathematical Handbook for Scientists and Engineers* (McGraw-Hill, New York, 1968).
8.81 E. Fortunato, G. Lavareda, R. Martins, F. Soares and L. Fernandes, Sensors & Actuators **A51**, 135 (1996).
8.82 G.E. Roberts and H. Kaufman: *Table of Laplace Transforms* (W.B. Saunders, Philadelphia, 1960).
8.83 G. Doetsch: *Guide of the application of the Laplace and Z-Transforms*, 2nd ed. (Van Nostrand-Reinhold, London, 1971).
8.84 G. Doetsch, H. Kniess and D. Voelker: *Tabellen zur Laplace Transformation* (Springer, berlin, 1947).
8.85 E. Fortunato, G. Lavareda, F. Soares and R. Martins, Thin Solid Films **272**, 148 (1996).
8.86 A. Kawasaki and M. Goto, Sensors & Actuators **A21–A23**, 534 (1990).
8.87 D. Qian, W. Wang, I. J. Busch-Vishniac and A. B. Buckman, IEEE Transaction and Measurement **42**, 14 (1993).
8.88 R. Williams and R.S. Crandall, Solar Cells **40**, 371 (1979).
8.89 H. Dersch, L. Schweitzer and J. Stuke, Phys. Rev. B **28**, 4678 (1983).
8.90 T.J. McMahon, B.G. Yacobi, K. Sadlon, J. Dick and A. Madan, J. Non-Cryst. Solids **6**, 375 (1984).
8.91 B.G. Yacobi, T.J. McMahon and A. Madan, Solar Cells **12**, 329 (1984).
8.92 R. Martins and E. Fortunato, J. Appl. Phys. **75**, 3481 (1995).
8.93 E. Fortunato, G. Lavareda, R. Martins, F. Soares and L. Fernandes, SPIE Proc. **2397**, 259 (1995).
8.94 S.M. Sze: *Physics of Semiconductors Devices*, 2nd ed. (Wiley, New York, 1981).
8.95 G. Hanke, F.A.G. Kugelfischer Georg Schafer, KGaA, Sensor Review **10** (1), 30 (1990).
8.96 F. Soares and R. Martins (private communication).

Symbols and Abbreviations

List of Symbols

Symbol	Description	Unit
A	Area	(cm^2)
a	Half length of the incident light line	(cm)
α	Fall-off parameter	(cm^{-1})
α'	Fall-off parameter with the recombination losses	(cm^{-1})
α_0	Fall-off parameter neglecting losses	(cm^{-1})
α_1	ac lateral fall-off parameter	(cm^{-1})
a_n	attenuation factor	(cm^{-1})
α_T	Temperature coefficient	(K^{-1})
α_x	Fall-off parameter ascribrd to the xx' direction	(cm^{-1})
α_y	Fall-off parameter ascribed to the yy' direction	(cm^{-1})
B	Bandwidth	(Hz)
c	Constant	(V)
c'	Constant	(V)
C_0	Junction capacitance	(F/cm^2)
C	Capacitance	(F)
C_{rmd}	Capacitance associated with the doped layer	(F/cm^2)
C_1	Capacitance per unit length	(F/cm)
Δ	Device with	(mm)
d	Thickness	(Å)
δ	Position detection error	–
D^*	Detectivity	(cm Hz$^{1/2}$/W)
Δa	Illuminated area	(cm^2)
ΔE	Activation energy	(eV)
$\Delta \phi$	Distortion potential	(V)
D_{na}	Ambipolar diffusion coefficient	(cm^2/s)
dP	Positional resolution	(μm)
ΔV	Lateral photovoltage	(V)

Symbols and Abbreviations

Symbol	Description	Unit
E	Electric field	(V/cm)
ε	Relative dielectric constant of a-Si:H	–
E_0	Electric field (dc component)	(V/cm)
ε_0	Dielectric vacuum constant	(F/cm)
E_1	Electric field (ac component)	(V/cm)
E_c	Equivalent lateral electric field	(V/cm)
F	Full scale	–
f	Frequency	(Hz)
$\phi(y)$	Lateral potential distribution	(V)
$\phi_0(y)$	Static spatial lateral potential distribution	(V)
f_λ	Photon flux	$(\text{cm}^{-3}\,\text{s}^{-1})$
f_m	Modulated frequency	(Hz)
G	Total channel conductance	(Ω^{-1})
γ	Losses parameter	–
g	Generation rate	$(\text{cm}^{-3}\,\text{s}^{-1})$
G_0	Lateral conductance	(Ω^{-1})
$G_0(y)$	Channel conductance along yy' axis	(Ω^{-1})
η	Diode quality factor	–
$I(t), I(t)$	Time dependent current	(A)
I_0	Total photocurrent	(A)
$I_{01}\ldots I_{on}$	Currents collected in a LTFPSD	(A)
I_1	Photocurrent at electrode 1	(A)
I_2	Photocurrent at electrode 2	(A)
I_{DR}	Dark reverse current	(A)
I_{in}	Input current at	(A)
$I_{fl}(x,y)$	Current components along the x,y surface	(A)
I_L	Lateral photocurrent	(A)
I_{L1}	Lateral photocurrent at electrode 1	(A)
I_{L2}	Lateral photocurrent at electrode 2	(A)
I_n	Noise current	
$I_{0,i}$	Static current component at xx' boundary	(A)
$I_{0,j}$	Static current component at yy' boundary	(A)
$I_{0,i,j}$	Static current surface components	(A)
I_{ph}	Total photocurrent	(A)
I_s	Saturation current	(A)
$I_{x1,x2}$	Current at xx' direction	(A)
$I_{y1,y2}$	Current at yy' direction	(A)
J_0	Pre-exponential factor of J_s	(A/cm^2)
J_L	Lateral photocurrent density	(A/cm^2)

Symbol	Description	Unit
J_{ph}	Photocurrent density	(A/cm^2)
J_{ph0}	Photocurrent density (dc component)	(A/cm^2)
J_{ph1}	Photocurrent density (ac component)	(A/cm^2)
J_{R}	Reverse current density	(A/cm^2)
J_{R0}	Reverse saturation current density ($T = 0\,\text{K}$ and $V = -1\,\text{V}$)	(A/cm^2)
J_{s}	Saturation current density	(A/cm^2)
J_x	Transverse current density	(A/cm^2)
J_y	Lateral current density	(A/cm^2)
$J_{y\text{d}}$	Lateral current density collected by collecting resistive layer	(A/cm^2)
$J_{y\text{l}}$	Lateral current density ascribed to the saturation current	(A/cm^2)
k	Boltzmann's constant	(eV/K)
L	Device length	(cm)
λ	Wavelength	(nm)
L^{-1}	Laplace inverse transform	–
l_{a}	Ambipolar diffusion length	(cm)
μ_{a}	Ambipolar mobility	(cm^2/Vs)
$\mu\tau E$	Shubweg – drift carriers distance under the action of an electric field	(cm)
n	Integer (varies from 1 to 127)	–
N	maximum number of detectable light beams	–
n'	Conduction carriers	(cm^{-3})
n_0	Free carriers under steady state condition	(cm^{-3})
n_1	Free electrons under light illumination	(cm^{-3})
n_{a}	Excess of generated carriers	(cm^{-3})
P	Spatial position coordinates	(cm)
p_1	Free holes under light illumination	(cm^{-3})
$P_{\max}$	Maximum power due to each light beam	(W)
P_{n}	Spatial coordinates of a line on a LTF-PSD	(cm)
P_{s}	Power deliverd by the light sources	(W)
P_x	x coordinates of the incident light beam position	(cm)
P_y	y coordinates of the incident light beam position	(cm)
q	Electron charge	(C)
Q_{a}	Ambipolar charge concentration	(C·cm^{-3})
ρ	Resistivity	(Ω cm)
R_0	Parallel resistance	(Ω)

Symbol	Description	Unit
ρ_0	Resistivity of the i layer	($\Omega\,\text{cm}$)
r_1	Resistance beween the incident point and the el. 1	(Ω)
r_2	Resistance beween the incident point and el. 2	(Ω)
R_d	Equivalent resistance due to the doped coll. layer	(Ω)
R_0	Load resistance at $x=0$ of the device	(Ω)
ρ_d	Transverse resistivity of the i-layer	($\Omega\,\text{cm}$)
R_{eq}	Circuit sensor equivalent resistance	(Ω)
R_L	Load resistance at $x=L$ of the device	(Ω)
R_l	Recombination losses	($\text{cm}^{-3}\text{s}^{-1}$)
R_λ	Spectral response	(A/W)
R_{lr}	longitudinal discrete resistance of the thin collecting layer	(Ω)
R_r	Rectification ratio	—
ρ_s	sheet resistivity of the TCRL	($\Omega/\square$)
ρ_{sd}	sheet resistivity of doping collecting layer	($\Omega/\square$)
$r_{x1,x2}$	Distributed resistances along the xx' direction	(Ω)
$r_{y1,y2}$	Distributed resistances along the yy' direction	(Ω)
σ	Root mean square deviation	(rms)
σ_0	i layer conductivity	($\Omega^{-1}\,\text{cm}^{-1}$)
σ_α	average conductivity	($\Omega^{-1}\,\text{cm}^{-1}$)
s_d	Static detection limits	(cm)
σ_d	Doped layer conductivity	($\Omega^{-1}\,\text{cm}^{-1}$)
s_r	Spatial resolution	(μm)
σ_r	Collecting resistive layer conductivity	($\Omega^{-1}\,\text{cm}^{-1}$)
T	Temperature	(K)
t	Time	(s)
τ	Delay time	(s)
T_a	Ambient temperature (300 K)	(K)
τ_a	Ambipolar recombination time	(s)
τ_d	Dynamic fall-off parameter	(s)
t_d	Decay time	(s)
t_r	Rise time	(s)
T_s	Sampling time	(s)
V	Voltage	(V)
V_0	Potential difference across the structure	(V)
v_d	Carriers diffusion velocity	(cm/s)
V_{inA}	Input voltage at terminal A	(V)

Symbol	Description	Unit
$V_{\text{in}B}$	Input voltage at terminal B	(V)
V_{out}	Output voltage	(V)
V_{R}	Reverse bias polarization	(V)
W	i layer thickness	(Å)
ω_0	Signal speed propagation within the device	(cm/s)
W_{d}	hickness of the doping collecting layer	(Å)
W_{r}	Thickness of the collecting layer	(Å)
x	Spatial dimension in the xx' direction	(cm)
$X_{1,2}$	Electrodes at the xx' direction	–
x_{G}	coordinates of the centre of the gravity of 2D TFPSD	(cm)
y	Spatial dimension in the yy' direction	(cm)
$y_{1,2}$	Electrodes at the yy' direction	–

Acronyms and Abbreviations

1D	One dimensional
2D	Two dimensional
A/D	Analog to digital converters
a-Si:H	Hydrogenated amorphous silicon
CCD	Charge-coupled devices
Cds-CdSe	Cadmium suphide-cadmium seleneide
CRT	Cathode ray tube
c-Si	Crystalline silicon
$Cu-Cu_2O$	Copper-copper dioxide
FET	Field effect transistor
Ge–In	Germanium indium alloy
GPIB	General purpose interface bus
ITO	Indium tin oxide
I–V	Current-voltage characteristic
J–V	Current density-voltage characteristics
LCD	Liquid crystal display
LTFPSD	Linear array of thin film position sensitive detector
MIS	Metal–insulator structure
MLS	Multilayer system
Mo	Molybdenium
MOS	Metal-oxide semiconductor
NEP	Noise equivalent power
OPAM	Operational amplifier
PECVD	Plasma-enhanced chemical vapour deposition
Poly-Si	Policrystalline silicon
PSD	Position sensitive detector
R_1	Resistor 1
R_2	Resistor 2
R_3	Resistor 3
R_3	Resistor 4
S/N	Signal-to-noise (ratio)
SR	Shift register
TCO	Transparent conductive oxide
TCRL	Thin collecting resistive layer
TFPSD	Thin film position sensitive detector
Ti	Titanium

Index

a-Si:H
- atomic structure   3
- electrical caracteristics   372
- growth   254
- optical characteristics   372
- TFPSD   349

a-SiC   275, 277
a-SiC:H   333
a-SiGe   278, 281
absorption coefficient   49, 309, 334
activation energy   49
active matrix LCDs   207
additional capacitance (C_{add})   14
additive noise   195, 199
adjustable threshold three color detector   323
aliasing   197, 199
ambipolar transport   360
amorphization   110
amplifier bandwidth   398
amplifier noise   188
analyzer   10
anisotropic dielectric constants   64
aperture ratio   12, 15, 24, 29, 33
autopolarization   328
avalanche gain   148

back reflector   270
back-channel-etched (BCE) TFT   41
back-to-back diode   60
backlight   11, 12
band diagram   260
band gap profiling   279
bend   27, 67, 68, 73
birefringence   66, 77, 89
black matrix   15
book scanner   215
boron doping   275

boundary conditions   360

capacitance–voltage characteristics   56
capacitance   45, 172
- of the TFT   158
carrier lifetime   150
CdSe   159, 344
channel length   28, 59
channel passivated (CHP) a-Si:H TFT   41
channel width   28, 46, 59
charge amplifiers   396
charge collection   150
charge trapping   170, 194
chemical annealing   297
chiral pitch   89
cholesteric   62
CMOS   121
- sensors   178
co-planar   60
collection efficiency   150, 174
color filter   12, 15, 23
color imaging   215
color sensitivity   306
color sensors   307
- current–voltage characteristics   318, 325
- response time   322
- spectral response   318, 320
component cells   281
composition profiling   279
conduction band   53
conductivity, thermal   238
contrast ratio   72, 80
cost of electronics   223
crosstalk   13, 171
CsI   201
CsI:Tl   193, 202

Subject Index

current matching 281
current stability 155
current–voltage characteristics 60, 325
– of a color sensor 325
– OLED 232
curvature strain 66
cyano-biphenyl 63

dangling bond 150
dangling bond defects 266
data acquisition 182
data line capacitance 186, 190
defect-creation kinetics 265
defects 321
degradation of solar cell 265, 282
density of states 4
depletion length 156
deposition parameters 256
deposition rate 296
design rule 225
detective quantum efficiency 195, 198
detectivity 386
deuterium dilution 297
device stability 130
diborane 39
dielectric constant 23, 26, 66
dielectric function 312
dielectric permittivity 27
differential capacitance 336
diode addressed arrays 175, 202
director orientation 72
disclination 15
disilane 278
document scanning 214
doped polysilicon, sheet resistance 115
doping 111, 275
double correlated sampling 181, 186
double diode addressing 177
DQE model 202
dual dielectric gate insulator 138
duolateral PSD 349
duolateral TFPSD 351

elastic constants 68, 83
electron cyclotron resonance 297
electronic circuits, large-area 222
electronic noise 185

ellipticity 77, 78
energy gap 53, 54
excimer laser 102
explosive crystallization 105
extented graphics array (XGA) 9, 31, 32
extended states 54
extrinsic noise 189

$1/f$ noise 185, 188
feed-through charge 168, 169, 176
fill factor 163, 171, 259, 282
flat-band voltage 55
flicker noise 23
fluoroscopy 208
frame frequency 13
frame time 162
free energy 73, 83
– of liquid crystal 68
frequency of rf plasma 257

gain and offset conection 183
gain-bandwidth product 44
gap state distribution 261
gas immersion laser doping (GILD) 113
gate capacitance 57
gate delay 17, 24, 31, 32
GdO_2S_2:Tb 193, 200, 206
generation-recombination 154
glow discharge 252, 254
gradual channel approximation 18, 43, 59
grain boundaries 98, 110
grain size 107

Hall mobility 99
heat flow analysis 103
high bandgap alloys 277
high fill factor array 171
homeotropic 71
hot-wire chemical vapor deposition 296
hydrogen 5, 266, 311
– dilution 257, 267, 278
– out-diffusion 136
– passivation 100, 101
hydrogenation 101, 128

ideality factor 153, 154

Subject Index   413

image contrast   205, 210
image intensifier tube   208
image lag   169, 183, 209
image sensor
– array size   216
– capacitance   168
– dynamic range   165
– integration time   166
– readout amplifiers   180
– readout time   165
– sensitivity   163, 174, 216
image spreading   195, 196
imaging systems   178
impurities   257, 266
in-plane-switching (IPS) mode   16, 81, 82
indium tin oxide (ITO)   36, 269, 349
instability   49, 155
inverted staggered-electrode   41
ionization rate   194
IPS (in-plane switching) mode   64, 85, 86
IR sensors   334
ITO (indium tin oxide)   14, 16, 40, 285

jet-printing   233
Johnson noise   368

kTC noise   187

Laplace transform   369
large area arrays   327
large lateral grain growth   105
laser crystallization   94–145
laser de-hydrogenation   141
laser doping   111
laser scribing   292
lateral grain growth   108
lateral photo-effect   342, 359, 361
layered structures   308
LCD
– diagonal size   14, 26
– fabrication   34
– gate delay   21
– normally black node   71
– normally white mode   72
– pixel capacitance   17
– pixel design   14
– scaling   26

– utility factor   12
lens efficiency   192
Leslie coefficients   69
liftoff   228
light emitting devices, organic   232
light scattering   273
light trapping   264
light-induced degradation (LID)   264, 266, 268
line-correlated noise   190
line-spread function   196
linear image sensor   352
linearity   350
liquid crystal   61
– anisotropic dielectric constant   23
– capacitance   76
– cholesteric   62
– dielectric constants   64
– elastic constants   66
– lyotropic   62
– nematic   62
– optical properties   76
– refractive indices   81
– smectic   62
– thermotropic   62
lithography   34
load-locked system   257
localized states   53, 54
low bandgap alloys   278
low noise   181
low-pressure chemical vapor deposition   97
low-temperature poly-Si   117

macroelectronics   222–251
mammography   207
manufacturing process   287
mask   228
– toner   228
matrix addressing   161
matrix array   328
Mauguin's limit   78
MBBA   81
MBMA   63
medical imaging   204
metal-insulator-metal (MIM) diodes   60
metastability   264
micro-doping   336

microcrystalline n-layer 283
microcrystalline p-layers 276
microcrystalline SiC 276
microcrystalline silicon 295
microstructure 267, 296
minority carrier lifetime 148
mobility 45
– TFT 121
mobility-lifetime product 150, 170
modulation transfer function 196, 201
module assembly 289
modulus
monomolecular recombination 383
multijunction solar cell 269
multilayer PSD 353
multiplexing 179

n-i-p-i-n detector 316
nematic 62
neutron imaging 214
node potential 18–22, 45
noise 197, 216
noise equivalent power (NEP) 387
noise power spectrum 198, 201
non-destructive evaluation 211
nucleation sites 109
numerical modeling 261
Nyquist frequency 197

OLED 232
– current-voltage characteristics 235
open-circuit voltage 259, 277, 282
optical absorption 270
optical design 308, 312
optical gap 309
– of a-$Si_{1-x}C_x$:H 311
optical inspection 392
optical path length 271
optical properties of amorphous silicon 308
– alloys 310
orientational order 62
output characteristics 44, 56

p-azoxyanisole (PAA) 63
p-CVD 40
p-i-n based TFPSD 351
p-i-n photodiodes 148
– leakage current 210

p-i-n-i-p photodetector 317
parasitic capacitance 22, 121
PbI_2 174, 194
PbO 174
phosphine 39
phosphor 191, 200, 206
phosphorus doping 275
photoconductivity 47
photoconductors 193, 344
photocurrent 15, 48, 49, 149
photodiode 148
– forward current 153
– leakage current 151, 152, 166
– p-i-n 148, 210
– reverse bias current 151
– thick photodiode sensors 155
photoionization 191
photoresist 34
phototransistor 323
pixel 13, 160, 161, 171
pixel amplifier 191
pixel capacitance 127
pixel circuit 161, 176, 187
pixel size 163, 206
plasma chemical vapor deposition 34, 39
plasma chemistry 254
plasma enhanced chemical vapor deposition 40, 97, 289
polarizer 10
poly(N-vinylcarbazole) (PVK) 232, 248
polycrystalline silicon 94, 159
polyimide 10, 70
polymeric substrates 292
polysilicon 96, 344
– grain boundaries 98
– grain growth 105
– grain size 138
– growth of polysilicon 96
– mobility 95
– peripheral circuits 95
position detection resolution 373
position sensitive detector (PSD)
– advantages of a-Si:H 345
presampled MTF 197, 200
pretilt angle 10, 86, 88, 89
printer

– ink-jet 232
– laser 228
printing 222, 227, 228
– direct 224
– mask 228
– speeds 227
production processes 349
projection aligner 34
PSD
– 2-dimensional 367
– curves 383
– dependence on wavelength 381
– detection limits 371
– duolateral 349
– equivalent electric circuit 347
– fabrication 349, 352, 354, 357
– fall-off parameter 362
– illumination 388
– large area 354
– linear array 357, 393
– linearity 372, 376
– principles of operation 346
– recombination losses 362
– resolution 395
– response time 379
– response to multiple light beams 374
– signal processing 389, 396
– spatial resolution 376
– spectral response 386
– static behaviour of E_y and ϕ_y 364
– temperature dependence 385
– tetralateral 349, 354
– theory 359
– transient response time 368
PVK, photoluminescence spectra 233

quantum efficiency 149, 215
– of solar cell 291
– of UV sensor, see also spectral response 334

radiation therapy 210
radiography 205
radius
– bending 241
– curvature of 242
readout 182, 329
readout amplifiers 179

readout pixel 169
refractive index 64, 66, 308
registration 225
resolution 225
response time 75
– of color sensor 325
retention 30, 32
– characteristics 29
roll-to-roll manufacturing 287
rotational viscosity 74

saturation current 43
scan lines 21, 33
Schottky based TFPSD 350
Schottky diodes 61
Schottky-barrier PSD 351
Se 174
secondary reactions 254
seeded grain growth 125
selective laser crystallization 133
selenium 194, 201
– dark current 175
sensitive amplifier 180
sensitivity 164
sensor leakage 167
short-circuit current 259, 272
shot noise 386
silane 39, 40
silicon nitride 40, 157
SiN 50
single-junction solar cells 258
smectic 62
solar cell 47
– efficiency 259, 270, 285
– substrate-type 259
– superstrate-type 259
solid-phase crystallization 97
space applications 294
space-charge-limited currents 58
spatial resolution 160, 373
spectral response
– PSD 386
– IR sensor 336
SPICE model 319
splay 27, 67, 68, 73
stability 294
Staebler–Wronski effect 51, 264
staggered electrode 53
steel foil 230

416 Subject Index

step-and-repeat aligners 34
STN 89
storage capacitance 14, 23, 24, 33
storage capacitor 160, 173
strain 243
– in thin films 240
strain tensor 67
stress-induced defects 130
structural changes 267
substrate 34, 236, 238
– compliant 242
– glass foil 236
– planarization 240
– plastic 244
– temperature 256
– thermal stability 35
substrate-type structure 287
subthreshold slope 45, 121
super-twisted nematic (STN) 64, 88
superstrate-type 292
supertwisted birefringent effect (SBE) 88
surface inspection 388
surface recombination 343
surface roughening 110
surface roughness 126

tape automated bonding (TAB) 10, 184
tetralateral PSD 349, 354
tetralateral TFPSD 351
TFPSD
– a-Si:H 349
– duolateral 351
– dynamic performance 376
– p-i-n based 351
– Schottky based 350
– tetralateral 351
thermal conductivity 238
thermal expansion, coefficient of 238
thermal generation 151
thermal noise 185, 188
thermionic emission 154
thin film position sensitive detectors 342–403
thin film transistor (TFT) 117, 157, 222
– back-channel-etched 14, 36
– bottom-gate 118

– channel length 41
– channel-passivated 14
– characteristics 41
– drain current 226
– fabrication 119, 135, 228, 245
– gate capacitance 43
– hybrid 133
– leakage 167
– leakage current 126, 158
– materials 159
– mobility 121, 123
– OLED integration 246
– on polyimide 244
– on steel foil 231
– output characteristics 38
– packing density 227
– parasitic capacitance 23
– retention 21
– self-aligned 118, 120
– simulation 53
– size 225
– stability 130
– structure 53
– threshold voltage 139
– top-gate 118
– transfer characteristics 22, 36, 38, 56, 122, 230, 247
– transient decay 158
– uniformity 124
three-color sensors 320
threshold voltage V_{th} 18, 27, 28, 38, 43, 45, 56, 59, 72, 74, 85, 86
– shift 49
– of liquid crystal 65
tilt angles 77
transconductance 38, 43
transfer characteristics 45, 158
triangulation 388
trimethyl boron 275
triple-cell structure 269
tunnel junction 282, 284
twist 27, 67, 68, 73
twisted nematic (TN) liquid crystal 69
two-color sensors 315
two-terminal devices 60

unloaded pixel 33
Urbach tail 310

UV sensors   332

V_t shift   49–51, 53
valence band   54
Vickers hardness   41
video graphics array (VGA)   9, 14, 31, 32, 53
viewing-angle characteristics   15, 81, 82, 87
viscous-stress tensor   69

X-ray attenuation   192
X-ray crystallography   212
X-ray dose   205, 209
X-ray photoconductor   173

yield of solar cell   291
Young's modulus   238, 242

ZnO   272

Springer Series in
MATERIALS SCIENCE

Editors: U. Gonser · R. M. Osgood, Jr. · M. B. Panish · H. Sakaki
Founding Editor: H. K. V. Lotsch

1. **Chemical Processing with Lasers***
   By D. Bäuerle

2. **Laser-Beam Interactions with Materials**
   Physical Principles and Applications
   By M. von Allmen and A. Blatter
   2nd Edition

3. **Laser Processing of Thin Films and Microstructures**
   Oxidation, Deposition and Etching of Insulators
   By I. W. Boyd

4. **Microclusters**
   Editors: S. Sugano, Y. Nishina, and S. Ohnishi

5. **Graphite Fibers and Filaments**
   By M. S. Dresselhaus, G. Dresselhaus, K. Sugihara, I. L. Spain, and H. A. Goldberg

6. **Elemental and Molecular Clusters**
   Editors: G. Benedek, T. P. Martin, and G. Pacchioni

7. **Molecular Beam Epitaxy**
   Fundamentals and Current Status
   By M. A. Herman and H. Sitter 2nd Edition

8. **Physical Chemistry of, in and on Silicon**
   By G. F. Cerofolini and L. Meda

9. **Tritium and Helium-3 in Metals**
   By R. Lässer

10. **Computer Simulation of Ion-Solid Interactions**
    By W. Eckstein

11. **Mechanisms of High Temperature Superconductivity**
    Editors: H. Kamimura and A. Oshiyama

12. **Dislocation Dynamics and Plasticity**
    By T. Suzuki, S. Takeuchi, and H. Yoshinaga

13. **Semiconductor Silicon**
    Materials Science and Technology
    Editors: G. Harbeke and M. J. Schulz

14. **Graphite Intercalation Compounds I**
    Structure and Dynamics
    Editors: H. Zabel and S. A. Solin

15. **Crystal Chemistry of High-T_c Superconducting Copper Oxides**
    By B. Raveau, C. Michel, M. Hervieu, and D. Groult

16. **Hydrogen in Semiconductors**
    By S. J. Pearton, M. Stavola, and J. W. Corbett

17. **Ordering at Surfaces and Interfaces**
    Editors: A. Yoshimori, T. Shinjo, and H. Watanabe

18. **Graphite Intercalation Compounds II**
    Editors: S. A. Solin and H. Zabel

19. **Laser-Assisted Microtechnology**
    By S. M. Metev and V. P. Veiko
    2nd Edition

20. **Microcluster Physics**
    By S. Sugano and H. Koizumi
    2nd Edition

21. **The Metal-Hydrogen System**
    By Y. Fukai

22. **Ion Implantation in Diamond, Graphite and Related Materials**
    By M. S. Dresselhaus and R. Kalish

23. **The Real Structure of High-T_c Superconductors**
    Editor: V. Sh. Shekhtman

24. **Metal Impurities in Silicon-Device Fabrication**
    By K. Graff

25. **Optical Properties of Metal Clusters**
    By U. Kreibig and M. Vollmer

26. **Gas Source Molecular Beam Epitaxy**
    Growth and Properties of Phosphorus Containing III–V Heterostructures
    By M. B. Panish and H. Temkin

27. **Physics of New Materials**
    Editor: F. E. Fujita 2nd Edition

28. **Laser Ablation**
    Principles and Applications
    Editor: J. C. Miller

29. **Elements of Rapid Solidification**
    Fundamentals and Applications
    Editor: M. A. Otooni

30. **Process Technology for Semiconductor Lasers**
    Crystal Growth and Microprocesses
    By K. Iga and S. Kinoshita

* The 2nd edition is available as a textbook with the title: *Laser Processing and Chemistry*

Printing (computer to plate): Mercedes-Druck, Berlin
Binding: Stürtz AG, Würzburg